Mathematical Cognition and Learning

Evolutionary Origins and Early Development of Number Processing

MATHEMATICAL COGNITION AND LEARNING

Series Editors

DANIEL B. BERCH
DAVID C. GEARY
KATHLEEN MANN KOEPKE

VOLUME 1

Evolutionary Origins and Early Development of Number Processing

Volume Editors

DAVID C. GEARY
DANIEL B. BERCH
KATHLEEN MANN KOEPKE

Mathematical Cognition and Learning

Evolutionary Origins and Early Development of Number Processing

Edited by

David C. Geary
Psychological Sciences
University of Missouri
Columbia, MO, USA

Daniel B. Berch
Curry School of Education
University of Virginia
Charlottesville, VA, USA

Kathleen Mann Koepke
Eunice Kennedy Shriver National Institute of Child
Health and Human Development (NICHD)
Bethesda, MD, USA

AMSTERDAM • BOSTON • HEIDELBERG • LONDON
NEW YORK • OXFORD • PARIS • SAN DIEGO
SAN FRANCISCO • SINGAPORE • SYDNEY • TOKYO
Academic Press is an imprint of Elsevier

Academic Press is an imprint of Elsevier
32 Jamestown Road, London NW1 7BY, UK
525 B Street, Suite 1800, San Diego, CA 92101-4495, USA
225 Wyman Street, Waltham, MA 02451, USA
The Boulevard, Langford Lane, Kidlington, Oxford OX5 1GB, UK

First edition 2015

British Library Cataloguing-in-Publication Data
A catalogue record for this book is available from the British Library

Library of Congress Cataloging-in-Publication Data
A catalog record for this book is available from the Library of Congress

ISBN: 978-0-12-420133-0

For information on all Academic Press publications
visit our web site at store.elsevier.com

Printed and bound in the United States of America

Contents

Part II
Number and Magnitude in Infants and Young Children

Part III
Number Judgments: Theoretical Perspectives and Evolutionary Foundations

Contributors

Numbers in Parentheses indicate the pages on which the author's contributions begin.

Christian Agrillo (3), Department of General Psychology, University of Padova, Italy

Joseph M. Baker (277), Stanford University School of Medicine, Department of Psychiatry and Behavioral Sciences, Stanford, CA, USA

Michael J. Beran (91), Language Research Center, Georgia State University, Decatur, GA, USA

Daniel B. Berch (335), Curry School of Education, University of Virginia, Charlottesville, VA, USA

Angelo Bisazza (3), Department of General Psychology, University of Padova, Italy

Elizabeth M. Brannon (123), Department of Psychology & Neuroscience and Center for Cognitive Neuroscience, Duke University, Durham, NC, USA

Jessica F. Cantlon (225), Department of Brain & Cognitive Sciences, University of Rochester, New York, NY, USA

Sara Cordes (253), Boston College, Chestnut Hill, MA, USA

Theodore A. Evans (91), Language Research Center, Georgia State University, Decatur, GA, USA

David C. Geary (335), Psychological Sciences, University of Missouri, Columbia, MO, USA

Justin Halberda (305), Department of Psychological and Brain Sciences, Johns Hopkins University, Baltimore, MD, USA

Kerry E. Jordan (277), Utah State University, Department of Psychology, Logan, UT, USA

Kathleen Mann Koepke (335), *Eunice Kennedy Shriver* National Institute of Child Health and Human Development (NICHD), Bethesda, MD, USA

Stella F. Lourenco (145), Emory University, Atlanta, GA, USA

Koleen McCrink (201), Barnard College, Columbia University, New York, NY, USA

Maria Elena Miletto Petrazzini (3), Department of General Psychology, University of Padova, Italy

Darko Odic (305), Department of Psychological and Brain Sciences, Johns Hopkins University, Baltimore, MD, USA

Audrey E. Parrish (91), Language Research Center and Department of Psychology, Georgia State University, Decatur, GA, USA

Irene M. Pepperberg (67), Department of Psychology, Harvard University, Cambridge, MA, USA

Tasha Posid (253), Boston College, Chestnut Hill, MA, USA

Ariel Starr (123), Department of Psychology & Neuroscience and Center for Cognitive Neuroscience, Duke University, Durham, NC, USA

Giorgio Vallortigara (35), Center for Mind/Brain Sciences, University of Trento, Italy

Kristy vanMarle (175), University of Missouri, Columbia, MO, USA

Foreword

Rochel Gelman and C.R. Gallistel

Rutgers University

It is a pleasure to introduce what is sure to become a prominent mark of progress in the scientific understanding of the origins of numerical cognition. This volume provides a comprehensive view of that progress, revealing the emergence of a broad consensus on the answers to several foundational questions, a consensus far removed from the views commonly entertained half a century ago. A half century or so ago it was commonly assumed that numerical cognition depended on language for its development and that it developed, therefore, only in humans. We now understand that numerical cognition is evolutionarily ancient. Many of the chapters in this volume review aspects of the empirical basis for this new understanding, but the reader is directed most particularly to the first four chapters and to the final chapter. We also now understand that the pre-linguistic system of numerical estimation and reasoning, which we share with most if not all vertebrates (and possibly with invertebrates), plays an important role in the development of verbalized mathematical reasoning. A majority of the chapters touch on the evidence for this latter conclusion, but the reader is referred most particularly to Chapters 5 through 10.

In this introduction, we begin with a brief history of contrasting ideas about what a number is. We do so to clarify for the reader some important distinctions, most importantly the distinction between numbers as symbols for discrete (countable) quantities and numbers as the players in the system of arithmetic reasoning. Gelman (1972) referred to this as the estimator-operator distinction.

Throughout most of the history of mathematics, numbers were defined by what they referred to, which is to say countable quantities. The traditional view is memorably summed up by a well known quote attributed to Kronecker, "God made the integers; all else is the work of man." (Quoted in an obituary by Weber, 1893.)

Only with the extensive formalization of mathematics in the latter part of the 19th century did an alternative view gain currency, the view that now dominates among mathematicians. This view is articulated by Knopp (1952, p. 5), who gives an axiomatization of arithmetic followed by:

"Once the validity of these fundamental laws has been established, it is unnecessary, in all further work with the literal quantities a, b,..., to make use again of the fact that these symbols denote rational numbers. ... From the important fact that the meaning *[that is, the reference] of the literal symbols need not be considered at all..., there results immediately the following extraordinarily significant consequence: If one has any other entities whatsoever besides the rational numbers, ..., but which obey the same fundamental laws, one can operate with them as with the rational numbers, according to exactly the same rules. Every system of objects for which this is true is called a* number system, *because, in a few words, it is customary to call all those objects* numbers *with which one can operate according to the fundamental laws we have listed." (italics in original).*

On this *formalist* view, numbers are defined in the same way that chess pieces are, that is, by what can be done with them. The empirical reasons for regarding the brain's non-verbal representation of discrete (countable) quantity as part of a more general system of arithmetic reasoning that represents *both* continuous *and* discrete quantity is the focus of Chapter 6 (see also Walsh, 2003; Gallistel, 2011). The fact that a system of symbols as simple as arithmetic can so effectively represent so much of experienced reality is astonishing (Wigner, 1960). The representational power of arithmetic reasoning probably explains why it emerged so early in evolution.

Our intuitive identification of number with *countable* quantity is evident already in the troublesome ambiguity about what exactly an author may mean when using 'number' in many contexts. Do they mean to refer to the numerosity of a set? Or do they mean to refer to a conventional symbol, such as 'two' or '11' or 'IV' or (5-2), that may (or may not) represent the numerosity of a set? The distinction here corresponds to the distinction in computer programming between, on the one hand, the bit-code representation of a number that enters into the arithmetic operations that a computer performs and, on the other hand, the bit-code representation of the textual symbol for that number. This distinction (between $x = 7$ and $x = $ '7') often flummoxes newcomers to the programming craft. It is now common in the technical literature to use 'numerosity' when the first meaning is intended. Gelman and Gallistel (1978) suggest 'numerlog' when reference to the conventional symbols is instead intended, but that coinage has not come into common use. In any event, failing to distinguish between the numerosities themselves, on the one hand, and the symbols for quantities in an arithmetic system, on the other, may lead some readers into confusion.

Some of the current authors use 'symbolic' to refer to conventional symbolizations of quantity, particularly when asserting that the evolutionarily ancient system is "non-symbolic" or that only humans have a "symbolic" representation of number. We would suggest that this usage invites confusion. It seems to imply that the evolutionarily ancient system does not employ symbols in the sense in which a computer scientist understands this term (see

Gallistel & King, 2010). Insofar as someone subscribes to a radical connec-
tionist view of the brain, a view in which brain activity is "sub-symbolic"
and does not in any true sense represent the experienced world, this might
be the intended meaning, but we do not believe it is the meaning intended
by any current author. To distinguish between conventional symbols and the
brain's symbols, we suggested 'numerons' to refer to the brain's symbols
for quantity (Gelman & Gallistel, 1978), but again that coinage has not gained
currency.

Symbols are essential elements of representations (Gallistel & King,
2010). A representation has two fundamental components, a component (set
of processes) that map from aspects of the represented system (for example,
from numerosities) to the symbols that thereby refer to them and a component
(set of processes) that operates on the symbols (ordering them, adding or subtracting them, multiplying or dividing them, concatenating or extracting them,
and so on). Symbols are the stuff of computation, and representations are constructed by computation. That is the essence of the computational theory of
mind, which is the central doctrine of cognitive science. If the brain represents
quantities—and, in our view, the chapters in this volume show beyond reasonable argument that it does—then it does so by means of symbols.

Because the meanings of numerical symbols were once commonly thought
to derive from their reference, there were doubts about whether negative numbers and irrational numbers really existed, even among professional mathematicians living into the 19th century (Kline, 1972). However, by then, the use of
numerical coordinates to represent different aspects of the physical and financial world was rapidly advancing. Because the origin of a useful coordinate
system is almost always arbitrary (the Greenwich meridian, for example)
and/or because many of the quantities represented by numbers are inherently
directed quantity (distance north and south, for example), it is all but impossible to avoid using negative numbers in practice. Moreover, in algebraic
reasoning, one often does not know until it is, so to speak, "too late," whether
one is or is not trafficking with negative quantities. Thus, some of what Kronecker regarded as the works of man (the negative numbers and fractions)
seemed unavoidable if one were to have a generally useful system for representing quantities, as even Kronecker conceded. Kronecker agreed that negative numbers and fractions could be admitted into the system made by God
(the counting numbers), provided that it could be proven that they could be
made to abide by the manipulation rules that God's creatures (the so-called
natural numbers) abided by. However, the enlargement of the system people
took to be the "natural" numbers to make it more generally useful led most
mathematicians to include among the numbers the so-called transcendental
numbers like π, whose very existence Kronecker disputed. (He insisted that
there was no such number!) One begins to see why the question whether a
symbol did or did not abide by the rules of arithmetic came to be an ever more
important consideration, emerging eventually as definitive of number itself.

Theories of the development of verbalized numerical reasoning divide along similar lines. Some authorities urge that the preschool child understands the meaning of 'one' if and only if s/he knows that this word refers to the numerosity of sets that have only one member, and similarly for the meaning of 'two' and 'three' (Carey, 2010). This view is similar to (and perhaps partially motivated by) Gottlob Frege's attempt to found mathematics on logic and set theory by defining '0' to *mean* (that is, denote or refer to) the numerosity of the empty set, '1' to *mean* the numerosity of sets that could be placed in 1:1 correspondence with the set that contains only the empty set, '2' to *mean* the numerosity of sets that can be placed in 1:1 correspondence with the set that contains only the empty set and the set that contains the empty set, and so on (Frege, 1884). On this first view of the development of verbal numerical reasoning, sets composed of the pointer symbols in the object tracking system (OTS) serve in place of Frege's recursively defined canonical sets. Reference to canonical pointer-sets in long term memory is what gives the first few number words their meaning in the mind of the very young child, on this view.

Other authorities urge that the non-verbal system of numerical reasoning (generally called the ANS, which is short for analog number system) is the principal foundation for the development of the human capacity to reason verbally about quantity (Gallistel & Gelman, 1992 and Chapters 5–10 of the current volume). This latter view stresses that the understanding of number words always involves an appreciation of two facts: (1) [ordinary] number words refer to countable quantities; (2) they also correspond to preverbal symbols in a system of preverbal arithmetic reasoning. On the first view, the child in the early stages of learning to count does not grasp that it is learning words for mental symbols that are subject to arithmetic reasoning, symbols that can be ordered, added and subtracted. On the second view, the child in the early stages of learning to count already understands that number words refer to symbols that are subject to arithmetic reasoning. In other words, the child just learning to count already understands that number words also correspond to players in the game of arithmetic, a game its brain already knows how to play. Several chapters in the current volume review the now extensive evidence that the brains of infants and toddlers already play the number game when thinking about both countable and continuous quantities (Chapters 5–10). Other chapters review the evidence that the brain's ability to play this game has ancient evolutionary origins (Chapters 1–4 and Chapter 13, see also Livingstone, 2014).

The question of the extent to which the development of numerical reasoning depends on the realization that number words correspond to entities that are subject to arithmetic reasoning bears crucially on proposals that there is more than one non-verbal system for representing numerosity. A profoundly important formal property of the number system is *closure*. A symbolic system is closed if the valid operations on its symbols always yield other symbols within that same system. When it is proposed that there is one non-verbal system for representing small numerosities (usually 4 or less) and a different non-verbal system for representing larger numerosities, one must ask why this does not lead to lack of closure in both systems. The problem is that 2+3, which is an operation on symbols for small numerosities, yields 5, which is not a symbol in the putative system for representing small numbers (i.e., the OTS). Similarly, 7–5, which is an operation on symbols for large numerosities, yields 2, which, at least on some stories, is a symbol that exists only in a different system (i.e., the OTS). It is as if a move in chess carried you out of the game of chess and into the game of checkers. But even a parrot can do 2+3=5 (see Chapter 3).

The closure consideration does not argue that, for example, the OTS does not exist. There is little doubt that it does exist. And, as Chapter 10 discusses at length, it does sometimes control behavior in tasks that the experimenter intended to tap an understanding of number. The closure consideration argues only that the OTS should not be regarded as in any sense a numerical representation, because a numerical representation requires both symbols that may refer to quantities and arithmetic operations on those symbols. In any such system, closure is a major consideration. Absent closure, the system will crash whenever it processes two symbols that cannot be combined arithmetically within the limits of that system. Chapter 7 offers a novel, and plausible way of dealing with this issue, while Chapters 8–10 review the extensive evidence for non-verbal arithmetic reasoning in infants and toddlers. (See also Zur & Gelman, 2004.)

As emphasized in most of the chapters in this volume and discussed at length in Chapter 12, a signature of the ANS is that it represents even discrete (countable) quantity with fixed percent resolution, that is, in accord with Weber's Law. This property is often explained on the assumption that the brain's symbols for quantities are noisy signals, with the noise scaling with signal magnitude. If the Weber Law property is really ascribable to scalar noise, then it is a bug or limitation of the system. Gallistel (2011) suggested that Weber's Law may be a feature, not a bug. It may reflect an automatic scaling mechanism such as is built into most contemporary electronic instruments to enable them to give useful measurements over many orders of magnitude. Following up on this suggestion, Wilkes (2014, under review) shows that any system for representing the computable numbers over many orders of magnitude that makes efficient use of limited physical resources must exhibit Weber's Law. Both the principle and the need for it are illustrated by scientific number notation, which uses some fixed number of digits to represent any number (hence fixed percent resolution), together with a fixed number of digits to represent the scale (order of magnitude). Livingstone et al. (2014) remind us that autoscaling (aka normalization) is a well established neurobiological mechanism (see also Brenner, Bialek, & de Ruyter van Steveninck, 2000).

Two other points of broad significance run through the chapters of this volume. The first is that while number is indeed a highly abstract, fundamentally amodal aspect of our experience (see particularly Chapter 11), it is not derived from concrete sensory experience through some process of abstraction and induction, as has traditionally been supposed in empiricist philosophy and in much psychological work on number. Rather, it is a foundational aspect of the brain's capacity to represent the experienced world. The second point, one particularly stressed by Vallortigara in Chapter 2 and adopted by vanMarle in Chapter 7, is that number knowledge is Exhibit A in the argument that there are core knowledge domains. These domains may be thought of as skeletal structures (Gelman, 1990; Gelman, 1993; Gelman & Williams, 1998) that organize sensory input and direct attention in such a way as to make possible the acquisition of further knowledge, such as, knowledge of how to count in one's native language.

We hope these introductory remarks will help the reader to appreciate the broad and deep importance of the rich empirical and theoretical results reported in the following chapters.

REFERENCES

Brenner, N., Bialek, W., & de Ruyter van Steveninck, R. (2000). Adaptive rescaling maximizes information transmission. *Neuron, 26*, 695–702.

Carey, S. (2010). *The origin of concepts.* New York: Oxford University Press.

Frege, G. (1884). *Die Grundlagen der Arithmetik: eine logisch-mathematische Untersuchung über den Begriff der Zahl.* Breslau: Wilhelm Koebner.

Gallistel, C. R. (2011). Mental magnitudes. In S. Dehaene, & L. Brannon (Eds.), *Space, time and number in the brain: Searching for the foundations of mathematical thought* (pp. 3–12). New York: Elsevier.

Gallistel, C., & Gelman, R. (1992). Preverbal and verbal counting and computation. *Cognition, 44*, 43–74.

Gallistel, C. R., & King, A. P. (2010). *Memory and the computational brain: Why cognitive science will transform neuroscience.* New York: Wiley/Blackwell.

Gelman, R. (1972). The nature and development of early number concepts. *Advances in Child Development and Behavior, 7*, 115–167.

Gelman, R. (1990). First principles organize attention to and learning about relevant data: Number and animate–inanimate distinction as examples. *Cognitive Science, 14*, 79–106.

Gelman, R. (1993). A rational-constructivist account of early learning about numbers and objects. *Learning and Motivation, 30*, 61–96.

Gelman, R., & Gallistel, C. R. (1978). *The child's understanding of number.* Cambridge, Mass: Harvard University Press.

Gelman, R., & Williams, E. M. (1998). Enabling constraints for cognitive development and learning: Domain specificity and epigenesis. In D. Kuhn, & R. S. Siegler (Eds.), *Cognition, perception and language: Vol. 2. Handbook of child psychology* (6th ed., pp. 575–630). New York: John Wiley and Sons.

Kline, M. (1972). *Mathematical thought from ancient to modern times.* New York: Oxford University Press.

Knopp, K. (1952). *Elements of the theory of functions.* New York: Dover.

Livingstone, M. S., Pettine, W. W., Srihasam, K., Moore, B., Morocz, I. A., & Lee, D. (2014). Symbol addition by monkeys provides evidence for normalized quantity coding. *Proceedings of the National Academy of Sciences, 111*, 6822–6827.

Walsh, V. (2003). A theory of magnitude: Common cortical metrics of time, space and quantity. *Trends in Cognitive Sciences, 7*, 483–488.

Weber, H. (1893). Leopold Kronecker. *Mathematische Annalen, 43*, 1–25.

Wigner, E. (1960). The unreasonable effectiveness of mathematics in the natural sciences. *Communications in Pure and Applied Mathematics, 13*(1).

Wilkes, J. T. (2014, under review). *Weber's law from first principles.*

Zur, O., & Gelman, R. (2004). Doing arithmetic in preschool by predicting and checking. *Early Childhood Quarterly Review, 19*, 121–137.

Preface

The current volume is the first in a series that will provide a broad survey of cutting edge research in mathematical cognition and learning. Three decades ago there were only a handful of scientists focusing on these topics, most of them studying school children and adults, but only a few working with other species. Since that time and the realization that non-verbal primates (including human infants) are capable of recognizing numerosities, the field has seen enormous growth in the number of researchers, the variety of organisms being studied, and the range of topics under investigation. The latter range from the sensitivity of various species of insect to relative magnitude to the brain systems supporting mathematical problem solving, to large-scale interventions focused on improving the achievement of students who struggle with mathematics. The depth and breadth of the field has accelerated our understanding of how people and other animals represent and understand numerical and mathematical information and how to best teach this information, but has come at the cost of the dissemination of findings across an exceedingly wide range of scientific journals. It is now nearly impossible to keep up with and integrate these advances. As a result, scholars, students, and practitioners are often aware of only specific subsets of knowledge within the broader field. This is where the current volume and the series more generally makes its contribution; specifically, to bring together and integrate some of the best work on key topics in the field of mathematical cognition and learning.

The evolution and early development of number systems seemed the logical place to start. To wit, we have assembled many of the most established and promising young scientists studying how non-human species and human infants and toddlers represent and understand non-symbolic (e.g., the relative magnitude of a collection of 3 items vs. a collection of 6 items) and symbolic (e.g., the quantity represented by the Arabic numeral '3') number and magnitude. Although edited books on animal cognition or early (human) cognitive development often include chapters on numerical cognition, none have pulled together the latest work in these areas into a single volume. We do so in this book. Across the chapters, the reader will find an integrative approach to the topic, providing not only a comprehensive treatment of relevant research with non-human species and human infants but also a thorough yet comprehensible coverage of the prevailing methodological approaches used by leading researchers in the field.

We anticipate that this volume and subsequent ones will be of interest to researchers, graduate students and undergraduates specializing in cognitive development, infant cognition, cognitive neuroscience, behavioral genetics, educational psychology, psycholinguistics, early childhood education, special education, and many other domains. In other words, although the chapters herein are focused on number, they represent a model approach for studying the evolution and here-and-now expression of brain and cognition, and extending this to the study of educational issues in the modern world. This volume could be used as a textbook for several kinds of courses taught in psychology (e.g., comparative cognition, animal cognition, and infant cognition), general education, and cognitive neuroscience programs.

As implied by the title, the scope is broad and deep, touching on the evolved functions of number systems, their potential evolutionary conservation across a wide variety of species, and how these competencies are expressed in non-human species and preverbal humans. Many of the chapters discuss the methods that are best used to study these competencies in infants, and several link these to specific brain regions and to children's learning of school-based mathematics. The latter is intriguing as it suggests that these evolved number systems might provide the foundation upon which some aspects of school-based mathematics are built and thus have broad implications not only for more fully understanding mathematical achievement but also for studying how non-evolved symbolic competencies can be built from evolved competencies more generally.

We have organized the volume into three sections. In the first section are chapters that review number and magnitude systems in non-human species. Across these chapters, the reader is provided with cutting edge reviews of number competencies in fish, birds, and primates. A striking feature of these studies is the cross-species similarities in number competencies, despite the clear diversity of species covered across these chapters. Perhaps more striking, or at least less intuitive to people unfamiliar with this research, are the similarities between the number competencies of non-human animals and those of preverbal humans; and, in some cases, the similarities in their brain systems that are sensitive to number and magnitude. The chapters in the second section focus on these similarities, as well as infants' and young children's sensitivity to and understanding of number and magnitude more generally, and tie these to the learning of symbolic mathematics (e.g., cardinality, understanding the quantities represented by Arabic numerals). The reader will be provided with up-to-date reviews of the different ways in which infants and young children can represent number and magnitude, the implicit competencies built into the organization of these systems (e.g., cardinal value and sensitivity to additions and subtractions from collections of items), and the cues (e.g., sequence of tones) that shift attention to the numerical and magnitude features of the world. The third section addresses related theoretical issues, including the nature of the core system for numerical representations and how to measure it, the contexts in which number systems evolved biologically and historically across human cultures, as well as a few thoughts on directions for future research.

We thank the Child Development and Behavior Branch of the *Eunice Kennedy Shriver* National Institute of Child Health and Human Development, NIH, for the primary funding of the conference on which this volume is based, Dr. Bruce Fuchs, former Director of the NIH's Office of Science Education, whose supplementary support for this meeting was critical to its success, and likewise Dr. Deborah Speece, former Commissioner of the Institute of Education Science's (IES) National Center for Special Education Research and Dr. John Easton, Director of IES, for their supplementary funding of the conference. We are also grateful to Marcy Baughman and her colleagues at Pearson for their generous support of the meals at the conference.

David C. Geary
Daniel B. Berch
Kathleen Mann Koepke

Part I

Number and Magnitude in Non-Human Animals

Chapter 1

At the Root of Math: Numerical Abilities in Fish

Christian Agrillo, Maria Elena Miletto Petrazzini and Angelo Bisazza
Department of General Psychology, University of Padova, Italy

INTRODUCTION

Laboratory and field studies have suggested that numerical abilities may be useful for mammals and birds in several ecological contexts. For instance, chimpanzees (*Pan troglodytes*) and dogs (*Canis lupus familiaris*) are more willing to enter social contexts when their group outnumbers that of opponents (Bonanni, Natoli, Cafazzo, & Valsecchi, 2011; Wilson, Britton, & Franks, 2002). Numerical abilities may also be important for anti-predator strategies. It has been shown that redshanks (*Tringa totanus*) are less likely to be predated when part of a larger flock, highlighting the importance of selecting the larger group of social companions (Cresswell, 1994). Foraging decisions represent another context in which numerical abilities may be useful, with robins (*Petroica australis*, Hunt et al., 2008) and parrots (*Psittacus erithacus*, Al Aïn, Giret, Grand, Kreutzer, & Bovet, 2009) being able to select the larger amount of food when different alternatives are available. Also, numerical skills can provide benefits in mate choice: the ability to count the number of conspecifics encountered seems to enable bank voles (*Myodes glareolus*) to adaptively adjust their reproductive strategy to the level of sperm competition (i.e., males produce more sperm when multiple other mates are present; Lemaitre, Ramm, Hurst, & Stockley, 2011). Indeed, laboratory studies controlling for non-numerical cues have demonstrated that many species of mammals (chimpanzees: Beran, McIntyre, Garland, & Evans, 2013, Beran et al., this volume; dogs: West & Young, 2002; rats [*Rattus norvegicus*]: Davis & Bradford, 1986) and birds (parrots: Pepperberg, 2012, this volume; chicks [*Gallus gallus domesticus*]: Rugani, Regolin, & Vallortigara, 2008; Vallortigara, this volume) can solve most of these tasks by using numerical information only.

Similar selective pressures in favor of processing quantitative information are likely to have also acted in species more distantly related to humans, such

Mathematical Cognition and Learning, Vol. 1. http://dx.doi.org/10.1016/B978-0-12-420133-0.00001-6

as fish. It has been demonstrated that several fish species tend to join the larger shoal when exploring a potentially dangerous environment (Hager & Helfman, 1991; Svensson, Barber, & Forsgren, 2000), a context in which the probability of being spotted by predators decreases with increasing shoal size (Brown, Laland, & Krause, 2011). The capacity to select the larger or smaller group may provide benefits in resource competition as well: for instance, when food is available, banded killifish (*Fundulus diaphanus*) form smaller shoals, probably to reduce the competition for food resources (Hoare, Couzin, Godin, & Krause, 2004). The successful rate of some predators (e.g., wolf-fish [*Hoplias malabaricus*], acara cichlids [*Aequidens pulcher*], and pike cichlids [*Crenicichla frenata*]) is related to the shoal size of their prey, and it was demonstrated that they often ignore a single near prey in order to attack a shoal containing a larger number of individuals (Botham, Kerfoot, Louca, & Krause, 2005). Lastly, in guppies (*Poecilia reticulata*) and mosquitofish (*Gambusia holbrooki*), the capacity to discriminate the sex-ratio of social groups is thought to allow males to adjust their reproductive strategies to the existing level of sperm competition (Lindstrom & Ranta, 1993; Smith & Sargent, 2006).

In the aforementioned examples, fish might use multiple strategies to discriminate the larger or smaller quantity. For example, they can base their discrimination on continuous variables that co-vary with numerosity such as the sum of the areas occupied by the items to enumerate, the overall space occupied by the entire set, or the density of objects in space; these are all cues to determine shoal size in natural contexts. Recent studies have been investigating these issues in teleost fish through controlled laboratory experiments (see Table 1-1).

OVERVIEW

In this chapter, we will first summarize methodological issues related to the study of numerical abilities in fish. In the second section, we will focus on the relation between numerical and non-numerical cues during numerical discrimination of fish (see also Lourenco, this volume). Debate surrounding the existence of single vs. multiple systems of numerical representation will be summarized in the third section, followed by an overview of the results of studies on the ontogeny of numerical abilities of fish. In the final section, we will examine inter-specific differences among teleost species and provide a comparison of the mathematical abilities of fish with warm-blooded vertebrates.

PROBLEMS AND METHODS OF STUDY

The investigation of numerical abilities in fish presents the same methodological problems that are encountered in the study of nonsymbolic numerical abilities in infants, mammals, and birds (Starr & Brannon, this volume).

TABLE 1-1 Summary of fish species in which some capacity to discriminate quantities (without experimental control of continuous quantities) is found, and species in which the use of numerical information has been demonstrated (with experimental control of continuous quantities)

Species (alphabetic order)	Quantity Discrimination (Number + Continuous quantities)	Numerical Discrimination
Angelfish	Gómez-Laplaza & Gerlai, 2011a, 2011b, 2012, 2013	Agrillo, Miletto Petrazzini, Tagliapietra, & Bisazza, 2012b
Banded killifish	Krause & Godin, 1994	—
Blue acara	Krause & Godin, 1995	—
Central mudminnow	Jenkins & Miller, 2007	—
Climbing perch	Binoy & Thomas, 2004	—
Fathead minnow	Hager & Helfman, 1991	—
Goldbelly topminnow	Agrillo & Dadda, 2007	—
Golden shiner	Reebs & Sauliner, 1997	—
Guppies	Agrillo, Piffer, Butterworth, & Bisazza, 2012c; Ledesma & McRobert, 2008; Piffer, Agrillo, & Hyde, 2012	Agrillo, Miletto Petrazzini, Tagliapietra, Bisazza, 2012b; Agrillo, Miletto Petrazzini, & Bisazza, 2014; Bisazza, Serena, Piffer, & Agrillo, 2010
Mosquitofish	Agrillo et al., 2007, 2008b	Agrillo, Dadda, Serena, & Bisazza, 2009; Agrillo, Piffer, & Bisazza, 2010; 2011, Agrillo, Miletto Petrazzini, Tagliapietra, & Bisazza, 2012b; Dadda, Piffer, Agrillo, & Bisazza, 2009
Red-bellied piranha	Queiroz & Magurran, 2005	—
Redtail splitfin	Stancher et al., 2013	Agrillo, Piffer, Bisazza, & Butterworth, 2012
Sailfin molly	Bradner & McRobert, 2001	—

Continued

TABLE 1-1 Summary of fish species in which some capacity to discriminate quantities (without experimental control of continuous quantities) is found, and species in which the use of numerical information has been demonstrated (with experimental control of continuous quantities)— Cont'd

Species (alphabetic order)	Quantity Discrimination (Number + Continuous quantities)	Numerical Discrimination
Siamese fighting fish	Snekser, McRobert, & Clotfelter, 2006	Agrillo, Miletto Petrazzini, Tagliapietra, & Bisazza, 2012b
Swordtail	Buckingham, Wong, & Rosenthal, 2007	—
Three-spined sticklebacks	Barber, Downey, & Braithwaite, 1998; Tegeder & Krause 1995	—
Two-spotted goby	Svensson et al. 2000	—
Wolf fish	Botham & Krause, 2005	—
Zebrafish	Pritchard, Lawrence, Butlin, & Krause, 2001	Agrillo, Miletto Petrazzini, Tagliapietra, & Bisazza, 2012b

The most obvious problem is the lack of verbal abilities in fish, which forces researchers to infer the existence of numerical abilities from subjects' behavior. Furthermore, we have all been able to discriminate between two quantities without necessarily counting the number of objects in everyday life, such as when looking at fruit baskets or two queues at the airport. Numerosity typically co-varies with other physical attributes, and fish can use the relative magnitude of non-numerical cues (hereafter "continuous quantities"), such as the cumulative surface area of the stimuli, the overall space occupied by the sets, or their density, to estimate which set is larger/smaller (Agrillo et al., 2008b, 2009; Gómez-Laplaza & Gerlai, 2013). For this reason, controlled experiments are necessary to assess whether fish can use discrete numbers, continuous quantities, or both types of information to determine relative quantities.

Methodologies Adopted

Different methods have been adopted in comparative psychology to study numerical abilities of mammals (Beran, 2006; Beran & Parrish, 2013;

Jordan, MacLean, & Brannon, 2008) and birds (Al Aïn et al., 2009; Hunt et al., 2008; Rugani et al., 2008; Rugani, Fontanari, Simoni, Regolin, & Vallortigara, 2009), including spontaneous choice tests, training procedures, habituation–dishabituation procedures, and expectancy violation procedures. In fish research, only the first two methodological approaches have been used thus far. The spontaneous choice paradigm consists of presenting the subjects with two groups of biologically relevant collections of objects differing in numerosity, with the assumption that if they are able to discriminate the two quantities, they will spontaneously select the most profitable one. With a conditioning paradigm, subjects undergo extensive training in which neutral stimuli of a given numerosity are repeatedly associated with a reward, and where the capacity to learn a numerical rule is taken as evidence of numerical abilities.

Spontaneous Choice Tests

Studies of mammals and birds have commonly used the spontaneous preference for larger over smaller quantities of food to study numerical abilities. Preference for a different number of social companions has been generally studied in fish. Many fish live in social aggregates such as shoals, and research has shown that in these species a single individual that happens to be in an unknown environment tends to join other conspecifics and, if two shoals are present, it exhibits a preference for the larger one (Buckingham et al., 2007; Hager & Helfman, 1991; Pritchard et al., 2001). This is thought to be an antipredatory strategy aimed at reducing the chance of being captured by predators (Hamilton, 1971). Actually, the risk of an individual prey to be caught diminishes as the quantity of individuals in the shoal increases, a phenomenon known as the "dilution effect" (Foster & Treherne, 1978). Larger shoals may also become more effective at detecting predators ("many eyes effect") and make it more difficult for a predator to single out an individual prey ("confusion effect"; see Landeau & Terborgh, 1986). Several studies have taken advantage of this strong tendency to select the larger shoal to assess the limits of quantity discrimination (Buckingham et al., 2007; Gómez-Laplaza & Gerlai, 2011a, b). A subject is typically moved into an unfamiliar tank where it sees two groups of social companions, differing in number. The proportion of time spent near the larger shoal is taken as a measure of discrimination (Figure 1-1a).

Using this procedure, Agrillo, Dadda, Serena, and Bisazza (2008) demonstrated that mosquitofish can discriminate between shoals differing by one unit, up to 4 items (1 vs. 2, 2 vs. 3, and 3 vs. 4, but not 4 vs. 5). Tested with a similar procedure, a closely related species, the guppy, exhibited the same limit (Agrillo, Piffer, Bisazza, & Butterworth, 2012), while angelfish (*Pterophyllum scalare*) were found to discriminate 1 vs. 2 and 2 vs. 3 but not 3 vs. 4 fish (Gómez-Laplaza & Gerlai, 2011b). Quantities larger than 4 could also be discriminated but only by increasing the numerical distance (i.e., the

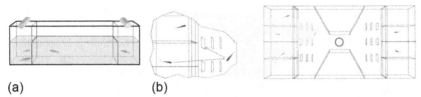

(a)   (b)

FIGURE 1-1   Methods used to assess spontaneous numerical abilities in fish. (a) Shoal-choice apparatus is composed of three adjacent tanks: subjects are singly inserted in the central tank while stimulus fish (e.g., 1 vs. 2) are placed into two opposite tanks. The proportion of time spent by the subjects near the larger shoal is taken as a measure of their quantity abilities (dependent variable). (b) A modification of the experimental apparatus is used to assess whether fish can use numerical information only: subjects are singly inserted in a central hourglass-shaped tank. Stimulus tanks are subdivided into several small compartments, each housing a stimulus fish. In each choice area, artificial vertical green plastic screens are aligned in a grid of 3 by 3, so that the subjects can see only one stimulus fish at a time from any position of its tank. In this way, fish cannot solve the task using non-numerical variables, such as cumulative surface area occupied by stimulus fish. *[Adapted by Dadda et al., 2009.]*

size ratio) between the shoals: mosquitofish could discriminate 8 vs. 16 (0.50 ratio) but not a smaller distance (8 vs. 12, 0.67 ratio) and their accuracy decreased with increasing numerical ratio, in agreement with Weber's law (i.e., discrimination is dependent on the ratio of the compared groups, not the absolute difference). Similarly, guppies (Agrillo, Piffer, Bisazza, & Butterworth, 2012) and swordtails (*Xiphophorus hellerii*, Buckingham et al., 2007) were able to discriminate 0.50 ratio, and their accuracy increased as the numerical ratio between the smaller and larger numbers increased. Ratio dependence for larger quantities has also been found in angelfish; this species, however, proved able to discriminate a 0.56 ratio (Gómez-Laplaza & Gerlai, 2011a).

Some differences between studies might be explained by differences in social motivation, as in some conditions fish might judge two large shoals as equivalent options even if they can perceive the small numerosity difference between them. We must note that no study investigated spontaneous quantity discrimination with visual contact of predators, a condition that might potentially increase subjects' motivation to select the larger shoal even when the numerical distance between the stimulus shoals is small.

A modification of the aforementioned procedure was used to test quantitative abilities with regard to mate choice. Tested in a binary choice, male mosquitofish were found to select the larger group of females in a comparison between 1 vs. 3 and 2 vs. 4 (Agrillo, Dadda, & Serena, 2008). When the same number of females was available, one with males and the other without, male mosquitofish preferred the latter. However, when the number of males per group was varied (2 males + 3 females vs. 4 males + 3 females), no preference was observed for the more favorable sex ratio, suggesting that male mosquitofish cannot simultaneously assess the quantity of males and females within a shoal.

In all the aforementioned studies, no attempt was made to control for continuous quantities that co-vary with number, and as a result, it was not possible to establish the exact mechanism adopted by fish to select the larger quantity. An experimental strategy to assess which mechanism is used by fish consists of controlling one continuous quantity at a time. For example, to assess the influence of cumulative surface area, Agrillo, Dadda, Serena, and Bisazza (2008) studied spontaneous preference of mosquitofish in 2 vs. 3 and 4 vs. 8 discriminations in which larger individuals were placed in the smaller shoals such that the overall area occupied by stimulus fish was the same for each numerical contrast. Interestingly, when the overall area of the stimuli was matched, mosquitofish had no preference for 2 vs. 3 or 4 vs. 8, suggesting the importance of cumulative surface area in their shoal choices. In 2013, Gómez-Laplaza and Gerlai studied whether the ability of angelfish to select the larger shoal (5 vs. 10 discrimination) is affected by the density of the two stimulus shoals. To keep this variable constant, the volume of the stimulus compartment was reduced by 50% for the smaller of the two stimulus shoals: angelfish were unable to select the larger shoal in this condition, indicating a spontaneous use of density to discriminate shoals.

The quantity of movement of stimulus fish represents another factor that might indirectly reveal the numerosity of a shoal. To control for this cue, researchers have used two different procedures. The first one consists of varying the water temperature of the stimulus aquaria. Most fish can live in a wide range of temperatures, and their activity is directly influenced by water temperature such that higher temperatures typically result in increased activity. This technique was originally adopted by Pritchard et al. (2001) to study shoal choice in zebrafish (*Danio rerio*). The authors found that when the two stimulus shoals were in water of the same temperature, subjects generally preferred the larger shoal in a 4 vs. 2 discrimination. However, this preference declined when colder water was used to reduce the activity of the larger shoal. Agrillo, Dadda, Serena, and Bisazza (2008) adopted the same technique with mosquitofish and showed that the total activity of conspecifics affects shoal choice in a 4 vs. 8 but not in a 2 vs. 3 discrimination. Subsequently, Gómez-Laplaza and Gerlai (2012) found similar results in angelfish: total activity of conspecifics influences the capacity to discriminate in the large (5 vs. 10) but not in the small (2 vs. 3) number range (see Section 3 for a definition of small and large number ranges; vanMarle, this volume).

A second procedure to control for the quantity of movements consists of confining each stimulus fish in single sectors that restricts movement so that the quantity of movement in the two shoals can be equated. Using this procedure, Gómez-Laplaza and Gerlai (2012) found that angelfish can discriminate not only 2 vs. 3 but also 5 vs. 10. The two studies on angelfish thus found opposite outcomes depending on which strategy was used to control for total movement of the shoal. This result highlights the importance of using multiple experimental strategies to control for continuous quantities before drawing a firm conclusion about the influence of continuous quantities on performance in numerical tasks.

Although the preceding examples highlight the possibility of controlling for continuous quantities in spontaneous choice experiments, this procedure may introduce other confounding factors. For example, when one controls for body size of stimuli, one must consider that fish often prefer to shoal with similar-sized individuals in order to minimize their distinctiveness in case of predatory attack (Landeau & Terborgh, 1986; Wong & Rosenthal, 2005). Similarly, controlling for movement may decrease the attractiveness of larger shoals, because in some species immobility is a strategy for avoiding predator detection ("freezing"; Chivers & Smith, 1994, 1995). Thus, many static individuals may be interpreted as an alarm cue on that side of the tank. In this sense, albeit interesting, the lack of preference for the larger shoals reported in these studies cannot permit us to draw firm conclusions about fish ability to process numerical information nor about the exact role of each continuous quantity.

Another possible strategy to reduce the use of continuous variables consists of simultaneously presenting two groups of conspecifics differing in numerosity, but—at the time of choice—the two groups are made visually unavailable because two opaque partitions are lowered in front of them. Using this procedure, Stancher, Sovrano, Potrich, and Vallortigara (2013) found that males of redtail splitfin (*Xenotoca eiseni*) are able to spontaneously select the larger number of females in 1 vs. 2, 2 vs. 3, and 3 vs. 4. No preference was found with larger numerosities. Interestingly, similar performance was reported with a 5-second delay and with a 30-second delay, showing surprisingly good working memory in these fish. However, this procedure does not entirely exclude the possibility that animals use continuous quantities, as subjects can potentially compare the different areas occupied by stimuli when they are visible and then remember the position occupied by the larger amount prior to its disappearance.

Some of the preceding limitations can be overcome by using another experimental strategy called "item-by-item presentation." This technique, previously adopted with mammals (e.g., chimpanzees: Beran et al., 2013; rhesus monkeys [*Macaca mulatta*]: Hauser, Carey, & Hauser, 2000; elephants [*Loxodonta africana*]: Perdue, Talbot, Stone, & Beran, 2012) and birds (New Zealand robins: Hunt, Low, & Burns, 2008; domestic chicks: Rugani et al., 2009), is based on the sequential presentation of items within each set so that subjects can never have a global view of the entire contents of the groups. To solve the task, animals have to attend to each item and to build a representation of the contents of the set on the basis of the items that come sequentially into view. Then they have to repeat the process for the second set and, finally, compare the two representations (Beran, 2004; Beran & Beran, 2004).

The technique was recently applied to assess whether mosquitofish have a spontaneous number representation when choosing between two groups of social companions (Dadda et al., 2009). A modified version of the shoal-choice test was used: each stimulus fish was confined in a separate compartment of the stimulus tanks, and several opaque screens were inserted in the subject tank to prevent the possibility of subjects seeing more than one stimulus fish at a time (Figure 1-1b). In this way, subjects were required to add the number of stimuli on one side, make the same operation on the other side, and then compare the two numerosities in order to select the optimal shoal. Results showed that mosquitofish spent more time near the larger shoal in 2 vs. 3 and 4 vs. 8. Further experiments controlled for density of individuals and the overall space occupied by the stimulus shoals ruled out the possibility that fish used these cues to choose the larger shoal.

Training Procedure

Spontaneous choice tests present two main limitations. First, motivation plays a key-role and null results do not necessarily imply a lack of discrimination. For instance, if a fish does not show a significant preference in a 12 vs. 16 discrimination, this might be due to the fact that subjects are simply not motivated to select the larger shoal because both groups provide seemingly equivalent protection from predators. Second, although several experimental procedures have been used to control for continuous quantities, such controls remain difficult when living organisms (such as social companions) are used as stimuli. As an alternative, classical and operant conditioning have often been adopted in fish research, as previously done with mammals (Beran, 2006; Davis, 1984) and birds (Koehler, 1951; Rugani et al., 2008).

Studies using training procedures commonly require that the subjects learn a numerical rule in order to receive a reward. In fish studies, both social and food reward were used. Agrillo et al. (2009) trained mosquitofish to discriminate between 2 and 3 bidimensional figures using social reward. Subjects were singly inserted into an unfamiliar environment provided with two doors, one associated with three objects and the other associated with two objects, placed at two opposite corners. To rejoin social companions, subjects were required to discriminate between the two numerosities and select the door associated with the reinforced numerosity (Figure 1-2a). In the first experiment, fish were trained to discriminate between 2 and 3 with no control for continuous quantity. Subjects were then retested while controlling for one continuous quantity at a time: their performance dropped to chance level when stimuli were matched for the cumulative surface area or for the overall space occupied by the arrays, indicating that these latter cues had been spontaneously used by the fish to solve the task. In a second experiment, where all continuous quantities were controlled, mosquitofish proved able to learn the discrimination, suggesting a pure use of numerical information.

A follow-up study (Agrillo et al., 2010) using the same methodology investigated whether there is an upper limit in fish numerical discrimination and whether the performance of trained fish is influenced by numerical ratio.

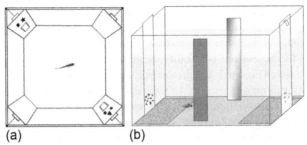

(a)   (b)

FIGURE 1-2   Two training procedures adopted by numerical cognition studies in fish. (a) Training procedure using social reward: subjects are singly placed in the middle of a test chamber provided with two doors—one associated to the reinforced numerosity (e.g., 3) and one to the nonreinforced one (2)—placed at two opposite corners. Only the reinforced door allows subjects to rejoin shoal mates in the outer tank (not shown). Proportion of correct choices is taken as dependent variable. (b) Training procedure using food reward: each subject is housed in an experimental tank for the entire experiment. At intervals, stimuli (two groups of figures differing in numerosity) are presented on the two short walls, and food is delivered near the reinforced stimulus. Learning is measured as a proportion of time spent near the reinforced stimulus in probe trials done in extinction.

To achieve the first goal, researchers initially trained fish to discriminate between quantities with a 0.5 numerical ratio and similar total numerosity (4 vs. 8, 5 vs. 10, and 6 vs. 12) and then, in the test phase, presented with three numerical contrasts, again with a 0.50 ratio, but differing in the total numerosity (4 vs. 8, 15 vs. 30, and 100 vs. 200 figures). To assess the influence of numerical ratio, researchers again trained mosquitofish to discriminate a 0.5 ratio; in the test phase, they were presented with three novel numerical contrasts that had similar total numerosity (7 vs. 14, 8 vs. 12, and 9 vs. 12) but different numerical ratios (0.50, 0.67, and 0.75, respectively). Results showed that fish have virtually no upper limit of numerical discrimination, at least with a 0.5 ratio, as they discriminate 4 vs. 8 as well as 100 vs. 200, while the numerical ratio had a clear effect on performance (that is, fish successfully discriminated 0.50 and 0.67 ratio, but they were less accurate in the latter condition; no discrimination was found instead in 0.75 ratio).

Recently, a different training procedure was devised using food as reward. Agrillo, Miletto Petrazzini, Piffer, Dadda, and Bisazza (2012) placed mosquitofish individually in rectangular tanks. Two stimuli (sets of bidimensional figures of different numerosity) were simultaneously introduced at opposite ends and food was provided near the stimulus to be rewarded. The proportion of time spent in probe trials near the previously rewarded stimulus was taken as a measure of discrimination performance (Figure 1-2b). Results obtained with the new procedure paralleled those reported by Agrillo et al. (2010): numerical discrimination of mosquitofish had no upper limit while their accuracy was influenced by the numerical ratio of the stimuli (Agrillo, Miletto Petrazzini, Piffer, Dadda, & Bisazza, 2012). Because the two studies used a different type of reward—social (Agrillo et al., 2010) vs. food (Agrillo, Miletto Petrazzini, Piffer, Dadda, & Bisazza, 2012)—we conclude that the limits exhibited by mosquitofish in both studies are likely to reflect the general limits of their numerical abilities and not the result of motivational factors associated with the use of one type of reward. Recently, this procedure has been adopted to study numerical abilities in other fish species (Agrillo, Miletto Petrazzini, Tagliapietra, & Bisazza, 2012b; Agrillo et al., 2014) and in very young guppies (Miletto Petrazzini, Agrillo, Piffer, & Bisazza, 2014).

Concerning the control of continuous quantities, different strategies have been adopted with the two types of procedures. In some experiments (Agrillo et al., 2009, 2010), cumulative surface area was controlled by using proportionally smaller figures in the more numerous set. However, as a by-product of this type of control, smaller-than-average figures were more frequent in the larger group, and mosquitofish could have used this non-numerical cue to discriminate between the two groups. To rule out this possibility, at the end of the experiment, researchers presented a control set in which figures had the same size for all stimuli. In other studies, a different type of control for cumulative surface area was performed (Agrillo, Miletto Petrazzini, Piffer, Dadda, & Bisazza, 2012; Agrillo, Miletto Petrazzini, Tagliapietra, & Bisazza, 2012; Agrillo et al., 2014). For one-third of the stimuli used in the training phase, the two numerosities were 100% equated for cumulative surface area; for another third, the cumulative surface area was controlled to 85%; and, in the remaining third, it was controlled to 70%. The biggest figure within each pair was shown half the time in the larger set and half in the smaller set. In each probe trial, cumulative surface area was matched to 100%. As a consequence, neither individual figure size (an unreliable cue in the training phase) nor cumulative surface area (matched to 100% in probe trials) could be used to discriminate between the two quantities. Overall, the results from these training studies indicate that while fish often use continuous quantity to make magnitude-based choices, they are also able to make these choices based on discrete quantity.

NUMBER VS. CONTINUOUS QUANTITIES: IS NUMBER MORE COGNITIVELY DEMANDING?

As described in the previous section, experiments done with spontaneous-choice tests and training procedures showed that fish can use both discrete (numerical) and continuous quantities when required to make relative numerosity judgments. Some of these studies suggest that when both numerical and continuous quantities are available, continuous quantities are often used as primary cues, as reported in other vertebrates (Gebuis & Reynvoet, 2012; Kilian, Yaman, Fersen, & Güntürkün, 2003; Pisa & Agrillo, 2009). Shoal-choice experiments have shown that mosquitofish easily select the larger shoal in 2 vs. 3 and 4 vs. 8, but, as soon as the cumulative surface area was controlled, their performance dropped to chance level, suggesting a spontaneous use of the area (Agrillo, Dadda, Serena, & Bisazza, 2008). Similarly, when the quantity of movement of social companions was controlled, both mosquitofish (Agrillo, Dadda, Serena, & Bisazza, 2008) and angelfish (Gómez-Laplaza & Gerlai, 2012) were unable to select the larger shoal in 4 vs. 8 (mosquitofish) and 5 vs. 10 (angelfish).

The scenario is similar for studies using the training procedure. The performance of mosquitofish, initially trained to discriminate between two sets of figures (2 vs. 3) that did not control for continuous quantity, dropped to chance level when stimuli were matched for the cumulative surface area and the overall space occupied by the figures. These two continuous quantities are likely to have been spontaneously used during the learning process. A similar result was observed with larger numbers, with mosquitofish being unable to discriminate between 4 and 8 figures when cumulative surface area was equated (Agrillo et al., 2010). In both studies, fish learned to use numerical information only when all continuous variables were simultaneously controlled for from the beginning of the training (Agrillo et al., 2009, 2010).

These results apparently indicate that, as previously advanced in other species (Davis & Memmott, 1982; Davis & Perusse, 1988; Kilian et al., 2003), fish may use numerical information as a "last resort" strategy, raising the question of whether processing numerical information is more cognitively demanding than processing continuous quantities. However, it must be noted that these studies involved comparisons of trials in which number was confounded with continuous quantities to trials in which continuous quantities were controlled (Agrillo et al., 2009; Gómez-Laplaza & Gerlai, 2012). Adopting such an experimental design, these studies are comparing subjects that could use one type of information only (numerical) with subjects that could use numerical information plus one or more continuous attributes of the stimulus. To specifically test if number is more cognitively demanding, Agrillo, Piffer, and Bisazza (2011) used a training procedure to compare the number of trials required for mosquitofish to learn a discrimination based on number only as compared to one based on continuous quantities only. In short, fish were trained to discriminate 2 from 3 in three different conditions. In one condition, fish could use only numerical information to distinguish between the quantities (2 vs. 3 figures); in a second condition, fish could use only continuous quantities (1 vs. 1 discrimination, 2/3 ratio between the areas), while in the third condition both number and continuous information were available (2 vs. 3 figures and 2/3 ratio between the areas). If number was more difficult to process, more learning trials would be needed when only numerical information was available compared to the conditions in which continuous quantities were available.

Mosquitofish allowed to simultaneously use numerical and continuous information needed fewer trials to reach the criterion and were generally more accurate compared with the condition in which only continuous or only

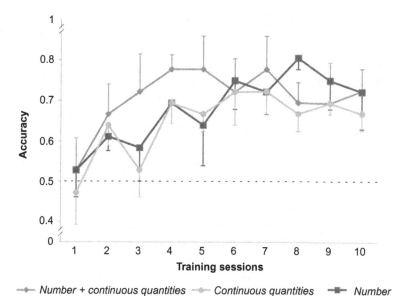

FIGURE 1-3 Results of an experiment measuring learning rate in three conditions, "number only," "continuous quantities only," and "number+continuous quantities" (Agrillo et al., 2011). Fish learned to discriminate more quickly when both number and continuous quantities were provided simultaneously. No difference in the learning rate was observed between "number only" and "continuous quantities only," suggesting that number is not more cognitively demanding than other non-numerical attributes of the stimuli. *[Adapted by Agrillo et al., 2011.]*

numerical information could be used (Figure 1-3). In line with this finding, a recent study in another species showed that newborn guppies could not discriminate a 0.5 ratio either when only numerical information was available or when only continuous quantities could be used, while they successfully discriminated when both types of information were available (Miletto Petrazzini et al., 2014). Therefore a combination of number and continuous quantities seems to be the easiest condition; consistent with previous literature showing that redundant information presented either in the same sensory modality or across modalities facilitates learning in human and nonhuman species (Lickliter, Bahrick, & Markham, 2006; Neil, Chee-Ruiter, Scheier, Lewkowicz, & Shimojo, 2006; Suanda, Thompson, & Brannon, 2008). After all, as outlined by Gebuis and Reynvoet (2012), the combination of number and continuous quantities is unlikely to be violated in nature: more pieces of food normally occupy more area and larger shoals typically have larger cumulative surfaces. In this sense, it is not surprising that fish, as well as other species, have evolved mechanisms that exploit multiple cues to estimate quantities.

Interestingly, Agrillo et al. (2011) found no difference in the learning rate between mosquitofish trained to use numerical information only and those trained to use continuous quantity only, suggesting that processing number is not more complex than processing continuous quantities for mosquitofish. This result apparently conflicts with previous evidence showing that mosquitofish rely on number only after continuous quantities were controlled (Agrillo, Dadda, Serena, & Bisazza, 2008; Agrillo et al., 2009, 2010). If number and continuous quantities require the same cognitive effort, why do fish often exhibit a spontaneous use of continuous quantities? A potential explanation is that, while there is just one numerical dimension, stimuli may differ for several continuous quantities (overall space occupied by the array, cumulative area or contour of stimuli, density, brightness, etc.), and there is a larger probability that the most salient cue for a given type of stimuli will be a non-numerical one. Agrillo et al. (2009) found in mosquitofish that individuals that attended to the cumulative surface area were not affected by the overall space occupied by the arrays and vice versa, suggesting that each individual preferentially relies on one cue, probably the one that is particularly salient for it.

ONE OR MULTIPLE SYSTEMS OF NUMERICAL REPRESENTATION?

Several studies in cognitive psychology have demonstrated that humans use two types of processes to enumerate without verbal counting (Chesney & Haladjian, 2011): subitizing and estimation (vanMarle, this volume). The former is a fast and accurate process (Revkin, Piazza, Izard, Cohen, & Dehaene, 2008) that appears to be based on an evolutionarily ancient mechanism, the object tracking system (OTS), that allows individuals to track some (usually up to 4) objects in parallel, even if these are moving in space and undergo brief periods of occlusion (Assad & Maunsell, 1995; Scholl & Pylyshyn, 1999; Trick & Pylyshyn, 1994). For this reason, our capacity to subitize does not usually exceed 3–4 items. The second process permits us to estimate the numerosity of larger quantities (Kaufman, Lord, Reese, & Volkmann, 1949). This process is thought to be supported by an approximate number system (ANS) that is fast but less accurate (Nieder & Dehaene, 2009). Its accuracy depends on the numerical ratio between the numerosities to be compared. The lack of a ratio effect is a distinctive signature that allows experimental differentiation of the OTS from the ANS (Feigenson, Dehaene, & Spelke, 2004): our performance is very similar when discriminating 3 vs. 4 or 1 vs. 4 objects, whereas we are much more accurate in discriminating 5 from 20 objects than 15 from 20 objects.

To date, comparative psychologists disagree as to whether in nonhuman species a single system, the ANS, accounts for discrimination over the whole numerical range, or a distinct system, the OTS, operates over the small number range. Several studies support the single-system hypothesis (ANS over the whole numerical range) in mammals (Barnard et al., 2013; Beran, 2004, 2007;

Cantlon & Brannon, 2007) and birds (Al Aïn et al., 2009; Rugani, Regolin, & Vallortigara, 2013), but other studies indirectly support the two-system hypothesis (mammals: Bonanni et al., 2011; Hauser et al., 2000; Utrata, Virányi, & Range, 2012; birds: Hunt et al., 2008).

Recently, this issue was also addressed with regard to fish. In support of the two-system hypothesis, Agrillo, Dadda, Serena, and Bisazza (2008) found that mosquitofish discriminate between small shoals that differ in numerosity up to 3:4 ratio (3 vs. 4), but they succeed with only up to a 1:2 numerical ratio (4 vs. 8 or 8 vs. 16 but not 6 vs. 8) when they have to discriminate between large numerosities. Similarly, two separate studies (Gómez-Laplaza & Gerlai, 2011a, b) showed that angelfish can discriminate up to 2 vs. 3 while the same ratio is not discriminated in the large number range (4 vs. 6). In short, those studies showed that numerical acuity of fish is different within and outside the subitizing range. A recent study provides another confirmation of this phenomenon (Agrillo, Piffer, Bisazza, & Butterworth, 2012). Guppies were tested in shoal choice with the same five numerical ratios (0.25/0.33/0.50/0.67/0.75) both within the small number range (1–4) and beyond it. The discrimination of two large numbers increased in accuracy as the difference in the ratio between them increased, but the performance was the same for all contrasts involving small numbers.

Some researchers have argued that a different ratio dependence does not necessary imply the existence of two separate systems (Gallistel & Gelman, 1992; Van Oeffelen & Vos, 1982; Vetter, Butterworth, & Bahrami, 2008). Gallistel and Gelman (1992), for example, pointed out that the representation of larger numbers would be more variable and, as a consequence, representation of nearest values may overlap in the large number range, leading to a ratio effect only in the large number range. Another explanation for the different performance in the two numerical ranges is that, unlike large sets, small sets of items usually generate recognizable geometric patterns (i.e., 1 item = a dot; 2 items = a line; 3 items = a triangle), the so-called pattern recognition hypothesis (Mandler & Shebo, 1982; Neisser, 1967).

An alternative way to test the existence of multiple numerical systems is by looking at factors other than ratio dependency in the two numerical ranges. For example, a potential prediction of the single-system hypothesis is that manipulation of physical properties of the stimuli should have similar effects on the estimation of small and large numbers, whereas the alternative hypothesis predicts that the two systems may be differentially affected by the type of stimuli to be enumerated. In humans, moving versus static items represent one of the variables that seem to affect numerical estimation differently. For example, Trick, Audet, and Dales (2003) found that even very slow motion reduced enumeration speed for stimuli containing 6–9 items, while the enumeration of 1–4 items was unaffected. Similarly, Alston and Humphreys (2004) found a faster and more accurate enumeration in the subitizing range in the presence of items in motion compared to static items.

Agrillo et al. (2014) investigated if the motion of the items had an influence on large- and small-number discrimination in fish. Here, guppies were initially trained to discriminate an easy numerical ratio (0.50), both within (2 vs. 4) and outside (6 vs. 12) the subitizing range: half of the subjects were trained in the presence of moving stimuli; the other half was trained in the presence of static stimuli. In the test phase, fish were presented with a more difficult numerical ratio (0.75), in both the small- (3 vs. 4) and the large-number (9 vs. 12) ranges. Guppies successfully discriminated 0.50 in both conditions (static items/items in motion). When the ratio became more difficult (0.75), items in motion were successfully discriminated in the subitizing range (3 vs. 4) but not with the large-number discrimination (9 vs. 12). This is in line with the idea that the OTS might be particularly activated by items in motion in the subitizing range. These results also challenge the "pattern recognition" hypothesis, as general configuration of those items was constantly changing, and no stable pattern could be recognized.

Other data are also consistent with distinct cognitive systems to process small and large numbers in fish. In shoal-choice experiments on mosquitofish, small- and large-number discriminations are differentially affected by continuous quantities (2 vs. 3: cumulative surface area; 4 vs. 8: cumulative surface area and quantity of movements of stimulus fish, Agrillo et al., 2008b). The continuous quantities used by trained mosquitofish (Agrillo et al., 2009, 2010) to discriminate between sets of figures differed in the two numerical ranges (2 vs. 3: cumulative surface area and overall space occupied by the two sets; 4 vs. 8: cumulative surface area only). In guppies, the developmental trajectory was found to differ for small- and large-number discrimination (Bisazza et al., 2010): the spontaneous ability to discriminate between small quantities of conspecifics is displayed at birth, while this ability appears at 20–40 days old for quantities beyond 4 units (see Section 4). Another prediction of the two-system hypothesis is interference and thus reduced performance when comparing numerosities across the small and large ranges. This has been confirmed by studies with infants (Feigenson & Carey, 2005; Xu, 2003). Recently, this hypothesis has been tested in guppies using the shoal-choice paradigm. The study found that guppies also fail to discriminate across the small number/large number divide (Piffer et al., 2012).

In sum, data from fish research generally support differences in performance below and above 4 units, suggesting the existence of two separate systems for processing small and large numbers, even though we must acknowledge that none of these studies provide direct evidence in favor of a subitizing-like process in fish. This evidence, however, raises the intriguing question about why the results from fish studies support the two-system hypothesis while most of the studies on mammals and birds are more in line with the single-system hypothesis. The possibility exists that there are differences across species in the numerical systems devoted to small number processing: the precision of OTS in some species, for instance, might result from different evolutionary pressures, such as the different need to track social companions in their ecological niches (Vonk & Beran, 2012). Alternatively, it was recently hypothesized that the ANS may be recruited in some cases to represent numbers in the OTS range, and the context (nature of stimuli, sensory modality, procedure, expertise in magnitude estimation) in which the representation is elicited may determine which system is activated in the small number range (Agrillo & Piffer, 2012; Cordes & Brannon, 2008; Hyde, 2011; vanMarle, this volume). If activation of either OTS or the ANS in the 1 to 4 range depends on contextual variables, the different results reported among mammals, birds, and fish might be at least partially ascribed to the different experimental methods adopted.

ONTOGENY OF NUMERICAL ABILITIES

There is evidence that our nonsymbolic numerical abilities are present at birth and increase in precision during development, before the emergence of language (reviewed in Cordes & Brannon, 2008). Newborns are able to discriminate a 0.33 ratio (4 vs. 12) but not a 0.5 ratio (4 vs. 8, Izard, Sann, Spelke, & Streri, 2009); 6-month-old infants can discriminate numerosities with a 0.5 ratio (such as 8 vs. 16) but not a 0.67 ratio (Xu & Spelke, 2000), whereas 10-month-old infants are able to discriminate numerosities with a 0.67 but not a 0.8 ratio (Xu & Arriaga, 2007). The resolution of numerical systems continues to increase throughout childhood, with 6-year-olds being able to discriminate a 0.83 ratio and adults even a ratio of 0.9 (Halberda, Mazzocco, & Feigenson, 2008). However, it is always difficult to compare the results of different studies on this topic: very different paradigms are used across the different ages, and devising experimental paradigms that are simultaneously applicable to newborns, infants, toddlers, and adults is problematic. In addition, practical and ethical reasons prevent us from manipulating experience during development in primates, which precludes the possibility of disentangling the relative contribution of maturation and experience.

Little is known about numerical abilities in nonadult animals, with the notable exception of research on domestic chickens and, of course, human infants. Studies conducted using both spontaneous-choice tests (Rugani et al., 2009; Rugani, Regolin, & Vallortigara, 2011) and training procedures (Rugani, Regolin, & Vallortigara, 2007; Rugani et al., 2008) showed that chicks already display numerical abilities in the first week of life (for further information, see the chapter by Vallortigara in this book). However, because no study has been done on the adults of this species, we have no information about the development of numerical abilities in nonhuman species.

Recently, the ontogeny of numerical abilities has been investigated in fish. Bisazza et al. (2010) used guppies as a model: this species is a livebearer with a relatively short lifespan giving birth to fully developed offspring that are independent and display a full social repertoire. In this sense, they represent a suitable model for examining the developmental trajectories of numerical abilities. Experiments were done using a small-scale version of the experimental apparatuses adopted to study spontaneous quantity discrimination in adult fish (Gómez-Laplaza & Gerlai, 2011a; Piffer et al., 2012). At birth, the capacity of guppies to discriminate between shoals differing by one individual included all numerical contrasts in the range 1–4 (1 vs. 2, 2 vs. 3, 3 vs. 4), but not contrasts involving larger quantities such as 4 vs. 5 and 5 vs. 6. That is, the capacity of newborn guppies to discriminate between small numerosities is the same as that of adults (Agrillo, Piffer, Bisazza, & Butterworth, 2012). Conversely, unlike adults, one-day-old fish were unable to select the larger group in 4 vs. 8 (0.50 ratio) and 4 vs. 12 (0.33) comparisons, suggesting that large quantity discrimination is absent or highly approximate at birth.

In a subsequent experiment, Bisazza et al. (2010) investigated the development of large quantity discrimination (e.g., 4 vs. 8). To assess the role of social experience, in one experimental condition, the authors reared fish in large shoals with the possibility to experience naturally occurring subgroups of variable numerosities, while in another condition, fish were reared in pairs, thus preventing them from seeing more than one fish at a time. The authors found that, at 40 days (approximately the onset of secondary sexual characteristics), fish from both treatments discriminated 4 from 8 fish, while at 20 days this capacity was observed only in fish reared in large shoals.

This study also included an experiment similar to that reported in adult fish using the "item by item" procedure (Dadda et al., 2009). Newborn guppies proved able to discriminate 2 vs. 3 by using numerical information only; similarly, juvenile fish (40-day-old guppies) discriminated 4 vs. 8 when prevented from using continuous quantities (Bisazza et al., 2010).

In a recent study, Miletto Petrazzini et al. (2014) investigated whether newborn fish can be trained to discriminate between different groups of inanimate objects by using numerical information only. Using a modification of the procedure adopted for adults (Agrillo, Miletto Petrazzini, Piffer, Dadda, & Bisazza, 2012, see Figure 1-2b), the authors trained 4- to 9-day-old fish with food reward to discriminate between groups of dots differing in numerosity comparing two conditions: number and continuous quantities or number only. Subjects rapidly learned to discriminate 2 from 4 items under the first condition, but failed to discriminate in the number-only condition even with a very easy discrimination (1 vs. 4 items). The comparison with the results of the shoal-choice test (Bisazza et al., 2010) indicates that newborns' capacity to use number is specific to social context. Although the possibility exists that attentional bias may partially explain the different results reported in the two studies, developmental differences in the capacity to make numerical discriminations in social contexts vs. collections of inanimate objects suggest separate systems underlying these discriminations in social vs. nonsocial contexts. In this sense, data on guppies are suggestive of the existence of multiple quantification mechanisms in fish that are domain-specific and serve to solve a limited set of problems, in agreement with the idea advanced by Penn, Holyoak, and Povinelli (2008) according to which several cognitive abilities in nonhuman animals would be mainly restricted to the ecological contexts in which they evolved.

SIMILARITIES IN NUMERICAL ABILITIES AMONG FISH AND BETWEEN FISH AND OTHER VERTEBRATES

After two decades of comparative research, it is now clear that all vertebrates possess one or more numerical abilities, and it seems time to take stock and ask whether all species share the same numerical systems, or if numerical abilities have independently evolved multiple times during vertebrate evolution (termed "convergent evolution"). Unfortunately, despite the large amount of published data, the answer is still unclear.

Some studies report similar performance in distantly related species. First, numerical discrimination obeys Weber's law in most species, especially in the large number range (see "One or Multiple Systems of Numerical Representation?" earlier). The universality of Weber's law reinforces the hypothesis of a similar numerical system. Additional evidence in favor of shared numerical abilities among vertebrates is a similar numerical acuity reported in animals as diverse as macaques (Hauser et al., 2000), New Zealand robins (Hunt et al., 2008), guppies (Agrillo, Piffer, Bisazza, & Butterworth, 2012), and mosquitofish (Agrillo, Dadda, Serena, & Bisazza, 2008), with all of these species being spontaneously able to discriminate between 1 vs. 2, 2 vs. 3, and 3 vs. 4, but not 4 vs. 5. Other studies, however, have highlighted inter-specific differences. For example, horses (*Equus caballus*), domestic chicks, salamanders (*Plethodon cinereus*), and angelfish differ from these other species, being able to discriminate up to 2 vs. 3, with no discrimination of 3 vs. 4 (Gómez-Laplaza & Gerlai, 2011b; Rugani et al., 2008; Uller, Jaeger, Guidry, & Martin, 2003; Uller & Lewis, 2009). Such differences are potentially compatible with the idea of similar numerical systems that vary in their precision across species.

In other studies, however, differences in performance are more evident. Trained chimpanzees are able to discriminate 8 vs. 10 items (Tomonaga, 2008); trained pigeons (*Columba livia*), 6 vs. 7 items (Emmerton & Delius, 1993); and trained dogs, 6 vs. 8 (MacPherson & Roberts, 2013); while, with similar procedures, mosquitofish, guppies, and domestic chicks showed a limit of 2 vs. 3 items (Agrillo, Miletto Petrazzini, Piffer, Dadda, & Bisazza, 2012; Agrillo et al., 2014; Rugani et al., 2008). Sometimes, wide differences have been reported even between closely related species; the accuracy of African elephants in spontaneous choice tasks, for example, is affected by the numerical ratio (Perdue et al., 2012) while the accuracy of Asian elephants (*Elephas maximus*) appears to be insensitive to this variable (Irie-Sugimoto, Kobayashi, Sato, & Hasegawa, 2009).

The inconsistencies reported in literature might be attributed in part to differences in methods, such as different procedures, stimuli, numerical contrasts, and/or assessed sensory modality. In some cases, there is direct evidence that different methods lead to different results. For instance, gold-belly topminnows (*Girardinus falcatus*) have been shown to discriminate up to 2 vs. 3 with one experimental procedure while they were unable to solve the same task using a different procedure (Agrillo & Dadda, 2007). Similarly, mosquitofish can discriminate between 3 and 4 social companions in free-choice tests (Agrillo, Dadda, Serena, & Bisazza, 2008); whereas they are unable to discriminate the same numerical contrast when geometrical figures are used as stimuli (Agrillo, Miletto Petrazzini, Piffer, Dadda, & Bisazza, 2012).

To circumvent these problems, we need cross-species studies using the same methodology: the kinds of studies that have been done almost exclusively on primates (Beran, 2006; Cantlon & Brannon, 2006, 2007; Hanus & Call, 2007). Agrillo, Miletto Petrazzini, Tagliapietra, and Bisazza (2012) investigated whether different fish species have the same numerical acuity by testing guppies, zebrafish, angelfish, redtail splitfin, and Siamese fighting fish (*Betta splendens*) with the same apparatus, stimuli, and procedure. Subjects initially were trained on an easy numerical ratio (0.50: 5 vs. 10 and 6 vs. 12). Once they reached the learning criterion, their ability to generalize to more difficult ratios (0.67 and 0.75) or to a larger (25 vs. 50) or a smaller (2 vs. 4) total set size was tested. Results showed a similar pattern of performance across species. No species was able to discriminate the 0.75 ratio (9 vs. 12); all species but angelfish were able to discriminate the 0.67 ratio (8 vs. 12). The pattern of generalization of the numerical rule to a different set size was also very similar across the different species. Most species generalized the learned discrimination to a smaller set size (2 vs. 4), while no species was able to generalize to a larger set size (25 vs. 50). One species, zebrafish, showed a lower performance compared to other species. However, a control experiment showed that the observed difference resulted from the zebrafish's difficulty in learning the training procedure rather than from a cross-species variation in the numerical domain. On the whole, similarities were greater than differences, suggesting the existence of similar numerical systems. From a phylogenetic point of view, the five species studied are distantly related. According to recent estimates, some of these fish belong to groups that diverged more than 250 million years ago (Steinke, Salzburger, & Meyer, 2006). Since they also encompass a broad spectrum of ecological specializations, with some species being highly social and others essentially solitary, some inhabiting open areas and others densely vegetated shallow waters, some providing parental care and others not, the observation of so few differences seems more in accord with the existence of ancient quantification systems inherited from a common ancestor than by convergent evolution, that is, the independent evolution of numerical abilities in each species.

Other studies have tried to compare fish with distantly related vertebrates. Agrillo, Piffer, Bisazza, and Butterworth (2012) compared numerical abilities of fish with nonsymbolic numerical abilities of adult humans. College students and guppies were tested with five numerical ratios (Figure 1-4) for both small (1 vs. 4, 1 vs. 3, 1 vs. 2, 2 vs. 3, and 3 vs. 4) and large (6 vs. 24, 6 vs. 18, 6 vs. 12, 6 vs. 9, 6 vs. 8) numerical contrasts. Students were tested with a procedure commonly used to measure nonsymbolic numerical abilities in adults, a computerized numerical judgment with sequential presentation of the stimuli (e.g., a group of 6 dots followed by a group of 12 dots), executed under verbal suppression (verbally repeating "abc"). Fish were tested in a classical shoal-choice test. Interestingly, the results of fish aligned with that of humans; in both species, the ability to discriminate between large numbers (>4) was approximate and strongly dependent on the ratio between the numerosities. Conversely, in both fish and humans, discrimination in the small number range was ratio-independent (i.e., discriminating 3 from 4 was as easy as discriminating 1 from 4).

In this study, however, different types of stimuli were used for the two species (dots for humans and social companions for fish). In another study, humans and fish were presented with the same stimuli (groups of figures differing in numerosity): undergraduates were required to estimate the larger of two groups of figures in a computerized task, and mosquitofish were trained to discriminate between the same stimuli (Agrillo et al., 2010). Again, the performance of mosquitofish was consistent with that of humans, with accuracy decreasing as ratio increased and with no influence of total set size in either humans or mosquitofish (i.e., 4 vs. 8 and 100 vs. 200 items were discriminated with the same cognitive effort).

The tasks in which the greatest differences between fish and warm-blooded vertebrates are observed are those involving training procedures, as previously advanced by Penn et al. (2008). In the aforementioned study, for

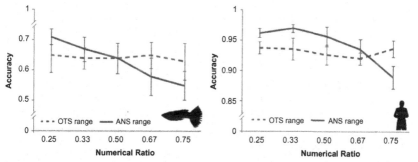

FIGURE 1-4 A comparison of the performance of guppies and college students tested with the same numerical ratios (Agrillo, Miletto Petrazzini, Piffer, Dadda, & Bisazza, 2012). Both species showed ratio insensitivity for large numbers (ANS range) and ratio insensitivity in the small number range (OTS range). *[Adapted by Agrillo, Miletto Petrazzini, Piffer, Dadda, & Bisazza, 2012.]*

example, mosquitofish did not significantly discriminate 9 vs. 12 elements while human capacities go well beyond this ratio. In experiments with operant conditioning, chimpanzees and macaques can easily discriminate 8 vs. 10 (Cantlon & Brannon, 2007; Tomonaga, 2008), and pigeons can discriminate up to 6 vs. 7 elements (Emmerton & Delius, 1993), while the limit thus far recorded for fish is 3 vs. 4 items, in spontaneous choice tests and with training procedures but only when using moving stimuli (Agrillo, Dadda, Serena, & Bisazza, 2008; Agrillo, Piffer, Bisazza, & Butterworth, 2012; Agrillo et al., 2014). It is worth noting, however, that studies done with fish usually employ few reinforced trials (typically 20–30) while studies done on other species often involve several hundred reinforced trials (Beran, 2006; Brannon, Wusthoff, Gallistel, & Gibbon, 2011). It is possible that at least part of our knowledge gap would be filled by studying fish with similarly extensive training procedures.

Of course, more investigation is required in this field, also encompassing other vertebrate species. The studies described here, nonetheless, have revealed interesting similarities among fish and between fish and humans tested nonverbally, opening the possibility of shared numerical systems even among distantly related species. The similar performance between human and fish is especially worth noting as it raises the intriguing possibility that the foundation of our numerical abilities might be evolutionarily more ancient than previously thought (see also Starr & Brannon, this volume), dating back at least as far as the divergence between fish and land vertebrates, which occurred approximately 450 million years ago. Of course, as with most comparative data, it is potentially possible that similarities in numerical abilities are the product of convergent evolution and that similar performance reflects dissimilar underlying mechanisms. In this sense, phylogenetic analyses of the species investigated are necessary to help to determine if these numerical abilities evolved independently or from a single common ancestor.

The literature so far examined indicates that surprisingly fish (which are provided with a much smaller brain than warm-blooded vertebrates) can often rival even our primate relatives in processing numerical information and solving complex numerical tasks. The results of this research are consistent with a recent discovery that bony fish possess several other cognitive abilities that were previously believed to be uniquely present in species with large, complex brains (i.e., mammalian and avian species). For instance, teleost fish have been shown to recognize up to 40 familiar individuals, cooperate to achieve a common goal, learn new habits from experienced conspecifics, use tools, and have cultural traditions (reviewed in Bisazza, 2010; Brown & Laland, 2003; Bshary, Wickler, & Fricke, 2002).

One possible explanation for this surprising finding could be related to a genetic advantage enjoyed by teleost fish during their early evolution. One of the major discoveries of the genomic era is the presence of thousands of ancient duplicated genes in teleost genomes that are present as single-copy genes in humans and in the other land vertebrates. There is now substantial evidence that these differences are due to a whole-genome duplication event that occurred between 400 and 450 million years ago after the separation of the ray-finned fishes (which are composed 99% by teleost fishes) and the lobe-finned fishes, the lineage leading to the tetrapods (amphibians, reptiles, birds, and mammals), but before the massive diversification of the teleosts (reviewed in Meyer & Schartl, 1999). Although many duplicated genes were secondarily lost, many others were freed up to evolve novel functions, and many authors have suggested that the increased genetic complexity of the teleosts might be the reason for their evolutionary success and astonishing biological diversity (Volff, 2005; Wittbrodt, Meyer, & Schartl, 1998). A high proportion of the duplicate genes of the teleosts were found to be somehow related to neural development or brain functions (Steinke et al., 2006; Winkler, Schafer, Duschl, Schartl, & Volff, 2003), thus raising the possibility that the high level of genomic plasticity also favored the evolution of novel and more complex cognitive functions in this group of vertebrates.

Recently, Schartl et al. (2013) have tested this hypothesis by studying in teleosts the rate of duplicate gene retention in a list of 190 genes that are assumed (from human and mouse studies) to be involved in core cognitive functions or in brain development. They found a much higher duplicate retention rate for putative cognition-related genes compared with the genes known to have other functions, thus supporting the idea that this early evolutionary event may have actually promoted the appearance in this vertebrate group of complex behaviors and sophisticated cognitive abilities, including the notable numerical abilities that have been so far demonstrated.

CONCLUSIONS AND FUTURE DIRECTIONS

Until a few years ago, no study had investigated the cognitive mechanisms underlying quantity discrimination in fish. During the past few years, nonetheless, a large body of experimental evidence has been accumulated, allowing us to draw at least preliminary conclusions about the numerical abilities and underlying cognitive mechanisms in these species. Through carefully controlled experiments, at least six diverse species (guppies, mosquitofish, zebrafish, angelfish, redtail splitfin, and Siamese fighting fish) have been shown to process numerical information. These capacities appear to be similar among fish, and in general they parallel those reported in other vertebrates. In partial contrast to the literature in other species, results of fish studies are suggestive of the existence of multiple systems of numerical representation which serve to solve a limited set of problems. Some of these seem to be displayed at birth, whereas others appear later, as a consequence of maturation and experience.

Of course, several topics still need to be investigated. First, even from a quick glimpse of the literature (Table 1–1), it appears clear that most studies have investigated quantity discrimination in tasks where number and continuous quantities were confounded. Further investigation is required to assess whether the capacity to use pure numerical information is restricted to a few species or instead—as suggested by the inter-specific study of Agrillo, Miletto Petrazzini, Tagliapietra, and Bisazza (2012)—is a cognitive skill shared by all fish. To better understand the evolution of the cognitive systems underlying numerical abilities of fish, we should conduct research on a wider range of distantly related species. To date, only near-shore, freshwater teleosts have been investigated; there have been no studies of the numerical abilities of saltwater or pelagic teleosts or of cartilaginous fish.

Again, so far all fish studies have been restricted to the visual modality. In our own species, most studies suggest a single amodal ANS. However, Tokita, Ashitani, and Ishiguchi (2013) reported performance differences in numerosity judgments tested in visual and auditory conditions, advancing the idea of multiple core number systems in which visual and auditory numerosities are mentally represented with different signal variabilities. The possibility exists that the numerical acuity of fish is modality-dependent. Future studies are needed to test this hypothesis. Partially related to this topic, studies investigating the capacity to match visual and auditory stimuli on the only basis of numerosity—the so-called cross-modal interaction of numerical information—are also needed, of the kind recently performed on primates (Jordan, Brannon, Logothetis, & Ghazanfar, 2005; Jordan et al., 2008). Finally, no study has investigated the relation between numerical and other magnitude abilities in fish. Several studies with humans (e.g., Agrillo & Piffer, 2012; Bueti & Walsh, 2009; Vicario, 2011) suggest a common magnitude system for nonsymbolic estimation of time, space, and number, the so-called ATOM (A Theory of Magnitude, Walsh, 2003; see also Lourenco, this volume). Recent evidence also supports the existence of a single magnitude system in nonhuman primates (Haun, Jordan, Vallortigara, & Clayton, 2010; Mendez, Prado, Mendoza, & Merchant, 2011; Merritt, Cassanto, & Brannon, 2010). Studies investigating the validity of ATOM in distantly related species, such as fish, may help us to assess whether this common magnitude system is a recent evolutionary development of the primate lineage or rather is a common feature among vertebrates.

As a last note, we believe that the study of numerical abilities of fish might play a key role in the near future by providing a broader framework for considering the factors underlying the acquisition of our own mathematical abilities. Recently, correlational studies (Furman & Rubinsten, 2012; Piazza et al., 2010) found that nonsymbolic numerical systems are less accurate in individuals with deficits in mathematics learning. Nonsymbolic and symbolic numerical abilities were also found to be causally related in a study by Park and Brannon (2013), which is in agreement with the idea that although the construction of symbolic numbers depends on processes that are culture-dependent, they are nevertheless rooted in nonsymbolic numerical systems

(Halberda et al., 2008). We are aware that the exact relation between symbolic and nonsymbolic numerical abilities is still not clear (see De Smedt, Noel, Gilmore, & Ansari, 2013). However, if the causal link between nonsymbolic and symbolic numerical abilities should be confirmed, animal models may become fundamental in increasing our knowledge of human cognitive systems. The whole genome of zebrafish and of a half dozen fish species is already sequenced, and several other species are currently being investigated. Considering the suggestions that vertebrates might share the same nonsymbolic numerical systems (Beran, 2008; Feigenson et al., 2004) and that dyscalculia might have a heritable origin (Butterworth & Laurillard, 2010), the use of a model such as zebrafish may help us to understand the biological origin of nonsymbolic numerical abilities. In this sense, studying the quantitative abilities of fish may help us not only to assess the evolutionary origin of nonsymbolic numerical abilities but also to better understand the foundation of our own mathematical skills.

ACKNOWLEDGMENTS

We would like to thank the editors for inviting us to make this contribution. Financial support was provided by FIRB grant *"Futuro in ricerca 2013"* (prot.: RBFR13KHFS) from Ministero dell'Istruzione, Università e Ricerca (MIUR, Italy) to Christian Agrillo.

REFERENCES

Agrillo, C., & Dadda, M. (2007). Discrimination of the larger shoal in the poeciliid fish, *Girardinus falcatus*. *Ethology Ecology & Evolution, 19*, 145–157.

Agrillo, C., Dadda, M., & Bisazza, A. (2007). Quantity discrimination in female mosquitofish. *Animal Cognition, 10*, 63–70.

Agrillo, C., Dadda, M., & Serena, G. (2008a). Choice of female groups by male mosquitofish (*Gambusia holbrooki*). *Ethology, 114*(5), 479–488.

Agrillo, C., Dadda, M., Serena, G., & Bisazza, A. (2008b). Do fish count? Spontaneous discrimination of quantity in female mosquitofish. *Animal Cognition, 11*(3), 495–503.

Agrillo, C., Dadda, M., Serena, G., & Bisazza, A. (2009). Use of number by fish. *PLoS One, 4*(3), e4786.

Agrillo, C., Miletto Petrazzini, M. E., & Bisazza, A. (2014). Numerical acuity of fish is improved in the presence of moving targets, but only in the subitizing range. *Animal Cognition, 17*(2), 307–316.

Agrillo, C., Miletto Petrazzini, M. E., Piffer, L., Dadda, M., & Bisazza, A. (2012a). A new training procedure for studying discrimination learning in fishes. *Behavioral Brain Research, 230*, 343–348.

Agrillo, C., Miletto Petrazzini, M. E., Tagliapietra, C., & Bisazza, A. (2012b). Inter-specific differences in numerical abilities among teleost fish. *Frontiers in Psychology, 3*(483). http://dx.doi.org/10.3389/fpsyg.2012.00483.

Agrillo, C., & Piffer, L. (2012). Musicians outperform non-musicians in magnitude estimation: Evidence of a common processing mechanism for time, space and numbers. *Quarterly Journal of Experimental Psychology, 65*(12), 2321–2332.

Agrillo, C., Piffer, L., & Bisazza, A. (2010). Large number discrimination by fish. *PLoS One*, *5*(12), e15232.

Agrillo, C., Piffer, L., & Bisazza, A. (2011). Number versus continuous quantity in numerosity judgments by fish. *Cognition*, *119*, 281–287.

Agrillo, C., Piffer, L., Bisazza, A., & Butterworth, B. (2012c). Evidence for two numerical systems that are similar in humans and guppies. *PLoS One*, *7*(2), e31923.

Al Aïn, S., Giret, N., Grand, M., Kreutzer, M., & Bovet, D. (2009). The discrimination of discrete and continuous amounts in African grey parrots (*Psittacus erithacus*). *Animal Cognition*, *12*, 145–154.

Alston, L., & Humphreys, G. W. (2004). Subitization and attentional engagement by transient stimuli. *Spatial Vision*, *17*, 17–50.

Assad, J. A., & Maunsell, J. H. R. (1995). Neuronal correlates of inferred motion in primate posterior parietal cortex. *Nature*, *373*, 518–521.

Barber, I., Downey, L. C., & Braithwaite, V. A. (1998). Parasitism, oddity and the mechanism of shoal choice. *Journal of Fish Biology*, *53*(6), 1365–1368.

Barnard, A. M., Hughes, K. D., Gerhardt, R. R., DiVincenti, L., Jr., Bovee, J. M., & Cantlon, J. F. (2013). Inherently analog quantity representations in olive baboons (*Papio anubis*). *Frontiers in Psychology*, *4*(253). http://dx.doi.org/10.3389/fpsyg.2013.00253.

Beran, M. J. (2004). Chimpanzees (*Pan troglodytes*) respond to nonvisible sets after one-by-one addition and removal of items. *Journal of Comparative Psychology*, *118*, 25–36.

Beran, M. J. (2006). Quantity perception by adult humans (*Homo sapiens*), chimpanzees (*Pan troglodytes*), and rhesus macaques (*Macaca mulatta*) as a function of stimulus organization. *International Journal of Comparative Psychology*, *19*, 386–397.

Beran, M. J. (2007). Rhesus monkeys (*Macaca mulatta*) enumerate large and small sequentially presented sets of items using analog numerical representations. *Journal of Experimental Psychology: Animal Behavior Processes*, *33*, 42–54.

Beran, M. J. (2008). The evolutionary and developmental foundations of mathematics. *PLoS Biology*, *6*, e19.

Beran, M. J., & Beran, M. M. (2004). Chimpanzees remember the results of one-by-one addition of food items to sets over extended time periods. *Psychological Science*, *15*, 94–99.

Beran, M. J., McIntyre, J. M., Garland, A., & Evans, T. A. (2013). What counts for "counting"? Chimpanzees (*Pan troglodytes*) respond appropriately to relevant and irrelevant information in a quantity judgment task. *Animal Behaviour*, *85*, 987–993.

Beran, M. J., & Parrish, A. E. (2013). Visual nesting of stimuli affects rhesus monkeys' (*Macaca mulatta*) quantity judgments in a bisection task. *Attention, Perception, & Psychophysics*, *75*, 1243–1251.

Binoy, W., & Thomas, K. J. (2004). The climbing perch (*Anabas testudineus Bloch*), a freshwater fish, prefers larger unfamiliar shoals to smaller familiar shoals. *Current Science*, *86*(1), 207–211.

Bisazza, A. (2010). Cognition. In J. Evans, A. Pilastro, & I. Schlupp (Eds.), *Ecology and evolution of poeciliid fishes* (pp. 165–173). Chicago: Chicago University Press.

Bisazza, A., Serena, G., Piffer, L., & Agrillo, C. (2010). Ontogeny of numerical abilities in guppies. *PLoS One*, *5*(11), e15516.

Bonanni, R., Natoli, E., Cafazzo, S., & Valsecchi, P. (2011). Free-ranging dogs assess the quantity of opponents in intergroup conflicts. *Animal Cognition*, *14*(1), 103–115.

Botham, M., Kerfoot, C., Louca, V., & Krause, J. (2005). Predator choice in the field: Grouping guppies, *Poecilia reticulata*, receive more attacks. *Behavioral Ecology and Sociobiology*, *59*, 181–184.

Botham, M. S., & Krause, J. (2005). Shoals receive more attacks from the wolf-fish (*Hoplias malabaricus Bloch*, 1794). *Ethology*, *111*(10), 881–890.

Bradner, J., & McRobert, S. P. (2001). The effect of shoal size on patterns of body colour segregation in mollies. *Journal of Fish Biology, 59*(4), 960–967.

Brannon, E. M., Wusthoff, C. J., Gallistel, C. R., & Gibbon, J. (2011). Subtraction in the pigeon: Evidence for a linear subjective number scale. *Psychological Science, 12*(3), 238–243.

Brown, C., & Laland, K. N. (2003). Social learning in fishes: A review. *Fish and Fisheries, 4,* 280–288.

Brown, C., Laland, K., & Krause, J. (2011). *Fish cognition and behavior*. London (UK): Wiley-Blackwell.

Bshary, R., Wickler, W., & Fricke, H. (2002). Fish cognition: A primate's eye view. *Animal Cognition, 5,* 1–13.

Buckingham, J. N., Wong, B. B. M., & Rosenthal, G. G. (2007). Shoaling decision in female swordtails: How do fish gauge group size? *Behaviour, 144,* 1333–1346.

Bueti, D., & Walsh, V. (2009). The parietal cortex and the representation of time, space, number and other magnitudes. *Philosophical Transactions of the Royal Society of London. Series B: Biological Sciences, 364,* 1831–1840.

Butterworth, B., & Laurillard, D. (2010). Low numeracy and dyscalculia: Identification and intervention. *ZDM Mathematics Education, 42*(6), 527–539.

Cantlon, J. F., & Brannon, E. M. (2006). Shared system for ordering small and large numbers in monkeys and humans. *Psychological Science, 17*(5), 401–406.

Cantlon, J. F., & Brannon, E. M. (2007). Basic math in monkeys and college students. *PLoS Biology, 5*(12), e328.

Chesney, D. L., & Haladjian, H. (2011). Evidence for a shared mechanism used in multiple-object tracking and subitizing. *Attention, Perception, & Psychophysics, 73,* 2457–2480.

Chivers, D. P., & Smith, R. J. F. (1994). The role of experience and chemical alarm signalling in predator recognition by fathead minnows, *Pimephales promelas. Journal of Fish Biology, 44,* 273–285.

Chivers, D. P., & Smith, R. J. F. (1995). Fathead minnows, *Pimephales promelas*, learn to recognize chemical stimuli from high risk habitats by the presence of alarm substance. *Behavioral Ecology, 6,* 155–158.

Cordes, S., & Brannon, E. M. (2008). Quantitative competencies in infancy. *Developmental Science, 11*(6), 803–808.

Cresswell, W. (1994). Flocking is an effective anti-predation strategy in Redshanks, *Tringa tetanus. Animal Behaviour, 47,* 433–442.

Dadda, M., Piffer, L., Agrillo, C., & Bisazza, A. (2009). Spontaneous number representation in mosquitofish. *Cognition, 112,* 343–348.

Davis, H. (1984). Discrimination of the number three by a raccoon (*Procyon lotor*). *Animal Learning and Behavior, 12,* 409–413.

Davis, H., & Bradford, S. A. (1986). Counting behaviour by rats in a simulated natural environment. *Ethology, 73,* 265–280.

Davis, H., & Memmott, J. (1982). Counting behavior in animals: A critical evaluation. *Psychological Bulletin, 92,* 547–571.

Davis, H., & Perusse, R. (1988). Numerical competence in animals: Definitional issues, current evidence and a new research agenda. *Behavioral and Brain Sciences, 11,* 561–579.

De Smedt, B., Noel, M. P., Gilmore, C., & Ansari, D. (2013). How do symbolic and non-symbolic numerical magnitude processing relate to individual differences in children's mathematical skills? A review of evidence from brain and behaviour. *Trends in Neuroscience and Education, 2,* 48–55.

Emmerton, J., & Delius, J. D. (1993). Beyond sensation: Visual cognition in pigeons. In H. P. Zeigler & H.-J. Bischof (Eds.), *Vision, brain, and behavior in birds* (pp. 377–390). Cambridge, MA: MIT Press.

Feigenson, L., & Carey, S. (2005). On the limits of infants' quantification of small object arrays. *Cognition, 97,* 295–313.

Feigenson, L., Dehaene, S., & Spelke, E. S. (2004). Core systems of number. *Trends in Cognitive Sciences, 8*(7), 307–314.

Foster, W. A., & Treherne, J. E. (1978). Dispersal mechanisms in an inter-tidal aphid. *Journal of Animal Ecology, 47*(1), 205–217.

Furman, T., & Rubinsten, O. (2012). Symbolic and non symbolic numerical representation in adults with and without developmental dyscalculia. *Behavioral and Brain Functions, 8,* 55.

Gallistel, C. R., & Gelman, R. (1992). Preverbal and verbal counting and computation. *Cognition, 44,* 43–74.

Gebuis, T., & Reynvoet, B. (2012). The role of visual information in numerosity estimation. *PLoS One, 7*(5), e37426.

Gómez-Laplaza, L. M., & Gerlai, R. (2011a). Can angelfish (*Pterophyllum scalare*) count? Discrimination between different shoal sizes follows Weber's law. *Animal Cognition, 14,* 1–9.

Gómez-Laplaza, L. M., & Gerlai, R. (2011b). Spontaneous discrimination of small quantities: Shoaling preferences in angelfish (*Pterophyllum scalare*). *Animal Cognition, 14,* 565–574.

Gómez-Laplaza, L. M., & Gerlai, R. (2012). Activity counts: The effect of swimming activity on quantity discrimination in fish. *Frontiers in Psychology, 3,* 484.

Gómez-Laplaza, L. M., & Gerlai, R. (2013). Quantification abilities in angelfish (*Pterophyllum scalare*): The influence of continuous variables. *Animal Cognition, 16,* 373–383.

Hager, M. C., & Helfman, G. S. (1991). Safety in numbers: Shoal size choice by minnows under predatory threat. *Behavioral Ecology, 29,* 271–276.

Halberda, J., Mazzocco, M. M. M., & Feigenson, L. (2008). Individual differences in nonverbal number acuity correlate with maths achievement. *Nature, 455,* 665–668.

Hamilton, W. D. (1971). Geometry for the selfish herd. *Journal of Theoretical Biology, 31,* 295–311.

Hanus, D., & Call, J. (2007). Discrete quantity judgments in the great apes: The effect of presenting whole sets vs. item-by-item. *Journal of Comparative Psychology, 121,* 241–249.

Haun, D. B. M., Jordan, F., Vallortigara, G., & Clayton, N. S. (2010). Origins of spatial, temporal and numerical cognition: Insights from comparative psychology. *Trends in Cognitive Sciences, 14,* 552–560.

Hauser, M. D., Carey, S., & Hauser, L. B. (2000). Spontaneous number representation in semi-free-ranging rhesus monkeys. *Proceedings of the Royal Society of London B: Biological Sciences, 267,* 829–833.

Hoare, D. J., Couzin, I. D., Godin, J. G. J., & Krause, J. (2004). Context-dependent group-size choice in fish. *Animal Behaviour, 67,* 155–164.

Hunt, S., Low, J., & Burns, C. K. (2008). Adaptive numerical competency in a food-hoarding songbird. *Proceedings of the Royal Society of London B: Biological Sciences, 10,* 1098–1103.

Hyde, D. C. (2011). Two systems of non-symbolic numerical cognition. *Frontiers in Human Neuroscience, 5*(150). http://dx.doi.org/10.3389/fnhum.2011.00150.

Irie-Sugimoto, N., Kobayashi, T., Sato, T., & Hasegawa, T. (2009). Relative quantity judgment by Asian elephants (*Elephas maximus*). *Animal Cognition, 12*(1), 193–199.

Izard, V., Sann, C., Spelke, E. S., & Streri, A. (2009). Newborn infants perceive abstract numbers. *Proceedings of the National Academy of Sciences of the United States of America, 106,* 10382–10385.

Jenkins, J. R., & Miller, B. A. (2007). Shoaling behavior in the central mudminnow (*Umbra limi*). *American Midland Naturalist, 158*(1), 226–232.

Jordan, K. E., Brannon, E. M., Logothetis, N. K., & Ghazanfar, A. A. (2005). Monkeys match the number of voices they hear to the number of faces they see. *Current Biology, 15*, 1034–1038.

Jordan, K. E., MacLean, E. L., & Brannon, E. M. (2008). Monkeys match and tally quantities across senses. *Cognition, 108*(3), 617–625.

Kaufman, E. L., Lord, M. W., Reese, T. W., & Volkmann, J. (1949). The discrimination of visual number. *American Journal of Psychology, 62*(4), 498–525.

Kilian, A., Yaman, S., Fersen, L., & Güntürkün, O. (2003). A bottlenose dolphin discriminates visual stimuli differing in numerosity. *Learning and Behaviour, 31*, 133–142.

Koehler, O. (1951). The ability of birds to count. *Bulletin of Animal Behaviour, 9*, 41–45.

Krause, J., & Godin, J. G. J. (1994). Shoal choice in banded killfish: Effects of predation risk, fish size, species composition and size of shoals. *Ethology, 98*, 128–136.

Krause, J., & Godin, J. G. J. (1995). Predator preferences for attacking particular prey group sizes: Consequences for predator hunting success and prey predation risk. *Animal Behaviour, 50*(2), 465–473.

Landeau, L., & Terborgh, J. (1986). Oddity and the confusion effect in predation. *Animal Behaviour, 34*, 1372–1380.

Ledesma, J. M., & McRobert, S. P. (2008). Shoaling in juvenile guppies: The effects of body size and shoal size. *Behavioural Processes, 77*(3), 384–388.

Lemaitre, J. F., Ramm, S. A., Hurst, J. L., & Stockley, P. (2011). Social cues of sperm competition influence accessory reproductive gland size in a promiscuous mammal. *Proceedings of the Royal Society B: Biological Sciences, 278*, 1171–1176.

Lickliter, R., Bahrick, L. E., & Markham, R. G. (2006). Intersensory redundancy educates selective attention in bobwhite quail embryos. *Developmental Science, 9*, 604–615.

Lindstrom, K., & Ranta, E. (1993). Social preferences by male guppies, *Poecilia reticulata*, based on shoal size and sex. *Animal Behaviour, 46*, 1029–1031.

MacPherson, K., & Roberts, W. A. (2013). Can dogs count? *Learning and Motivation, 44*(4), 241–251.

Mandler, G., & Shebo, B. J. (1982). Subitizing: An analysis of its component processes. *Journal of Experimental Psychology, 111*, 1–22.

Mendez, J. C., Prado, L., Mendoza, G., & Merchant, H. (2011). Temporal and spatial categorization in human and non-human primates. *Frontiers in Integrative Neuroscience, 5*(50). http://dx.doi.org/10.3389/fnint.2011.00050.

Merritt, D., Cassanto, D., & Brannon, E. M. (2010). Do monkeys think in metaphors? Representations of space and time in monkeys and humans. *Cognition, 117*, 191–202.

Meyer, A., & Schartl, M. (1999). Gene and genome duplications in vertebrates: The one-to-four (to-eight in fish) rule and the evolution of novel gene functions. *Current Opinion in Cell Biology, 11*(6), 699–704.

Miletto Petrazzini, M. E., Agrillo, C., Piffer, L., & Bisazza, A. (2014). Ontogeny of the capacity to compare discrete quantities in fish. *Developmental Psychobiology, 56*(3), 529–536.

Neil, P. A., Chee-Ruiter, C., Scheier, C., Lewkowicz, D. J., & Shimojo, S. (2006). Development of multisensory spatial integration and perception in humans. *Developmental Science, 9*, 454–464.

Neisser, U. (1967). *Cognitive psychology*. Englewood Cliffs, NJ: Prentice-Hall.

Nieder, A., & Dehaene, S. (2009). Representation of number in the brain. *Annual Review of Neuroscience, 32*, 185–208.

Park, J., & Brannon, E. M. (2013). Training the approximate number system improves math proficiency. *Psychological Science. 24*(10). http://dx.doi.org/10.1177/0956797613482944, online first.

Penn, D. C., Holyoak, K. J., & Povinelli, D. J. (2008). Darwin's mistake: Explaining the discontinuity between human and nonhuman minds. *Behavioral and Brain Sciences, 31*(2), 109–129.

Pepperberg, I. M. (2012). Further evidence for addition and numerical competence by a Grey parrot (*Psittacus erithacus*). *Animal Cognition, 15*, 711–717.

Perdue, B. M., Talbot, C. F., Stone, A., & Beran, M. J. (2012). Putting the elephant back in the herd: Elephant relative quantity judgments match those of other species. *Animal Cognition, 15*, 955–961.

Piazza, M., Facoetti, A., Trussardi, A. N., Berteletti, I., Conte, S., Lucangeli, D., et al. (2010). Developmental trajectory of number acuity reveals a severe impairment in developmental dyscalculia. *Cognition, 116*, 33–41.

Piffer, L., Agrillo, C., & Hyde, D. C. (2012). Small and large number discrimination in guppies. *Animal Cognition, 15*, 215–221.

Pisa, P. E., & Agrillo, C. (2009). Quantity discrimination in felines: A preliminary investigation of the domestic cat (*Felis silvestris catus*). *Journal of Ethology, 27*(2), 289–293.

Pritchard, V. L., Lawrence, J., Butlin, R. K., & Krause, J. (2001). Shoal choice in the zebrafish, *Danio rerio*: The influence of shoal size and activity. *Animal Behaviour, 62*, 1085–1088.

Queiroz, H., & Magurran, A. E. (2005). Safety in numbers? Shoaling behaviour of the Amazonian red-bellied piranha. *Biology Letters, 1*(2), 155–157.

Reebs, S. G., & Sauliner, N. (1997). The effect of hunger on shoal choice in golden shiners (Pisces: Cyprinidae, *Notemigonus crysoleucas*). *Ethology, 103*(8), 642–652.

Revkin, S. K., Piazza, M., Izard, V., Cohen, L., & Dehaene, S. (2008). Does subitizing reflect numerical estimation? *Psychological Science, 19*, 607–614.

Rugani, R., Fontanari, L., Simoni, E., Regolin, L., & Vallortigara, G. (2009). Arithmetic in newborn chicks. *Proceedings of the Royal Society B: Biological Sciences, 276*, 2451–2460.

Rugani, R., Regolin, L., & Vallortigara, G. (2007). Rudimental competence in 5-day-old domestic chicks: Identification of ordinal position. *Journal of Experimental Psychology: Animal Behavior Processes, 33*(1), 21–31.

Rugani, R., Regolin, L., & Vallortigara, G. (2008). Discrimination of small numerosities in young chicks. *Journal of Experimental Psychology: Animal Behavior Processes, 34*(3), 388–399.

Rugani, R., Regolin, L., & Vallortigara, G. (2011). Summation of large numerousness by newborn chicks. *Frontiers in Psychology*. http://dx.doi.org/10.3389/fpsyg.2011.00179, 2(179).

Rugani, R., Regolin, L., & Vallortigara, G. (2013). One, two, three, four, or is there something more? Numerical discrimination in day-old domestic chicks. *Animal Cognition, 16*(4), 557–564.

Schartl, M., Walter, R. B., Shen, Y., Garcia, T., Catchen, J., Amores, A., et al. (2013). The genome of the platyfish, *Xiphophorus maculatus*, provides insights into evolutionary adaptation and several complex traits. *Nature Genetics, 45*, 567–572.

Scholl, B. J., & Pylyshyn, Z. W. (1999). Tracking multiple items through occlusion: Clues to visual objecthood. *Cognitive Psychology, 38*, 259–290.

Smith, C. C., & Sargent, R. C. (2006). Female fitness declines with increasing female density but not male harassment in the western mosquitofish, *Gambusia affinis*. *Animal Behaviour, 71*, 401–407.

Snekser, J. L., McRobert, S. P., & Clotfelter, E. D. (2006). Social partner preferences of male and female fighting fish (*Betta splendens*). *Behavioural Processes, 72*, 38–41.

Stancher, G., Sovrano, V. A., Potrich, D., & Vallortigara, G. (2013). Discrimination of small quantities by fish (redtail splitfin, *Xenotoca eiseni*). *Animal Cognition, 16*, 307–312.

Steinke, D., Salzburger, W., & Meyer, A. (2006). Novel relationships among ten fish model species revealed based on a phylogenomic analysis using ESTs. *Journal of Molecular Evolution, 62*, 772–784.

Suanda, S. H., Thompson, W., & Brannon, E. M. (2008). Changes in the ability to detect ordinal numerical relationships between 9 and 11 months of age. *Infancy, 13*, 308–337.

Svensson, P. A., Barber, I., & Forsgren, E. (2000). Shoaling behaviour of the two-spotted goby. *Journal of Fish Biology, 56*, 1477–1487.

Tegeder, R. W., & Krause, J. (1995). Density dependence and numerosity in fright stimulated aggregation behaviour of shoaling fish. *Philosophical Transactions of the Royal Society of London. Series B: Biological Sciences, 350*(1334), 381–390.

Tokita, M., Ashitani, Y., & Ishiguchi, A. (2013). Is approximate numerical judgment truly modality-independent? Visual, auditory, and cross-modal comparisons. *Attention, Perception, & Psychophysics, 75*, 1852–1861.

Tomonaga, M. (2008). Relative numerosity discrimination by chimpanzees (*Pan troglodytes*): Evidence for approximate numerical representations. *Animal Cognition, 11*, 43–57.

Trick, L. M., Audet, D., & Dales, L. (2003). Age differences in enumerating things that move: Implications for the development of multiple-object tracking. *Memory & Cognition, 31*(8), 1229–1237.

Trick, L. M., & Pylyshyn, Z. W. (1994). Why are small and large number enumerated differently: A limited-capacity preattentive stage in vision. *Psychological Review, 101*(1), 80–102.

Uller, C., Jaeger, R., Guidry, G., & Martin, C. (2003). Salamanders (*Plethodon cinereus*) go for more: Rudiments of number in a species of basal vertebrate. *Animal Cognition, 6*, 105–112.

Uller, C., & Lewis, J. (2009). Horses (*Equus caballus*) select the greater of two quantities in small numerical contrasts. *Animal Cognition, 12*, 733–738.

Utrata, E., Virányi, Z., & Range, F. (2012). Quantity discrimination in wolves (*Canis lupus*). *Frontiers in Psychology. 3*(505). http://dx.doi.org/10.3389/fpsyg.2012.00505.

Van Oeffelen, M. P., & Vos, P. G. (1982). A probabilistic model for the discrimination of visual number. *Perception & Psychophysics, 32*, 163–170.

Vetter, P., Butterworth, B., & Bahrami, B. (2008). Modulating attentional load affects numerosity estimation: Evidence against a pre-attentive subitizing mechanism. *PLoS One, 3*(9), e3269.

Vicario, C. M. (2011). Perceiving numbers affects the subjective temporal midpoint. *Perception, 40*, 23–29.

Volff, J. N. (2005). Genome evolution and biodiversity in teleost fish. *Heredity, 94*, 280–294.

Vonk, J., & Beran, M. J. (2012). Bears "count" too: Quantity estimation and comparison in black bears (*Ursus americanus*). *Animal Behaviour, 84*, 231–238.

Walsh, V. (2003). A theory of magnitude: Common cortical metrics of time, space and quantity. *Trends in Cognitive Sciences, 7*, 483–488.

West, R. E., & Young, R. J. (2002). Do domestic dogs show any evidence of being able to count? *Animal Cognition, 5*, 183–186.

Wilson, M. L., Britton, N. F., & Franks, N. R. (2002). Chimpanzees and the mathematics of battle. *Proceedings of the Royal Society of London B: Biological Sciences, 269*, 1107–1112.

Winkler, C., Schafer, M., Duschl, J., Schartl, M., & Volff, J. N. (2003). Functional divergence of two zebrafish midkine growth factors following fish-specific gene duplication. *Genome Research, 13*, 1067–1081.

Wittbrodt, J., Meyer, A., & Schartl, M. (1998). More genes in fish? *BioEssays, 20*, 511–515.

Wong, B. B. M., & Rosenthal, G. G. (2005). Shoal choice in swordtails when preferences conflict. *Ethology, 111*(2), 179–186.

Xu, F. (2003). Numerosity discrimination in infants: Evidence for two systems of representations. *Cognition, 89*(1), B15–B25.

Xu, F., & Arriaga, R. I. (2007). Number discrimination in 10-month-old infants. *British Journal of Developmental Psychology, 25*, 103–108.

Xu, F., & Spelke, E. S. (2000). Large number discrimination in 6-month-old infants. *Cognition, 74*, B1–B11.

Chapter 2

Foundations of Number and Space Representations in Non-Human Species

Giorgio Vallortigara
Center for Mind/Brain Sciences, University of Trento, Italy

INTRODUCTION

Studies on the ontogenetic origins of human knowledge suggest that cognition is not built on a blank slate. On the contrary, the human mind appears to be built on a set of innate systems that represent significant aspects of the environment such as physical objects, living beings, spatial relationships, and number. This idea, known as theory of core knowledge (Carey, 2009; Spelke, 2000), has an evolutionary foundation (Geary, 2005). The evolution of core systems means that at least some of them will be shared with other species, either because the systems were present in the last common ancestor (homologous systems) or because the associated competencies are so advantageous they evolved independently in many species (analogous systems; see Vallortigara, 2012a). Insects are certainly a case in point, and we discuss some examples later in this chapter, but further research is needed to establish to what extent vertebrates and invertebrates may share core knowledge mechanisms (Chittka & Niven, 2009; Srinivasan, 2009).

Two aspects are crucial to support the core knowledge theory: basic cognitive systems should be shared in different species, and they should not be acquired through experience (although they can be modified by experience). Here, we want to stress that in this regard animal models are relevant *per se*, and not simply as substitutes (replacements) in experiments that cannot be carried out on humans for practical or ethical reasons. First, animal models are useful for testing the generality of core knowledge systems and to prove the existence (or nonexistence) of certain computational capabilities independent of language or other symbolic abilities. Comparative psychologists and biologists are now in agreement that humans share some core systems with non-human animals, but there is also vigorous debate on the extent of these

Mathematical Cognition and Learning, Vol. 1. http://dx.doi.org/10.1016/B978-0-12-420133-0.00002-8

continuities for other systems (Penn, Holyoak, & Povinelli, 2008; Premack, 2007). Among the systems that show at least some degree of continuity across species are object, number, and space representation (e.g., apes and monkeys: Beran, Decker, Schwartz, & Schultz, 2011; Brannon, 2006; Call, 2007; Cantlon & Brannon, 2006; dogs: Kundey, De Los Reyes, Taglang, Baruch, & German, 2009; birds: Pepperberg, 1994, 2006a,b, 2012; Pepperberg & Carey, 2012; Regolin, Rugani, Stancher, & Vallortigara, 2011; Rugani, Regolin, & Vallortigara, 2011, 2013a,b; Vallortigara, Regolin, Chiandetti, & Rugani, 2010; see for general reviews: Haun, Jordan, Vallortigara, & Clayton, 2010; Vallortigara, 2004, 2006; Vallortigara, Chiandetti, Sovrano, Rugani, & Regolin, 2010).

Second, precocial species—those that are mobile and able to feed on their own soon after birth—seem to be a better model to investigate the innateness of core knowledge systems than altricial species, whose young are born in a relatively immature and highly dependent state. Precocial species can be tested behaviorally soon after birth because they are mobile and independent; consequently, they are ideal candidates to explore the degree to which a mechanism is available at birth or arises as a result of experience with the environment.

Availability of core knowledge at birth is clearly difficult to test in human infants, due to the unavoidable limitations in the control of early experience. The well-known example of preferential orientation of human neonates to faces (Johnson, 2005) makes this difficulty clear. It is not feasible to study human newborns maintaining them in a complete absence of face stimulation, but a similar attempt has been carried out on monkeys (Sugita, 2008). Infant Japanese monkeys reared with no exposure to faces for 6–24 months nevertheless showed a preference for human and monkey faces. However, in this experiment, monkeys were separated from their mother within a few hours after birth, making it possible, in principle, that very fast learning of face characteristics occurred. Furthermore, the monkeys had extensive visual experience with other (nonface) objects. Thus, it cannot be excluded that responses to face-like stimuli were in some significant ways shaped by experience with other types of complex objects. It could also be noted that monkeys (and other altricial animals) deprived from exposure to faces for several months might show an abnormal development of face processing. In contrast, testing monkeys soon after birth would require taking into account their motor immaturity, even in simple behavioral tasks.

The use of precocial species might simplify the study of core knowledge systems. The "God's organisms"—using the words of neuroscientist Steven Rose (2000)—to address the scientific problems related to the origins of knowledge could be provided by those species that are extremely precocial in their motor development, making possible quite sophisticated behavioral analyses at an early age (i.e., soon after birth or hatching) combined with a very precise control of early environmental stimulation (even *in utero* or *in ovo*), including a complete lack of some sensory experiences. Here, we champion the use of the young domestic chick (*Gallus gallus domesticus*) as a convenient model to investigate the origins of knowledge. In particular, we discuss evidence of precocious abilities in very young chicks to represent basic mechanical properties of physical objects (e.g., solidity); cardinal and ordinal/sequential aspects of numerical cognition (e.g., judgments of "more" and "less," and "third" and "fourth" in ordered numerosities), rudimentary arithmetic (addition and subtraction) with small numbers of discrete objects [e.g., $(4-1)>(1+1)$], and the geometrical relationships (e.g., to the left of the closest surface) among these objects. Through controlled rearing studies, it was possible to show that these abilities are observed in young chicks in the absence of (or with very reduced) experience (review in Vallortigara, 2012a).

CORE KNOWLEDGE OF OBJECTS

I start the exploration of the core knowledge systems with simple object mechanics, focusing on the properties of solid objects (see Baillargeon, Spelke, & Wasserman, 1985, for evidence in human infants). We tested some basic knowledge of mechanics with young chicks, taking advantage of the well-known phenomenon of filial imprinting. Through this learning process, the young of some animals, usually of precocial species, learn to recognize their mother/siblings/social partners by simply being exposed to them for a short period of time soon after hatching (Andrew, 1991; Bateson, 1990; Bateson & Gluckman, 2011). Chicks then follow the imprinting object. In my laboratory, soon after their hatching, we exposed chicks individually to a red cylindrical object for about 24 hours. With this procedure chicks became imprinted on the red cylinder. Then we confined the 1-day-old chicks behind a transparent partition while they saw the red cylinder "mother" moving and disappearing behind either one of two identical opaque screens tilted 45 degrees (Chiandetti & Vallortigara, 2011; see Figure 2-1).

This procedure was repeated several times to habituate the animal. We then assessed the knowledge of core-system mechanisms possessed by chicks. During the test, the mother moved toward the screen, and before the chick could see her going behind the right or left screen, we occluded the chick's view with an opaque screen. After the mother was hidden, the chick was presented with two slanted screens of the same size, but only one was slanted in a way that could fully hide the mother. We wondered whether the chick was able to take into account at the same time the shape and size of the object (retained in its memory) and the principle of solidity with respect to the slant. Chicks chose to approach preferentially the highest screen. This choice is consistent with the possession of the solidity concept but could also reflect a preference for higher screens. To discriminate between these effects, we ran a control experiment using a smaller mother, which could be hidden behind both screens. In this case chicks did not exhibit any significant preference

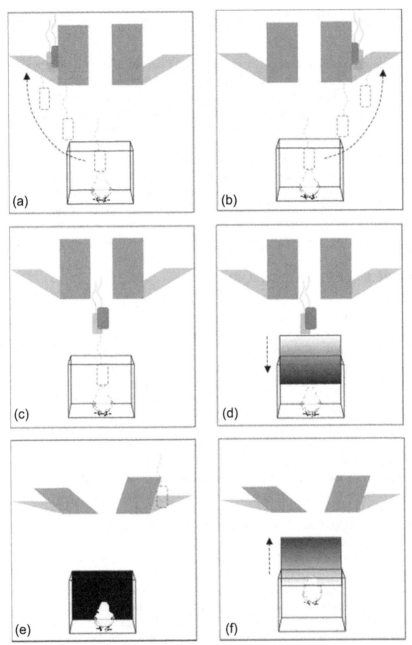

FIGURE 2-1 Schematic representation of the experimental paradigm used to investigate core knowledge of object mechanics in newborn chicks. Chicks were first allowed and habituated to rejoin an imprinting object moving and disappearing behind either of two identical opaque screens (a, b). At test, chicks were first shown the object moving along a straight trajectory midline toward the screens while they were confined behind a transparent partition (c) and then an opaque partition was lowered (d), whereas the slant (or, in other experiments, the height or the width) of the screens was changed in such a way to make one of the screens compatible and the other not compatible with the presence of the imprinting object hidden beneath them (e). After removal of the partitions, chicks were allowed to choose between the two screens (f). *(From Chiandetti & Vallortigara, 2011.)*

for the highest screen, suggesting that in the original condition they chose according to the physical properties of objects, in spite of their minimal experience with objects and occluding surfaces.

To exclude a role of the experience that chicks may have had with the imprinting object during the "exposition period" (during which chicks were reared with the red cylinder imprinting object), we repeated the experiment rearing chicks behind a transparent partition that divided them from the imprinting object. In the absence of direct interactions (touching and pecking) with the visible imprinting objects, we obtained identical results, showing that the interaction with the imprinting object is not necessary to show some knowledge of the basic mechanics of physical objects.

The same imprinting paradigm was used to explore other aspects of chicks' intuitive physics (Chiandetti & Vallortigara, 2011). Different groups of chicks were imprinted on objects of different size such as a slim or a large mother (see Figure 2-2a,b). At test, chicks were probed with screens that differed in their occluding properties. Each chick was held in a transparent box and could see the imprinting object moving along a straight midline toward two identical screens. Then the view was obscured by an opaque partition and the imprinting object removed. The screens were replaced with two other screens that were no longer identical. For example, the height or width or slant of the two screens was changed (see Figure 2-2c,d,e) so that only one screen was large/tall/slanted enough (in different experiments) to hide completely the imprinting object. The partition was then removed, and the chicks were allowed to search for the object. The chicks chose to search preferentially behind the screen that could hide the object (Figure 2-2d). It is possible to interpret this outcome as evidence that chicks possess certain knowledge about the physical properties of occluding and occluded objects even in the absence of significant experience with such objects.

CORE KNOWLEDGE OF NUMBER

A similar research strategy can be applied also to other domains such as number cognition. To increase protection against predators and avoid heat dissipation, young chicks show a tendency to approach larger numbers of social partners. Thus, by exposing chicks to different numbers of artificial imprinting objects and then observing the animals in free-choice tests with sets of different numbers of objects, we could explore their numerical abilities. In a series of experiments (Rugani, Regolin, & Vallortigara, 2010), separate groups of chicks were imprinted on three or one imprinting object, and then chicks' choices were tested between one and three objects. When chicks were reared with three objects, they preferred three objects at test; when they were reared with one object, they still preferred three objects at test. Apparently, the chicks had a preference for "more." When tested with two versus three, a more difficult task, the results were similar: reared with three, they preferred

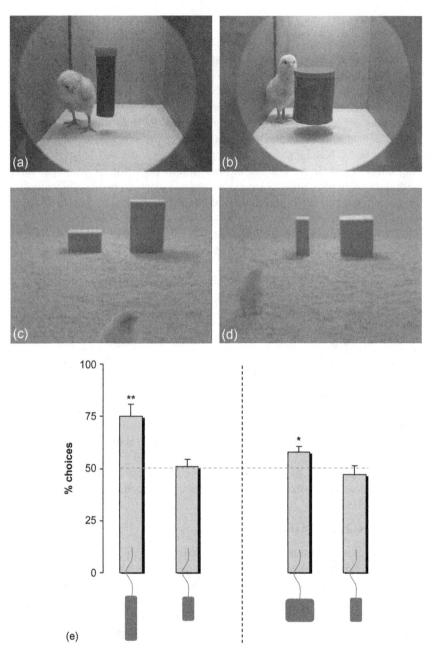

FIGURE 2-2 The procedure for investigating intuitive physics in newly hatched chicks. Different groups of chicks were reared with imprinting objects of different physical characteristics (a, b). At test, each chick is placed in a transparent pen while it is shown the imprinting object moving along a straight midline toward two identical screens. The screens are then replaced with two screens different in height (c) or width (d) so that one screen but not the other is a possible hiding location for the imprinting object. Figure (e) shows the percentage of choices for the highest (left bars) and the widest (right bars) screen as possible hiding locations depending on the size of the imprinting object during exposure (group means with SEM are shown). *p, 0.05; **p , 0.01. *(From Chiandetti & Vallortigara, 2011.)*

three; reared with two, they preferred three again. Choice in a group of chicks tested without any imprinting did not yield any preference for the larger number of objects, thus showing that preference for the larger number was related to imprinting. (Note that in natural contexts, there are typically multiple chicks in a clutch, and therefore they also imprint on their sibs as a mechanism to keep all of them together and following their mother.)

However, it could be that chicks were not using number in this task but were simply estimating the overall volume or area or some other continuous physical variables. Reliance on such continuous magnitudes has been shown in human infants in some circumstances. For instance, when objects have similar properties or belong to a domain in which physical extent can be particularly important (e.g., food; see Feigenson, Carey, & Hauser, 2002), infants favor extent over number. When, however, a task requires reaching for individual objects (e.g., Feigenson & Carey, 2005) or objects contrasting in color, pattern, or texture (Feigenson, 2005), then infants respond to number and not to spatial extent. Something similar occurred in chicks when they were imprinted on objects of different shape, color, area, and volume (Figure 2-3) and were then tested for choice with completely different novel objects (in size, color, and shape with respect to the exposure imprinting phase,

FIGURE 2-3 Chicks were exposed (imprinted) on either two or three objects with different characteristics (shape, color, size; see smaller photos on the left side). Then, at test, they could choose between pairs of objects of novel shape and color in which size (area or volume) was made identical. This was done to check whether chicks were able to respond only to the pure number of objects they were familiar with as a result of imprinting (see text).

Figure 2-3). These novel objects were controlled for their continuous extent (volume or surface area) such that they were the same in the three-element and in the two-element stimulus display. This time, when reared with three objects, at test chicks chose three objects; when reared with two objects, they chose two objects, showing their capacity to use number when continuous extent was equated (Rugani, Regolin, & Vallortigara, 2010).

Early availability of small numerosity discrimination by chicks strongly suggests that these abilities are in place at birth. Chicks are even sensitive to arithmetic-like change in number at birth. In some experiments (Rugani, Fontanari, Simoni, Regolin, & Vallortigara, 2009), chicks were reared with five identical imprinting objects. At test, each chick was confined to a holding pen, behind a transparent partition, from where it could see two identical opaque screens. The chick saw two sets of imprinting objects—a set of three objects and a set of two objects; each set disappeared (either simultaneously or one by one) behind one of the two screens. Immediately after the disappearance of both sets, the transparent partition was removed, and the chick was left free to move around and search within the arena. Chicks spontaneously inspected the screen occluding the larger set of three (and did so even when continuous physical variables were controlled for). In further experiments, after an initial disappearance of two sets (i.e., four objects disappeared behind a screen and one behind the other), some of the objects were visibly transferred, one by one, from one screen to the other before the chick was released (Figure 2-4a). Even in this case, chicks spontaneously chose the screen hiding the larger number of elements, and they did so irrespective of the directional cues provided by the initial and final displacements. Thus, chicks were sensitive and responded to, at their very first experience, a series of subsequent additions or subtractions of objects that appeared and disappeared over time (see Figure 2-4b). This evidence suggests the presence of early proto-arithmetic capacities. It is important to note that chicks did not have any experience of addition and subtraction of imprinting objects (and of their possible outcomes following a search behind occluding objects) before the test. In contrast, chicks' innate possession of such abilities appears to be highly adaptive for additions, and subtractions would be common in their natural environment. A mother and a group of chicks would, as a result of natural movements, experience (within their view) additions and subtractions from their group as their sibs move to and fro.

Human developmental research has identified different mechanisms to process small (1, 2, 3) and large (larger than 3) quantities (e.g., Coubart, Izard, Spelke, Marie, & Streri, 2014; Hyde & Spelke, 2009). A striking piece of evidence came from research showing that 12- to 14-month-old infants are able to discriminate between two small quantities (e.g., 1 vs. 3) but not between two quantities that cross the divide between the small and the large numerosity systems (e.g. 1 vs. 4, 1 vs. 5, or 2 vs. 4; Posid & Cordes, this volume). Since we did not observe these difficulties in chicks tested for arithmetic transitions

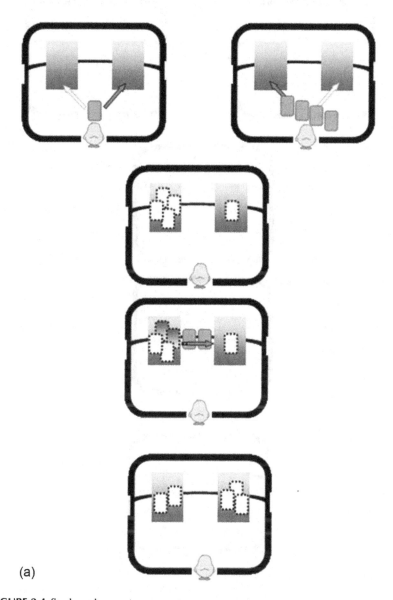

(a)

FIGURE 2-4 See legend on next page.

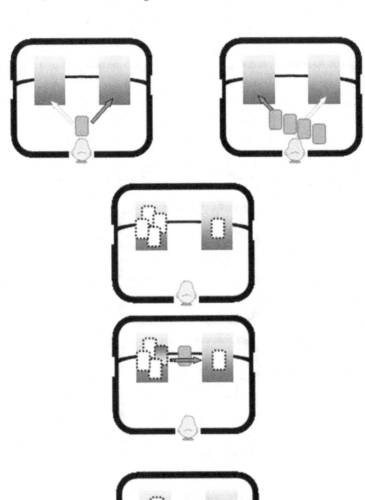

(b)

FIGURE 2-4 Schematic representation of the experiments to investigate sensitivity to arithmetic changes in young chicks. Newly hatched chicks were imprinted on five identical objects, and then one ball was hidden behind one screen and four balls were hidden—either one by one or as a group—behind the other screen. The sequence of events and the directions were randomized between trials. At the end of the first displacement event, therefore, either four balls or one ball was hidden behind each screen. Then in condition (a) two balls moved from the screen hiding four to the one hiding a single ball. At test, chicks approached the screen hiding the larger number of imprinting balls, even though it was not behind the screen where the larger number of balls had initially disappeared. In condition (b), only one ball moved from the screen hiding four to the one hiding a single ball. At test, chicks approached the screen hiding the larger number of imprinting balls, which was not behind the screen where the final hiding of balls had been observed. *(From Rugani et al., 2009.)*

(Rugani et al., 2009), we wondered whether their enhanced capability is a specialization to deal with social objects that are particularly relevant for chicks. To test this hypothesis, we ran similar experiments using imprinting objects (social stimuli), mealworms (food objects unrelated to social aspects), and two-dimensional imprinting objects of different sizes (as control for continuous physical variables such as area or contour length; see Figure 2-5). Results showed that chicks are able to deal with discriminations that crossed the boundaries between small and large numerousness (e.g., 1 vs. 4, 2 vs. 4, 1 vs. 5), even when continuous physical variables were controlled for and irrespective of whether social or palatable stimuli were used (Rugani, Cavazzana, Vallortigara, & Regolin, 2013; Rugani, Vallortigara, & Regolin, 2013a).

It has been suggested that the modality of stimulus presentation triggers activation of one or the other numerical systems (review in Mou & vanMarle, 2013; vanMarle, this volume). The current view is that the small numerosity system would be an object-based attention mechanism in which each element present in a scene is represented by a unique symbol—an "object file"—stored in working memory. Such an object file system (OFS) would have evolved to

Testing Comparisons	Rearing	Experimental Group	n	p
1 vs. 4		Social attractor (*SAG*) (*3D-NCG*)	8	<.001
		Food attractor (*FAG*)	8	<.001
1 vs. 5		Social attractor (*SAG*) (*3D-NCG*)	8	<.001
2 vs. 4		Social attractor (*SAG*) No Control (*3D-NCG*)	8	<.001
		Social attractor (*SAG*) No Control (*2D-NCG*)	8	
		Social attractor (*SAG*) Perimeter Control (*2D-PCG*)	8	<.001
		Social attractor (*SAG*) Area Control (*2D-ACG*)	8	

FIGURE 2-5 Chicks were tested for their ability for numerousness discrimination using pairs of social (imprinted) or nonsocial (food) objects that cross the boundaries between the small (<3) and the large (>3) number systems, i.e., 1 vs. 4, 1 vs. 5, and 2 vs. 4. Two-dimensional imprinted objects were also used in order to control of non-numerical continuous variables (area, contour length). Chicks appeared able to perform all the depicted numerousness discriminations. *(From Rugani et al., 2013b.)*

individuate in working memory each new object that is introduced into a scene. Spatio-temporal information and property/kind changes should thus be tracked by the object file system (for evidence of object individuation in animals, see Fontanari, Rugani, Regolin, & Vallortigara, 2011, 2014; Mendes, Rakoczy, & Call, 2008; Phillips & Santos, 2007). The OFS is therefore not specific to quantity representation, because numbers would be only implicitly represented in it. In other words, the OFS did not evolve to represent quantities per se, although simple quantitative tasks can sometimes be supported by this system, but only within the limits of working memory, that is, about three object files. The signature of the system would therefore be associated with a limit of working memory; that is, no more than about three object files can be simultaneously tracked and stored in working memory (Trick & Pylyshyn, 1994).

Discriminations of large numerousness (i.e., larger than 3) would be accomplished by the analog magnitude system (AMS), that in accord with Weber's law, is ratio dependent (e.g., Rugani, Regolin, & Vallortigara, 2011). (Note that human researchers tend to use the term "approximate number system," and animal researchers use the term "analog magnitude system"; Lourenco, this volume.) The larger the ratio between the numerousness to be discriminated, the faster the response times and higher the accuracy (Gallistel & Gelman, 1992). When attention is directed to the identity of each separate element, the OFS would be activated, whereas when attention is directed to the whole collection of elements, the AMS would be activated (Hyde & Spelke, 2011).

According to this hypothesis, the rearing experience of our chicks with a whole group of imprinting elements could have elicited subsequent processing of those elements by AMS, even if they were presented one by one during the testing session. In conditions in which attention was directed by pecking single separate elements, chicks did show evidence of an OFS with a set limit of about three elements (Rugani, Regolin, & Vallortigara, 2008). In contrast, in experiments with human infants, participants observed elements (crackers) being placed successively in one of two opaque containers (Feigenson & Carey, 2005; Feigenson, Carey, & Hauser, 2002; Feigenson, Dehaene, & Spelke, 2004). The infants thus lacked experience with an overall group of crackers and the fact that the crackers were presented one by one could, instead, have triggered processing by the OFS (and thus could have produced the difficulty in managing with quantities that spanned the boundary between OFS and AMS such as 4 vs. 1, see above). Clearly, further research work is needed to confirm these hypotheses. Note, however, that the ability to discriminate between numerousness that crosses the boundary between the small and the large numerosity systems has been confirmed in at least one species of fish, the redtail splitfin (*Xenotoca eiseni*), under conditions that would favor use of the AMS with small quantities (Stancher, Sovrano, Potrich, & Vallortigara, 2013).

As with many other species (e.g., Chittka & Geiger, 1995; Dacke & Srinivasan, 2008; Davis & Bradford, 1986), chicks possess the ability to identify an object on the exclusive basis of its position in a series of identical objects. When trained to peck at, for example, the third, fourth, or sixth position in a series of ten identically spaced locations (Figure 2-6), chicks learned to accurately identify the correct position (Rugani, Regolin, & Vallortigara, 2007).

Control experiments ruled out the possibility that they relied on distances as a cue. For instance, Rugani et al. (2007) trained chicks to peck at the fourth position and then tested them with a new sequence in which the distance from the starting point to each position was such that ordinal position and distance from the starting point differed from those used during the training. During the test, chicks pecked at the fourth serial position, even if that position was now located much farther away than during the training (Figure 2-7).

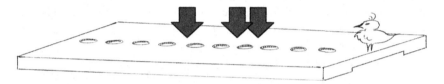

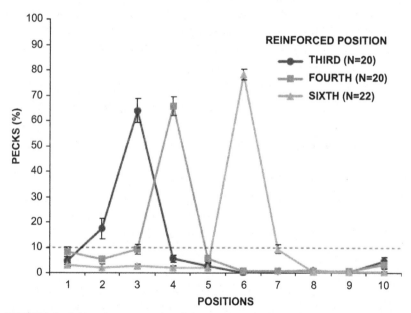

FIGURE 2-6   Schematic representation of the apparatus used for studying representation of ordinal position in chicks. The animals were faced with a series of containers arranged along a runway, and different groups of subjects have to learn that food was available in a particular position (i.e., the third, the fourth, or the sixth). Arrows (above) indicate the positions reinforced in the three experimental conditions; the results are shown below (mean values with SEM for the pecks directed to each position in the series). The dotted line represents chance level. *(After Rugani et al., 2007.)*

While investigating the abilities of chicks to deal with serial ordering and ordinal properties of numbers, we observed a bias for the left side. Chicks were trained to peck a selected position in a sagittally oriented series of identical elements (Figure 2-8a). During the test, the array was rotated by 90 degrees and oriented perpendicular to the chick's starting point; thus, the correct position could not be located on the basis of absolute distances from the starting point. Chicks selected only the fourth position from the left end, whereas the fourth from the right end was chosen at chance level (Figure 2-8b).

In subsequent work, newborn domestic chicks were compared in a similar task. In this case they were trained to peck at either the fourth or the sixth element in a series of 16 identical and aligned positions. Again, when the array was rotated 90 degrees, chicks showed a bias for the correct position from the left but not from the right end. The same results were observed in Clark's nutcrackers (*Nucifraga columbiana*) (Rugani, Kelly, Szelest, Regolin, & Vallortigara, 2010). The birds' bias is reminiscent of the human mental number line phenomenon. As early as 1880, Galton showed that humans describe and think of numbers as being represented on a mental number line, which is usually oriented from left to right (see also Dehaene, 1997). The mainstream interpretation of the spatial orientation of the human mental number line is that it is linked to writing and reading habits. We wondered, however, if this phenomenon may at least partially depend on a biological bias in the allocation of attention in extracorporeal space. In fact, a preference for targets located

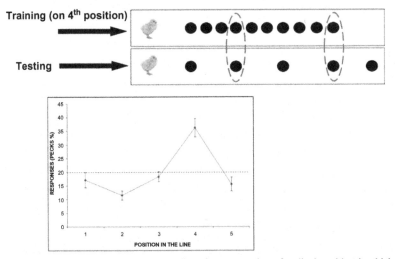

FIGURE 2-7   A control experiment in studies of representation of ordinal position in chicks. After chicks were trained to find food in the fourth position, the containers were spaced so that distance from the starting point would now correspond to the second ordinal position (top). Chicks, however, searched in the correct ordinal position (the fourth), ignoring spatial distance (bottom). *(From Rugani et al., 2007.)*

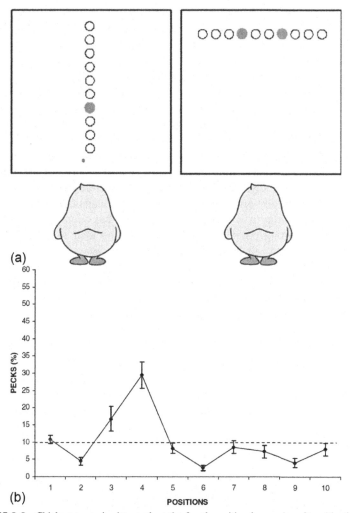

(a)

(b)

FIGURE 2-8 Chicks were trained to peck at the fourth position in a series of ten identical and sagittally oriented locations. Chicks were then required to respond to a different condition, where the correct position had to be identified within a series identical to the one used at training, but rotated by 90 degrees (a). Chicks identified as correct only the fourth position from the left end, and not the fourth position from the right end, which was chosen at chance level. *(From Rugani, Kelly et al., 2010; Rugani, Regolin et al., 2010).*

on the left hemispace may depend on a bias in the allocation of attention, somewhat similar to what is known in humans as "pseudoneglect" (e.g., Jewell & McCourt, 2000; and see also Loetscher et al., 2008). "Pseudoneglect" phenomena have been described in birds with selective allocation of attention to the left hemifield during free foraging (Diekamp, Regolin, Güntürkün, & Vallortigara, 2005).

Somewhat similar phenomena in which the left hemifield is favored in tongue strike responses associated with feeding have been described in several species of amphibian, such as the European toad (*Bufo bufo*) and cane toad (*B. marinus*) (Vallortigara, Rogers, Bisazza, Lippolis, & Robins, 1998; and for general reviews, see MacNeilage, Rogers, & Vallortigara, 2009; Rogers, Vallortigara, & Andrew, 2013; Vallortigara, 2000; Vallortigara, Chiandetti, & Sovrano, 2011; Vallortigara & Rogers, 2005). Direct evidence that the left bias in the ordinal task is due to hemispheric specialization comes from monocular tests. Chicks show complete cross-over of retinal fibers at the optic chiasma coupled with lack of a corpus callosum (Vallortigara, Rogers, & Bisazza, 1999). As a result, information entering each eye is processed mainly (although not completely) by contralateral brain structures. After chicks were trained with a sagittally oriented array of identical elements in binocular conditions, when the array was rotated by 90 degrees and oriented perpendicular to the chick's starting point, left-eyed chicks (using their right hemisphere) showed a bias to the left hemispace, whereas right-eyed chicks (using their left hemisphere) showed a bias to the right hemispace. The presence of a bias to the left side in the binocular condition clearly shows that the right hemisphere is in charge of control of this type of behavior when information is available to both eyes/hemispheres.

The issue of whether the bias observed in the ordinal number task is inherently numerical or spatial has been extensively studied (Rugani, Vallortigara, Vallini, & Regolin, 2011). When trained with systematic changes in inter-element distances, or when faced with similar distance changes at test, chicks identified as correct the target positions from both the left and the right end; that is, any bias disappeared. However, chicks encoded the ordinal position even when inter-element distances were kept fixed during the training, even if in this case the distances between elements would be sufficient to identify the target without any numerical computation. Hence, the ordinal numerical encoding is spontaneous, not triggered by a spatial training procedure. Right/left asymmetries in serial ordering tests can thus be explained with a simple model that assumes two differential encodings and different locations in the two hemispheres for spatial and numerical information (Rugani et al., 2011). The finding that, even after training with fixed inter-element distances, chicks encoded serial numerical ordering and not just distances between the elements suggests that two separate representations are formed during training. The first representation encodes serial ordering as such, without any association with the magnitude of distances on a spatial scale. Such a representation would encode something like "the target is in the fourth position"—that is, after the third container and before the fifth container—without any specification of the distances between the ordinal positions of the containers. The second representation would have instead a spatial nature, encoding magnitude of distances between containers in a relational fashion. Association of the purely ordinal and the purely spatial representation would provide the animal with a mental line of number.

The purely ordinal representation would be bilaterally represented in the left and right cerebral hemispheres. Activation of this representation, in contrast to a purely left-biased spatial attention task, would not produce any imbalance in the activity of the two hemispheres, giving rise to an identical allocation of attention toward the left and right visual hemifields. As evidence for this representation, when chicks are trained with stimuli arranged in a fronto-parallel line and distances are changed at test, they choose the correct target from both the left and right ends. In contrast, the purely spatial representation would be unilaterally represented in the right hemisphere. Such a role of the right hemisphere in spatial encoding is supported by independent evidence provided by both behavioral (Tommasi & Vallortigara, 2001; Vallortigara, Pagni, & Sovrano, 2004) and lesion studies (Tommasi, Gagliardo, Andrew, & Vallortigara, 2003; review in Vallortigara, 2009a). As a result, when at test chicks face a fronto-parallel line of elements whose relational spatial encoding matches the serial ordering, extra activation of the right hemisphere occurs, favoring the allocation of attention to the left hemispace. This in turn produces a bias to "count" selectively from left to right. When, however, because of a change in the spatial arrangement of the distances, chicks face a line of elements in which spatial properties do not match those acquired during training, the chicks rely on serial ordering alone. The outcome in this case is the bilateral activation of both cerebral hemispheres and the unbiased allocation of attention along both hemifields, resulting in chicks showing identical propensity "to count" from left to right and from right to left. Crucially, when a nonspatial property such as color of the elements is changed between training and test stimuli, a left bias is still observed.

CORE KNOWLEDGE OF GEOMETRY

Turning now to core knowledge of geometry, animals show very sophisticated knowledge of their spatial layout, with some understanding of geometric properties, presumably systems that have evolved to support navigation (Vallortigara, 2009b). This was originally shown in seminal studies by Cheng (1986) with rats and by Spelke (Hermer & Spelke, 1994, 1996) with children. Figure 2-9 illustrates a simple and extensively used setup, a rectangular room in which it is possible to (partly) disentangle between the different corners by using very simple geometrical properties: metric properties in the arrangement of surfaces such as the short/long length of a wall, and what is called "geometry sense," namely the distinction between right and left.

To understand the experimental setting, imagine you are located in an empty rectangular room, with identically colored walls and no external cues. In a corner, a visible goal is located. Then you are displaced from the room

FIGURE 2-9   Perspective view of the reorientation task in a rectangular enclosure (see text). The blue platforms indicate two geometrically equivalent positions. Imagine moving from the center of the enclosure toward one of the platforms: while facing the corner, you will have a longer wall on the left and a shorter wall on the right; the same would occur while remaining on the other blue platform. Partial reorientation in a rectangular enclosure is thus possible by combining elementary euclidian metric (short/long or close/far) and sense (right/left) properties.

and disoriented by being turned around passively a few times while your eyes are covered. Finally, you are reintroduced into the room. The goal object is no longer visible. Can you figure out in which corner the goal object was previously located? A partial solution is available by taking into account the rectangular shape of the room and the right/left geometric sense of the observer. In fact, only facing two of the four corners, you will observe the short wall on your right side and the long wall on your left side. Several species can solve this geometrical task, such as chicks of domestic fowl (Sovrano & Vallortigara, 2006; Vallortigara, Feruglio, & Sovrano, 2005; Vallortigara et al., 2004; Vallortigara, Zanforlin, & Pasti, 1990), pigeons (Kelly, Spetch, & Heth, 1998), monkeys (Gouteux, Thinus-Blanc, & Vauclair, 2001), fishes (Lee, Sovrano, & Vallortigara, 2012; Sovrano, Bisazza, & Vallortigara, 2002, 2003, 2005, 2007), and more recently ants (Wystrach & Beugnon, 2009; Wystrach, Cheng, Sosa, & Beugnon, 2011) and bumblebees (Sovrano, Potrich, & Vallortigara, 2013; Sovrano, Rigosi, & Vallortigara, 2012).

When newborn chicks are reared in enclosures of different shapes and metrical properties (Figure 2-10) it has been found that they need little, if any, experience to use geometric cues to locate previously seen objects (review in Vallortigara, Sovrano, & Chiandetti, 2009). While rectangular enclosures provided metrically distinct surfaces connected at right angles and two principal axes of symmetry, geometric information was not available in circular enclosures. In C-shaped enclosures, neither right angles nor differences in wall length were available, although the first principal axis could be

FIGURE 2-10   Schematic representations of rectangular (left), c-shaped (middle), and circular home cages (right) used for controlled-rearing experiments in chicks. In rectangular cages, geometric information is fully available because of the presence of metrically distinct surfaces connected at right angles and two principal axes of symmetry. In circular cages, this sort of geometric information is removed, and there is an infinite number of principal axes. In c-shaped cages, neither right angles nor differences in walls' length are available, but the first principal axis is still usable to encode the shape. Results (see text) showed that chicks used geometry in spatial reorientation irrespective of experience with different types of environment during rearing. *(From Vallortigara et al., 2009.)*

used to encode shape. Chicks proved to be equally capable of learning and performing navigational tasks based on geometric information irrespective of being reared in different environments (Chiandetti & Vallortigara, 2008, 2010; and see Brown, Spetch, & Hurd, 2007, for similar evidence in fish). Evidence for innateness in the encoding of space representations also comes from single-cell recording in the rat hippocampus and related regions. Different types of cells have been found: place cells, located in the hippocampus, that fire only when an animal is in a specific location within its environment; head direction cells, in the presubiculum, which is adjacent to the hippocampus, that fire only when an animal is facing a certain direction; grid cells, in the entorhinal cortex, which also lies next to the hippocampus, that fire when the animal is at multiple locations in its environment, boundary vector cells, which encode the organism's distance from geometric borders surrounding its environment. Place cells are present as early as postnatal day 15 soon after pups open their eyes (and the same seems to be true for head direction and border cells; see Langston et al., 2010; Wills, Cacucci, Burgess, & O'Keefe, 2010).

The issue of the axes is relevant with regard to the type of computations organisms use in encoding geometry. Geometric cues must be encoded in a way that allows for reliable orientation and, at the same time, that is not computationally demanding (Cheng, 2008). It has been argued that animals may encode the principal axes of an environment (the major principal axis passes lengthwise through the center of a shape, whereas the minor principal axis lies perpendicular to the major principal axis; Cheng, 1986). This hypothesis has been tested by first training chicks to locate food hidden in a rectangular arena and then modifying the shape of the arena in an L-shaped fashion to differentiate among possible encoding strategies (Figure 2-11). The chicks appeared to use mainly a local geometry strategy, and they did not show any behavior suggesting the use of principal axes. It is likely that although principal axes would allow for an economical means of computing heading, they do not

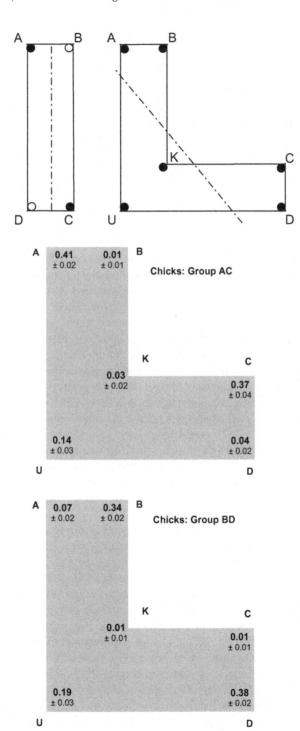

retain information about the environmental shape and thus they are not informative enough for use in complex environments (see Kelly, Chiandetti, & Vallortigara, 2011; Kelly, Durocher, Chiandetti, & Vallortigara, 2011).

The zebrafish (*Danio rerio*) is another species that has been used to test animals' geometric abilities—mostly for the prospects it offers for a possible genetic analysis of the mechanisms underlying geometric encoding. Lee, Vallortigara, Fiore, and Sovrano (2013) trained zebrafish in a variant of the classical rectangular task for reorientation by geometry. Test animals were located in the center while a conspecific was located in a corner. The test fish was rotated and the conspecific removed (Figure 2-12a). Based on the social motivation of this species to school as a means of predator avoidance, Lee et al. (2013) used the locations searched by the experimental fish to test their ability to orient. To determine which information the fish used to do so, researches manipulated the length and the walls and their distance. It was found that zebrafish use only distance and not length to solve this geometrical task (Figure 2-12b; note that vertical navigation based on water pressure has no role in this task because corners were all at the same vertical positions). Overall, the results support a model of spatial reorientation based on relative directions and positions computed with respect to the distance relationships among extended visible three-dimensional surfaces (see also Lee, Sovrano, & Spelke, 2012). Like human children (Lee, Sovrano, & Spelke, 2012), zebrafish failed to use either an equidistant array of walls of different lengths or a rectangular array of fragmented corners to guide their search for a hidden goal. These striking similarities in the spontaneous navigation of humans and fish are consistent with a domain-specific, evolutionarily ancient computation of surface distance and direction in navigation.

Given the specificity of the geometric reorientation capacity, it may be possible to gain insight into gene–behavior relationships in geometric navigation using zebrafish and mice as model organisms. This hypothesis is further motivated by reports of a specific navigational deficit in individuals with Williams syndrome (Lakusta, Dessalegn, & Landau, 2010). In many cognitive tasks, including most tests of spatial cognition that recruit attention to objects or other landmarks, adults with Williams syndrome perform like young children, showing overall deficits, but abilities qualitatively similar to those found early in human development (Landau, 2011; Landau & Lakusta, 2009). In

---

FIGURE 2-11   After training in a rectangular enclosure (one group of chicks trained with AC geometrically-equivalent corners as positive [rewarded], one group trained with BD geometrically-equivalent corners as positive, top), chicks were tested in an L-shaped enclosure (bottom). They showed clear reliance on local geometric information (i.e., angle formed by a long and short wall, with, for instance, the long wall to the right and the short to the left for AC) with no use of principal axis (numbers in bold represent percentages of choice with SEM indicated below). *(From Kelly, Chiandetti, & Vallortigara, 2011.)*

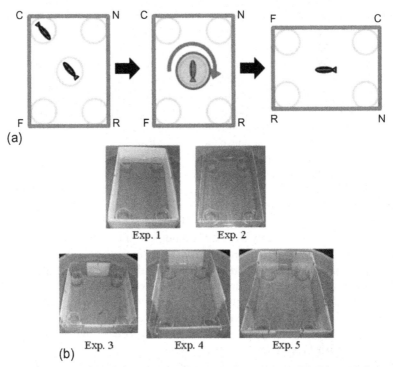

FIGURE 2-12 (a) A spatial reorientation task for zebrafish (see text). The fish could observe a conspecific in the goal location (C); then it was covered and rotated slowly in order to obtain spatial disorientation. When released into an empty tank to navigate freely, its approaches to the corner jars were coded as choices. Choosing the correct corner and its rotational geometric equivalent corner (R)—rather than the near (N) and far (F) incorrect corners—indicates successful use of geometrical information. (b) Photographs of the experimental set used to investigate encoding of distance and length information in the reorientation task by zebrafish. In Experiment 1, the ordinary rectangular enclosure with opaque walls was used replicate successful reorientation by geometry. In Experiment 2, a rectangular arena consisting of transparent surfaces was used as a control to verify whether fish were reorienting visually or using the lateral line (sensitivity to the pressure variation of the water). In Experiment 3, the geometric property of distance while equating for length was tested, using an array of four fragmented, opaque surfaces of identical length, within the same transparent rectangular arena. In Experiment 4, an array of four truncated corners was used so as to present the same area of opaque surfaces specifying the same rectangular configuration. In Experiment 5, four opaque surfaces within a square transparent arena were used so that pairs of surfaces whose length differed by a 2:1 ratio were presented at equal distances. Results showed that in this case zebrafish were unable to reorient themselves. *(From Lee et al., 2013.)*

contrast, when tested in spontaneous reorientation tasks, adults with Williams syndrome showed little sensitivity to surface layout geometry and perform far worse than normally developing toddlers (Lakusta et al., 2010). Through genetic modification of model species such as zebrafish or mice, it may be possible to clarify the role of specific genes in the emergence of specific spatial cognitive abilities, as well as their neural correlates (e.g., the hippocampus).

It is important to stress that comparative work is not only useful to establish commonalities of mechanisms as a result of true homology or convergent evolution (homoplasy). Similar behavioral performances may also arise from completely different mechanisms. Insects are relevant models for testing whether similar or different encoding strategies are used in species that are phylogenetically very distantly related to humans. The simplest model of insect navigation that has been proposed is based on view-based homing strategies (Collett & Collett, 2002; Collett, Graham, Harris, & Hempel de Ibarra, 2006). According to this theory, insects navigate to minimize the difference between the memorized panoramic image (a two-dimensional "snapshot" of the goal site) and the panorama perceived from the current location. Theoretical and computer modeling suggest that view-based strategies for reorientation could produce geometric errors in the rectangular enclosure task (Cheung, Stürzl, Zeil, & Cheng, 2008; Stürzl, Cheung, Zeil, & Cheng, 2008). Bumblebees *(Bombus terrestris),* for example, are able to reorient themselves using purely geometric cues using the same rectangular room test employed with the other species (Sovrano, Rigosi, & Vallortigara, 2012). Moreover, it has been shown that bumblebees distinguish between geometric equivalent locations when tested in the presence of featural information, both near and far from the goal (Sovrano et al., 2013).

However, recent work by Lee, Sovrano, and Vallortigara (2014) provided direct evidence that use of the same cues by insects and vertebrates is not evidence for the same processes underlying behavior; that is, the underlying navigation systems likely evolved independently. The authors studied the ability of bumblebees to learn a hidden escape route out of box-like enclosures, manipulating environmental cues such as landmark size and proximity, environment size, and boundary geometry. In contrast to vertebrates, which tend to be more proficient in navigating by boundaries in smaller environments, bumblebees appeared to learn boundary geometry more quickly in larger environments. Moreover, in contrast to vertebrate behavior that has been shown to be more strongly influenced by distal landmarks, the bumblebees' behavior was shown to depend mostly on factors such as the size of the featural landmark (i.e., the size of panels located at the corners of the rectangular enclosure) and its proximity or visibility from the goal location, consistent with image-matching accounts of navigation.

Spatial reorientation by layout geometry occurs in numerous species, but its underlying mechanisms are debated. Although some support for the image-matching theory has been provided in vertebrates (Pecchia, Gagliardo, & Vallortigara, 2011; Pecchia & Vallortigara, 2010a, 2010b, 2012), it is restricted to conditions of repeated testing that involve learning to use different cues, whereas in spontaneous reorientation a sense of place based on geometric computations over three-dimensional representations

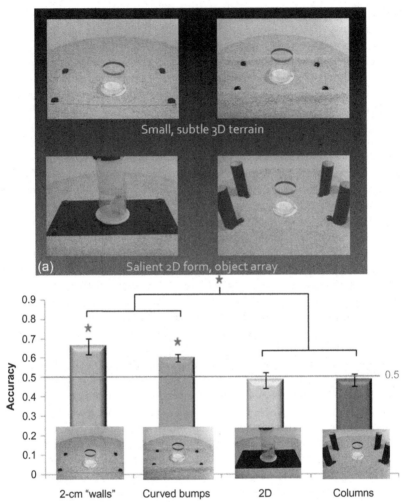

FIGURE 2-13 In all experimental conditions (a), a rectangular array of feeders was buried inside the sawdust such that only the covers were visible. In one condition, a 2 cm high rectangular frame surrounded the feeders such that each feeder was located directly at a corner of the rectangle (top left). In another condition, parallel bumps, made by pipes covered in sawdust, bordered each long side of the rectangular array (top right). The last two conditions presented a black two-dimensional rectangle at the level of the sawdust (bottom left) or four cylinders fixed directly behind each feeder such that there was a rectangular array of large freestanding columns (bottom right). The results (b) showed that chicks were able to reorient on the basis of perturbations to the three-dimensional terrain that produced only subtle image contrast borders but failed to reorient by visually salient two-dimensional forms or object arrays that produced more prominent contrast borders. *(From Lee, Spelke, & Vallortigara, 2012.)*

seems to provide a better account than panoramic image view matching (Lee, Tucci, Sovrano, &Vallortigara, 2014). Moreover, comparative studies with children (Lee & Spelke, 2011) and chicks (Lee, Spelke, & Vallortigara, 2012) showed that they both reorient by subtle three-dimensional perturbations of the terrain and fail to reorient using visually salient two-dimensional forms or object arrays that produced more prominent contrast borders (see Figure 2-13). The cue specificity for three-dimensional surface layouts provides evidence in favor of geometry-based navigation theories and against image-matching theories. It will be interesting to investigate invertebrate responses to these stimuli.

CONCLUSIONS

Summarizing, the evidence collected on different animal species—certainly vertebrates and likely invertebrates as well—shows that these organisms are naturally equipped with knowledge available in the absence of previous experience. Although I have primarily covered core physical and spatio-numerical knowledge, similar evidence is available for social core knowledge as well. This includes, for example, research on animacy and goal-directedness of certain types of objects conducted with domestic chicks (Mascalzoni, Regolin, & Vallortigara, 2010; Rosa Salva, Regolin, & Vallortigara, 2010; Vallortigara & Regolin, 2006; Vallortigara, Regolin, & Marconato, 2005; Wood, 2013) with direct parallels in research conducted with human infants (Mascalzoni, Regolin, Vallortigara, & Simion, 2013; Rosa Salva, Farroni, Regolin, Vallortigara, & Johnson, 2011; Simion, Regolin, & Bulf, 2008; for general reviews, see Vallortigara, 2009a; 2012b; and see Biro, Csibra, & Gergely, 2007, for human infants).

Humans' explicit knowledge of symbolic number and geometry may be our unique, abstract cognitive ability (Penn et al., 2008; Premack, 2007; Spelke, 2003), but it clearly depends, in part, on a system that is widely shared by other animals (Starr & Brannon, this issue). A system for geometry develops in many animals including chicks, fish, and rodents with no experience of a geometrically structured surface layout. And a system of approximate numerousness estimation and arithmetic develops in chicks with no experience of addition or subtraction of discrete items. Through use of a comparative approach, combined with neurobiological and genetical techniques, perhaps ancient hypotheses concerning the nature and origins of abstract concepts may now be amenable to study.

ACKNOWLEDGMENTS

I thank Elisabetta Versace and the editors for reading and commenting on a preliminary version of this manuscript. My work is funded by an ERC Advanced Grant (PREMESOR ERC-2011-ADG_20110406).

REFERENCES

Andrew, R. J. (Ed.), (1991). *Neural and behavioural plasticity: The use of the chick as a model.* Oxford: Oxford University Press.

Baillargeon, R., Spelke, E. S., & Wasserman, S. (1985). Object permanence in five-month-old infants. *Cognition, 20,* 191–208.

Bateson, P. (1990). Is imprinting such a special case? *Philosophical Transactions of the Royal Society of London Series B—Biological Sciences, 329,* 125–131.

Bateson, P., & Gluckman, P. (2011). *Plasticity, robustness, development and evolution.* New York: Cambridge University Press.

Beran, M. J., Decker, S., Schwartz, A., & Schultz, N. (2011). Monkeys (*Macaca mulatta* and *Cebus apella*) and human adults and children (*Homo sapiens*) compare subsets of moving stimuli based on numerosity. *Frontiers in Psychology, 2,* 61.

Biro, S., Csibra, G., & Gergely, G. (2007). The role of behavioral cues in understanding animacy, agency and goal-directed actions in infancy. In C. von Hofsten & K. Rosander (Eds.), *Progress in brain research: From action to cognition* (p. 303). New York: Elsevier.

Brannon, E. M. (2006). The representation of numerical magnitude. *Current Opinion in Neurobiology, 16,* 222–229.

Brown, A. A., Spetch, M. L., & Hurd, P. L. (2007). Growing in circles: Rearing environment alters spatial navigation in fish. *Psychological Science, 18,* 569–573.

Call, J. (2007). Apes know that hidden objects can affect the orientation of other objects. *Cognition, 105,* 1–25.

Cantlon, J., & Brannon, E. M. (2006). Shared system for ordering small and large numbers in monkeys and humans. *Psychological Science, 17,* 401–406.

Carey, S. (2009). *The origin of concepts.* New York: Oxford University Press.

Cheng, K. (1986). A purely geometric module in the rat's spatial representation. *Cognition, 23,* 149–178.

Cheng, K. (2008). Whither geometry? Troubles of the geometric module. *Trends in Cognitive Sciences, 12,* 355–362.

Cheung, A., Stürzl, W., Zeil, J., & Cheng, K. (2008). The information content of panoramic images II: View-based navigation arenas. *Journal of Experimental Psychology. Animal Behavior Processes, 34,* 15–30.

Chiandetti, C., & Vallortigara, G. (2008). Is there an innate geometric module? Effects of experience with angular geometric cues on spatial reorientation based on the shape of the environment. *Animal Cognition, 11,* 139–146.

Chiandetti, C., & Vallortigara, G. (2010). Experience and geometry: Controlled-rearing studies with chicks. *Animal Cognition, 13,* 463–470.

Chiandetti, C., & Vallortigara, G. (2011). Intuitive physical reasoning about occluded objects by inexperienced chicks. *Proceedings of the Royal Society of London B, 278,* 2621–2627.

Chittka, L., & Geiger, K. (1995). Can honey bees count landmarks? *Animal Behaviour, 49,* 159–164.

Chittka, L., & Niven, J. (2009). Are bigger brains better? *Current Biology, 19,* R995–R1008.

Collett, T. S., & Collett, M. (2002). Memory use in insect visual navigation. *Nature Reviews. Neuroscience, 3,* 542–552.

Collett, T. S., Graham, P., Harris, R. A., & Hempel de Ibarra, N. (2006). Navigational memories in ants and bees: Memory retrieval when selecting and following routes. *Advances in the Study of Behaviour, 36,* 123–172.

Coubart, A., Izard, V., Spelke, E. S., Marie, J., & Streri, A. (2014). Dissociation between small and large numerosities in newborn infants. *Developmental Science, 17*(1), 11–22.

Dacke, M., & Srinivasan, M. V. (2008). Evidence for counting in insects. *Animal Cognition, 11*, 683–689.

Davis, H., & Bradford, S. A. (1986). Counting behavior by rats in a simulated natural environment. *Ethology, 73*, 265–280.

Dehaene, S. (1997). *The number sense.* New York: Oxford University Press.

Diekamp, B., Regolin, L., Gunturkun, O., & Vallortigara, G. (2005). A left-sided visuospatial bias in birds. *Current Biology, 15*, R372–R373.

Feigenson, L. (2005). A double dissociation in infants' representation of object arrays. *Cognition, 95*, B37–B48.

Feigenson, L., & Carey, S. (2005). On the limits of infants' quantification of small object arrays. *Cognition, 97*, 295–313.

Feigenson, L., Carey, S., & Hauser, M. (2002). The representations underlying infants' choice of more: Object files versus analog magnitudes. *Psychological Science, 13*, 150–156.

Feigenson, L., Dehaene, S., & Spelke, E. (2004). Core systems of number. *Trends in Cognitive Sciences, 8*, 307–314.

Fontanari, L., Rugani, R., Regolin, L., & Vallortigara, G. (2011). Object individuation in 3-day-old chicks: Use of property and spatiotemporal information. *Developmental Science, 14*, 1235–1244.

Fontanari, L., Rugani, R., Regolin, L., & Vallortigara, G. (2014). *Use of kind information for object individuation in young domestic chicks. Animal Cognition,* in press.

Gallistel, C. R., & Gelman, R. (1992). Preverbal and verbal counting and computation. *Cognition, 44*, 43–74.

Galton, F. (1880). Visualised numerals. *Nature, 21*, 252–256.

Geary, D. C. (2005). *The origin of mind: Evolution of brain, cognition and general intelligence.* Washington, DC: American Psychological Association.

Gouteux, S., Thinus-Blanc, C., & Vauclair, J. (2001). Rhesus monkeys use geometric and nongeometric information during a reorientation task. *Journal of Experimental Psychology. General, 130*, 505–519.

Haun, D. B. M., Jordan, F., Vallortigara, G., & Clayton, N. (2010). Origins of spatial, temporal and numerical cognition: Insights from animal models. *Trends in Cognitive Sciences, 14*, 477–481.

Hermer, L., & Spelke, E. S. (1994). A geometric process for spatial reorientation in young children. *Nature, 370*, 57–59.

Hermer, L., & Spelke, E. S. (1996). Modularity and development: The case of spatial reorientation. *Cognition, 61*, 195–232.

Hyde, D. C., & Spelke, E. S. (2009). All numbers are not equal: An electrophysiological investigation of small and large number representations. *Journal of Cognitive Neuroscience, 21*, 1039–1053.

Hyde, D. C., & Spelke, E. S. (2011). Neural signatures of number processing in human infants: Evidence for two core systems underlying numerical cognition. *Developmental Science, 14*, 360–371.

Jewell, G., & McCourt, M. E. (2000). Pseudoneglect: A review and metanalysis of performance factors in line bisections tasks. *Neuropsychologia, 38*, 93–110.

Johnson, M. H. (2005). Subcortical face processing. *Nature Reviews. Neuroscience, 6*, 766–774.

Kelly, D. M., Chiandetti, C., & Vallortigara, G. (2011). Re-orienting in space: Do animals use global or local geometry strategies? *Biology Letters, 7*, 372–375.

Kelly, D. M., Durocher, S., Chiandetti, C., & Vallortigara, G. (2011). A misunderstanding of principal and medial axes? Reply to Sturz and Bodily (2011). *Biology Letters, 7*, 649–650.

Kelly, D. M., Spetch, M. L., & Heth, C. D. (1998). Pigeons' (*Columba livia*) encoding of geometric and featural properties of a spatial environment. *Journal of Comparative Psychology, 112,* 259–269.

Kundey, S. M. A., De Los Reyes, A., Taglang, C., Baruch, A., & German, R. (2009). Domesticated dogs' (*Canis familiaris*) use of the solidity principle. *Animal Cognition, 13,* 497–505.

Landau, B. (2011). The organization and development of spatial cognition: Insights from Williams syndrome. In J. Burack, R. M. Hodapp, G. Iarocci & E. Zigler (Eds.), *Handbook of intellectual disabilities and development* (2nd ed) (pp. 61–88). New York: Oxford University Press.

Landau, B., & Lakusta, L. (2009). Spatial representation across species: Geometry, language, and maps. *Current Opinion in Neurobiology, 19,* 12–19.

Lakusta, L., Dessalegn, B., & Landau, B. (2010). Impaired geometric reorientation caused by genetic defect. *Proceedings of the National Academy of Sciences, 107,* 2813–2817.

Langston, R. F., Ainge, J. A., Couey, J. J., Canto, C. B., Bjerknes, T. L., Witter, M. P., et al. (2010). Development of the spatial representation system in the rat. *Science, 328,* 1576–1580.

Lee, S. A., Sovrano, V. A., & Spelke, E. S. (2012). Navigation as a source of geometric knowledge: Young children's use of length, angle, distance, and direction in a reorientation task. *Cognition, 123,* 144–161.

Lee, S. A., Sovrano, V. A., & Vallortigara, G. (2012). Independent effects of geometry and landmark in a spontaneous reorientation task: A study of two species of fish. *Animal Cognition, 15,* 861–870.

Lee, S. A., Sovrano, V. A., & Vallortigara, G. (2014). *No evidence for homologies in vertebrate and invertebrate navigation.* Submitted for publication.

Lee, S. A., & Spelke, E. S. (2011). Young children reorient by computing layout geometry, not by matching images of the environment. *Psychonomic Bulletin and Review, 18,* 192–198.

Lee, S. A., Spelke, E. S., & Vallortigara, G. (2012). Chicks, like children, spontaneously reorient by three-dimensional environmental geometry, not by image matching. *Biology Letters, 8,* 492–494.

Lee, S. A., Tucci, V., Sovrano, V. A., & Vallortigara, G. (2014). *Spontaneous and learned representations of space: Dissociable systems of spatial memory and navigation in mice.* Submitted for publication.

Lee, S. A., Vallortigara, G., Fiore, M., & Sovrano, V. A. (2013). Navigation by environmental geometry: The use of zebrafish as a model. *Journal of Experimental Biology, 216,* 3693–3699.

Loetscher, T., Schwarz, U., Schubiger, M., & Brugger, P. (2008). Head turns bias the brain's internal random generator. *Current Biology, 18,* R60–R62.

MacNeilage, P. F., Rogers, L. J., & Vallortigara, G. (2009). Origins of the left and right brain. *Scientific American, 301,* 60–67.

Mascalzoni, E., Regolin, L., & Vallortigara, G. (2010). Innate sensitivity for self-propelled causal agency in newly hatched chicks. *Proceedings of the National Academy of Sciences, 107,* 4483–4485.

Mascalzoni, E., Regolin, L., Vallortigara, G., & Simion, F. (2013). The cradle of causal reasoning: Newborns' preference for physical causality. *Developmental Science, 16,* 327–335.

Mendes, N., Rakoczy, H., & Call, J. (2008). Ape metaphysics: Object individuation without language. *Cognition, 106,* 730–749.

Mou, Y., & vanMarle, K. (2013). Two core systems of numerical representations in infants in press, *Developmental Review.*

Pecchia, T., Gagliardo, A., & Vallortigara, G. (2011). Stable panoramic views facilitate snap-shot like memories for spatial reorientation in homing pigeons. *PLoS One, 6*(7), e22657.

Pecchia, T., & Vallortigara, G. (2010a). Re-orienting strategies in a rectangular array of landmarks by domestic chicks (*Gallus gallus*). *Journal of Comparative Psychology, 124*, 147–158.

Pecchia, T., & Vallortigara, G. (2010b). View-based strategy for reorientation by geometry. *Journal of Experimental Biology, 213*, 2987–2996.

Pecchia, T., & Vallortigara, G. (2012a). Spatial reorientation by geometry with freestanding objects and extended surfaces: A unifying view. *Proceedings of the Royal Society of London B, 279*, 2228–2236.

Penn, D. C., Holyoak, K. J., & Povinelli, D. J. (2008). Darwin's mistake: Explaining the discontinuity between human and nonhuman minds. *Behavioral and Brain Sciences, 31*, 109–178.

Pepperberg, I. M. (1994). Evidence for numerical competence in an African Grey parrot (*Psittacus erithacus*). *Journal of Comparative Psychology, 108*, 36–44.

Pepperberg, I. M. (2006a). Grey parrot (*Psittacus erithacus*) numerical abilities: Addition and further experiments on a zero-like concept. *Journal of Comparative Psychology, 120*, 1–11.

Pepperberg, I. M. (2006b). Grey parrot numerical competence: A review. *Animal Cognition, 9*, 377–391.

Pepperberg, I. M. (2012). Further evidence for addition and numerical competence by a Grey parrot (*Psittacus erithacus*). *Animal Cognition, 15*, 711–717.

Pepperberg, I. M., & Carey, S. (2012). Grey parrot number acquisition: The inference of cardinal value from ordinal position on the numeral list. *Cognition, 125*, 219–232.

Phillips, W., & Santos, L. (2007). Evidence for kind representations in the absence of language: Experiments with rhesus monkeys (*Macaca mulatta*). *Cognition, 102*, 455–463.

Premack, D. (2007). Human and animal cognition: Continuity and discontinuity. *Proceedings of the National Academy of Sciences of the United States of America, 104*, 13861–13867.

Regolin, L., Rugani, R., Stancher, G., & Vallortigara, G. (2011). Spontaneous discrimination of possible and impossible objects by newly hatched chicks. *Biology Letters, 7*, 654–657.

Rogers, L. J., Vallortigara, G., & Andrew, R. J. (2013). *Divided brains: The biology and behaviour of brain asymmetries.* New York: Cambridge University Press.

Rosa-Salva, O., Farroni, T., Regolin, L., Vallortigara, G., & Johnson, M. H. (2011). The evolution of social orienting: Evidence from chicks (*Gallus gallus*) and human newborns. *PLoS One, 6*(4), e18802.

Rosa-Salva, O., Regolin, L., & Vallortigara, G. (2010). Faces are special for newly hatched chicks: Evidence for inborn domain-specific mechanisms underlying spontaneous preferences for face-like stimuli. *Developmental Science, 13*, 565–577.

Rose, S. (2000). God's organism? The chick as a model system for memory studies. *Learning & Memory, 7*, 1–17.

Rugani, R., Cavazzana, A., Vallortigara, G., & Regolin, L. (2013). One, two, three, four, or is there something more? Numerical discrimination in day-old domestic chicks. *Animal Cognition, 16*, 557–564.

Rugani, R., Fontanari, L., Simoni, E., Regolin, L., & Vallortigara, G. (2009). Arithmetic in newborn chicks. *Proceedings of the Royal Society B, 276*, 2451–2460.

Rugani, R., Kelly, D. M., Szelest, I., Regolin, L., & Vallortigara, G. (2010). Is it only humans that count from left to right? *Biology Letters, 6*, 290–292.

Rugani, R., Regolin, L., & Vallortigara, G. (2007). Rudimental numerical competence in 5-day-old domestic chicks: Identification of ordinal position. *Journal of Experimental Psychology: Animal Behavioural Processes, 33*, 21–31.

Rugani, R., Regolin, L., & Vallortigara, G. (2008). Discrimination of small numerosities in young chicks. *Journal of Experimental Psychology. Animal Behavior Processes, 34,* 388–399.

Rugani, R., Regolin, L., & Vallortigara, G. (2010). Imprinted numbers: Newborn chicks' sensitivity to number vs. continuous extent of objects they have been reared with. *Developmental Science, 13,* 790–797.

Rugani, R., Regolin, L., & Vallortigara, G. (2011). Summation of large numerousness by newborn chicks. *Frontiers in Psychology, 2,* 179.

Rugani, R., Vallortigara, G., & Regolin, L. (2013a). Numerical abstraction in young domestic chicks *(Gallus gallus). PLoS One, 8*(6), e65262.

Rugani, R., Vallortigara, G., & Regolin, L. (2013b). From small to large: Numerical discrimination by young domestic chicks *(Gallus gallus). Journal of Comparative Psychology, 128*(2), 163–171.

Rugani, R., Vallortigara, G., Vallini, B., & Regolin, L. (2011). Asymmetrical number-space mapping in the avian brain. *Neurobiology of Learning and Memory, 95,* 231–238.

Simion, F., Regolin, L., & Bulf, H. (2008). A predisposition for biological motion in the newborn baby. *Proceedings of the National Academy of Sciences of the United States of America, 105,* 809–813.

Sovrano, V. A., Bisazza, A., & Vallortigara, G. (2002). Modularity and spatial reorientation in a simple mind: Encoding of geometric and nongeometric properties of a spatial environment by fish. *Cognition, 85,* B51–B59.

Sovrano, V. A., Bisazza, A., & Vallortigara, G. (2003). Modularity as a fish views it: Conjoining geometric and nongeometric information for spatial reorientation. *Journal of Experimental Psychology. Animal Behavior Processes, 29,* 199–210.

Sovrano, V. A., Bisazza, A., & Vallortigara, G. (2005). Animals' use of landmarks and metric information to reorient: Effects of the size of the experimental space. *Cognition, 97,* 121–133.

Sovrano, V. A., Bisazza, A., & Vallortigara, G. (2007). How fish do geometry in large and in small spaces. *Animal Cognition, 10,* 47–54.

Sovrano, V. A., Potrich, D., & Vallortigara, G. (2013). Learning of geometry and features in bumblebees *(Bombus terrestris). Journal of Comparative Psychology, 127,* 312–318.

Sovrano, V. A., Rigosi, E., & Vallortigara, G. (2012). Spatial reorientation by geometry in bumblebees. *PLoS One, 7*(5), e37449.

Sovrano, V. A., & Vallortigara, G. (2006). Dissecting the geometric module: Different linkage for metric and landmark information in animals' spatial reorientation. *Psychological Science, 17,* 616–621.

Spelke, E. S. (2000). Core knowledge. *The American Psychologist, 55,* 1233–1243.

Spelke, E. S. (2003). What makes us smart? Core knowledge and natural language. In D. Gentner & S. Goldin-Meadow (Eds.), *Language in mind: Advances in the investigation of language and thought* (pp. 277–311). Cambridge, MA: MIT Press.

Srinivasan, M. V. (2009). Honeybees as a model for vision, perception and 'cognition'. *Annual Review of Entomology, 55,* 267–284.

Stancher, G., Sovrano, V. A., Potrich, D., & Vallortigara, G. (2013). Discrimination of small quantities by fish (redtail splitfin, *Xenotoca eiseni). Animal Cognition, 16,* 307–312.

Stürzl, W., Cheung, A., Zeil, J., & Cheng, K. (2008). The rotational errors and the similarity of views in a rectangular experimental arenas. *Journal of Experimental Psychology. Animal Behavior Processes, 34,* 1–14.

Sugita, Y. (2008). Face perception in monkeys reared with no exposure to faces. *Proceedings of the National Academy of Sciences of the United States of America, 105,* 394–398.

Tommasi, L., Gagliardo, A., Andrew, R. J., & Vallortigara, G. (2003). Separate processing mechanisms for encoding geometric and landmark information in the avian hippocampus. *European Journal of Neuroscience, 17*, 1695–1702.

Tommasi, L., & Vallortigara, G. (2001). Encoding of geometric and landmark information in the left and right hemispheres of the avian brain. *Behavioral Neuroscience, 115*, 602–613.

Trick, L. M., & Pylyshyn, Z. W. (1994). Why are small and large numbers enumerated differently? A limited-capacity preattentive stage in vision. *Psychological Review, 101*, 80–102.

Vallortigara, G. (2000). Comparative neuropsychology of the dual brain: A stroll through left and right animals' perceptual worlds. *Brain and Language, 73*, 189–219.

Vallortigara, G. (2004). Visual cognition and representation in birds and primates. In L. J. Rogers & G. Kaplan (Eds.), *Vertebrate comparative cognition: Are primates superior to non-primates?* (pp. 57–94). New York: Kluwer Academic/Plenum Publishers.

Vallortigara, G. (2006). The cognitive chicken: Visual and spatial cognition in a non-mammalian brain. In E. A. Wasserman & T. R. Zentall (Eds.), *Comparative cognition: Experimental explorations of animal intelligence* (pp. 41–58). Oxford, UK: Oxford University Press.

Vallortigara, G. (2009a). Original knowledge and the two cultures. In E. Carafoli, G. A. Danieli, & G. O. Longo (Eds.), *The two cultures: Shared problems* (pp. 125–145). Springer Verlag.

Vallortigara, G. (2009b). Animals as natural geometers. In L. Tommasi, M. Peterson, & L. Nadel (Eds.), *The biology of cognition* (pp. 83–104). Cambridge, MA: MIT Press.

Vallortigara, G. (2012a). Core knowledge of object, number, and geometry: A comparative and neural approach. *Cognitive Neuropsychology, 29*, 213–236.

Vallortigara, G. (2012b). Aristotle and the chicken: Animacy and the origins of beliefs. In A. Fasolo (Ed.), *The theory of evolution and its impact* (pp. 189–200). New York: Springer.

Vallortigara, G., Chiandetti, C., & Sovrano, V. A. (2011). Brain asymmetry (animal). *Wiley Interdisciplinary Reviews: Cognitive Science, 2*, 146–157.

Vallortigara, G., Chiandetti, C., Sovrano, V. A., Rugani, R., & Regolin, L. (2010). Animal cognition. *Wiley Interdisciplinary Reviews: Cognitive Science, 1*, 882–893.

Vallortigara, G., Feruglio, M., & Sovrano, V. A. (2005). Reorientation by geometric and landmark information in environments of different size. *Developmental Science, 8*, 393–401.

Vallortigara, G., Pagni, P., & Sovrano, V. A. (2004). Separate geometric and non-geometric modules for spatial reorientation: Evidence from a lopsided animal brain. *Journal of Cognitive Neuroscience, 16*, 390–400.

Vallortigara, G., & Regolin, L. (2006). Gravity bias in the interpretation of biological motion by inexperienced chicks. *Current Biology, 16*, R279–R280.

Vallortigara, G., Regolin, L., Chiandetti, C., & Rugani, R. (2010). Rudiments of mind: Insights through the chick model on number and space cognition in animals. *Comparative Cognition & Behavior Reviews, 5*, 78–99.

Vallortigara, G., Regolin, L., & Marconato, F. (2005). Visually inexperienced chicks exhibit spontaneous preference for biological motion patterns. *PLoS Biology, 3*, 1312–1316.

Vallortigara, G., & Rogers, L. J. (2005). Survival with an asymmetrical brain: Advantages and disadvantages of cerebral lateralization. *Behavioral and Brain Sciences, 28*, 575–589.

Vallortigara, G., Rogers, L. J., & Bisazza, A. (1999). Possible evolutionary origins of cognitive brain lateralization. *Brain Research Reviews, 30*, 164–175.

Vallortigara, G., Rogers, L. J., Bisazza, A., Lippolis, G., & Robins, A. (1998). Complementary right and left hemifield use for predatory and agonistic behaviour in toads. *NeuroReport, 9*, 3341–3344.

Vallortigara, G., Sovrano, V. A., & Chiandetti, C. (2009). Doing Socrates experiment right: Controlled rearing studies of geometrical knowledge in animals. *Current Opinion in Neurobiology, 19*, 20–26.

Vallortigara, G., Zanforlin, M., & Pasti, G. (1990). Geometric modules in animal's spatial representation: A test with chicks. *Journal of Comparative Psychology, 104*, 248–254.

Wills, T. J., Cacucci, F., Burgess, N., & O'Keefe, J. (2010). Development of the hippocampal cognitive map in preweanling rats. *Science, 328*, 1573–1576.

Wood, J. N. (2013). Newborn chickens generate invariant object representations at the onset of visual object experience. *Proceedings of the National Academy of Sciences of the United States of America, 110*(34), 14000–14005.

Wystrach, A., & Beugnon, G. (2009). Ants learn geometry and feature. *Current Biology, 19*, 61–66.

Wystrach, A., Cheng, K., Sosa, S., & Beugnon, G. (2011). Geometry, features, and panoramic views: Ants in rectangular arenas. *Journal of Experimental Psychology. Animal Behavior Processes, 37*, 420–435.

Chapter 3

Numerical Concepts: Grey Parrot Capacities

Irene M. Pepperberg
Department of Psychology, Harvard University, Cambridge, MA, USA

INTRODUCTION

The study of numerical abilities is confounded by many issues. Even for humans, some researchers still disagree on what constitutes various stages of numerical competence; which are the most complex, abstract stages; what mechanisms are involved; and even what is enumerated (for a detailed review, see Carey, 2009). And considerable discussion exists as to the extent to which language—or at least symbolic representation—is required for numerical competence, not only for preverbal children but also for people in traditional cultures without formal schooling and for nonhuman species (e.g., Frank, Everett, Fedorenko, & Gibson, 2008; Gordon, 2004; Watanabe & Huber, 2006). If language and number skills require the same abstract cognitive capacities, then animals lacking human language and, for the most part, symbolic representation, should not succeed on abstract number tasks. An alternate view is that humans and animals have similar simple, basic number capacities but that only humans' language skills enable development of symbolic numerical representation and thus abilities such as verbal counting, addition, etc. (see Carey, 2009; Pepperberg, 2006b; Pepperberg & Carey, 2012). But what if a nonhuman had already acquired a certain level of abstract, symbolic representation that might assist in other cognitive tasks (e.g., Premack, 1983)? Could such abilities be adapted to the study of numerical competence?

The issue was particularly intriguing for someone like myself, studying avian abilities, because at the time I began my research, others had already demonstrated certain levels of avian numerical competence using subjects that lacked symbolic representation (e.g., canaries, chickens, corvids, parrots). Most animal studies prior to the 1970s avoided the issue of symbolic labeling by having subjects (1) choose a particular set of items from among several competing arrays, presumably based on quantity; (2) choose between observable arrays or their symbolic representation with respect to "more" versus

Mathematical Cognition and Learning, Vol. 1. http://dx.doi.org/10.1016/B978-0-12-420133-0.00003-X

"less"; (3) perform match-to-sample problems for various quantities and denote quantities in separate, simultaneously presented arrays as "same" or "different"; or (4) respond with a specific behavior to a particular quantity of sequential events. Differences among the preceding tasks, and why they might fail to demonstrate precise number comprehension, can best be clarified through examples; note that such tasks are still being used at present to demonstrate that most nonhumans do indeed have an approximate number system (i.e., one that is nonexact but able to discriminate magnitudes; see Carey, 2009).

In the first set of experiments noted previously, subjects such as canaries (Pastore, 1961) ignored color, shape, size, brightness, pattern, and so forth to select one quantity from simultaneously presented arrays of other quantities. If the designated quantity was three, subjects supposedly responded based on "threeness." Given a choice between, for example, two trays with either three or four chess pieces, or two trays with either two or three chess pieces, the birds choose the tray with three to get their reward. Once trained, the birds could also choose the third pill in a row among other objects on the tray to get a reward, but could not transfer to choosing the third of any other homogeneous set. Animals in these studies, moreover, did not respond to different quantities based on different cues presented at random, responded to only one quantity at a time (i.e., could be trained to choose three and *retrained* to choose two, but were not trained so that, for example, blue trays signaled that they should choose three, and red trays signaled they should choose two, with red and blue trays presented randomly in sequential trials), and could not (or were not given tests to) transfer their responses from simultaneous to sequential presentations (see Pepperberg, 1987b). A conservative interpretation of the results is that these animals recognized the particular aggregation as a single perceptual unit (Mandler & Shebo, 1982; von Glasersfeld, 1982) that was associated with a reward.

In the second set of experiments noted previously, subjects such as pigeons (*Columba livia*, Honig & Matheson, 1995; Honig & Stewart, 1989) likely used prenumerical abilities to choose between, for example, matrices of arrays of dots consisting of different proportion of colors, or of different numbers of dots, being rewarded for choosing "more" versus "less." Such choices did not require recognizing a specific quantity (note Koehler, 1950). Even when animals symbolically indicated "more" versus "less" and objects varied in size, brightness, etc., the task still did not require labeling specific quantities with abstract, unique symbols (Pepperberg, 1987b). More recently, Emmerton and Renner (2006) used a version of this task to demonstrate that pigeons (*Columba livia*) could distinguish between various amounts of small and large arrays in a manner consistent with Weber's Law; Scharf, Hayne, and Columbo (2011) have found similar effects. Bogale, Kamata, Mioko, and Sugita (2011) have shown that jungle crows (*Corvus macrorhynchos*) also perform relative quantity discriminations, again limited by Weber's law (i.e., the accuracy of discrimination of different quantities being based less with respect to their absolute differences than with the ratios between them). Ujfalussy, Miklosi, Bugnyar, and Kotrschal (2013) have demonstrated comparable behavior in jackdaws (*Corvus monedula*), as have Zorina and Smirnova (1996) in gray crows (*Corvus tristis*).

In the third set of studies, animals succeeded to different extents on simultaneous match-to-sample problems, indicating if two arrays were identical in quantity (see Pepperberg, 1987b). Grey parrots, ravens (*Corvus corax*), and jackdaws succeeded for quantities up to 8; pigeons reached 5 or 6; chickens managed 2 or 3 (Braun, 1952; Koehler, 1943, 1950; Lögler, 1959), with all tasks controlled for mass, density, contour, brightness, etc. Lögler (1959) showed that Grey parrots transferred such behavior to light flashes and flute notes up to 7, thus going from simultaneous visual representations to sequential visual and auditory ones. More recently, Smirnova, Lazareva, and Zorina (2000) showed that hooded crows (*Corvus cornix*) have performed match-to-sample tasks with numerical sets comparable to subjects in Koehler's (1950) studies. How the task relates to labeling of quantity or counting is, however, unclear: Koehler called the match-to-sample task "thinking in un-named numbers" or "non-numerical counting," because distinguishing same versus different requires only a one-to-one correspondence and not necessarily that the task be solved based on an understanding of, or perceiving, actual quantity (note Fuson & Hall, 1983; Gelman & Gallistel, 1986). Too, mechanisms involved in match-to-sample versus labeling are likely different: at least in humans, separate brain areas apparently mediate these two tasks (Geschwind, 1979).

In the fourth set of studies, animals produce a particular action in response to a certain number of events (review in Pepperberg, 1987b). Thus, a subject responds to auditory sequences of 3 but not 2 or 4 (Davis & Albert, 1986). According to Seibt (1982), however, these data indicate that animals merely form a one-to-one correspondence with some internal pattern such that external events are judged as to whether they match this pattern. Koehler's birds may have performed a more sophisticated version of this task: they learned to open boxes randomly containing 0, 1, or 2 baits until they obtained a fixed number (e.g., 4). Thus, the number of boxes needed to be opened to obtain the precise number of baits varied across trials, and the number being sought depended on independent visual cues (e.g., black box lids denoted 2 baits; green lids, 3; red lids, 4). Koehler claimed that his birds learned to respond correctly to four different cues simultaneously. He does not state, however, if different colored lids were presented randomly in a single series and thus whether colors may indeed have "represented" particular quantities (see Pepperberg, 1987b).

The question, at that time, was whether a parrot, named Alex, who had already acquired some level of symbolic representation using the sounds of English speech, could, as did Matsuzawa's (1985) chimpanzee (*Pan troglodytes*) Ai, for example, go beyond the number tasks described above and use numerical symbols as categorical labels (Pepperberg, 1987b). Specifically, could Alex learn to use vocal numerical labels to distinguish various numerical arrays and to denote number as an attribute of a collection, the same way he labeled its color? If successful, could he generalize (transfer) his behavior to any novel collection, including novel items placed in random patterns? And, finally, could he respond to a collection solely in terms of its quantity, that is, respond with equal accuracy to mixtures of different objects (e.g., balls and blocks)? If so, he would demonstrate numerical competence comparable to that of chimpanzees and young children.

ALEX'S NON-NUMERICAL CAPACITIES

When I first began numerical work with Alex in the 1980s, he had already achieved competence on various tasks once thought limited to young children or at least nonhuman primates (Pepperberg, 1999). Through the use of a modeling technique, roughly based on that of Todt (1975), Alex learned to use English speech sounds to referentially label a large variety of objects and their colors (Pepperberg, 1981); at the time he could also label two shapes ("3-corner" for triangles, "4-corner" for squares; later he identified various other polygons as "x-corner"; Pepperberg, 1983). He understood concepts of category: that the same item could be identified with respect to material, color, shape, and object name (e.g., "wood," "blue," "4-corner," and "block"; Pepperberg, 1983). He had functional use of phrases such as "I want X" and "Wanna go Y," X and Y being appropriate object or location labels. He was acquiring concepts of *same, different*, and *absence*—for any object pair he could label the attribute ("color," "shape," "matter") that was same or different, and state "none" if nothing was the same or different (Pepperberg, 1987a, 1988); he was also learning to view collections of items and state the attribute of the sole object defined by two other attributes—for example, in a set of many objects of which some were yellow and some were pentagonal, to label the material of the only one that was both yellow *and* pentagonal (Pepperberg, 1992). But could he form an entirely new categorical class consisting of quantity labels?

ALEX'S EARLY NUMERICAL ABILITIES

As noted previously, to succeed on number concepts, Alex would have to reorganize how he categorized objects in his world. He would have to learn that a new set of labels, "one," "two," "three," etc., represented a novel classification strategy; that is, one based on both physical similarity within a group and a group's quantity, rather than solely by physical characteristics of group members. He would also have to generalize this new class of number labels to sets of novel items, items in random arrays, heterogeneous collections, and eventually to more advanced numerical processes (Pepperberg, 1999, 2006b). If successful, he would demonstrate a *symbolic* concept of number, that is, vocally designate the *exact* quantity of a given array with an appropriate numerical, referential utterance in his repertoire (Pepperberg, 2012b).

Training and Testing Methods

Via our standard modeling technique that enabled Alex to produce labels for objects, colors, and shapes, he was initially trained to identify small number sets with English labels (note, however, that he initially used "sih" for six, because he had trouble pronouncing the final /s/; Pepperberg, 1987b). As we will see in detail later, this training was quite different from that experienced by young children. For example, unlike children (Carey, 2009), who learn numbers in the appropriate ordinal pattern (i.e., "one," "two," etc.), Alex was first trained on sets of three and four, because he already had those labels in his repertoire; he was then taught "five" and "two," then "six," and lastly "one." Training in such a manner also ensured that Alex was building his concept of number solely by forming one-to-one associations between specific quantities and their respective number labels (Pepperberg, 2006b). Unlike children, who seem to learn "one" fairly easily (i.e., "one" versus "many," Carey, 2009), "one" was actually rather difficult for Alex to acquire, because he already knew to label a single item with an object label and had to be trained for quite some time to add the number label. Training details are published elsewhere and will not be repeated (Pepperberg, 1987b). Training was limited to sets of a few familiar objects; testing involved transfer to sets of other familiar and novel exemplars. Various publications describe, again in great detail, testing procedures that ensured against myriad forms of possible external cuing, both with respect to inadvertent human cuing and cues based on non-number issues such as mass, brightness, density, surface area, odor, item familiarity, or canonical pattern recognition (Pepperberg, 1987b, 1994, 1999, 2006a,b,c).

Labeling of Basic Quantities and Simple Heterogeneous Sets

Initial studies demonstrated that Alex could use English labels to quantify small sets of familiar different physical items, up to six, exactly (78.9%, all trials; Pepperberg, 1987b); that is, he overall made few errors, and his data did not show a peak near a correct response with many errors of nearby numerals, which would have suggested only a general sense of quantity (i.e., an approximate number system). Rather, his most common errors across all sets was to provide the label of the object involved—to respond, for example, "key" rather than "four key," which accounted for almost 60% of his roughly 50 errors in ~250 trials (another ~20% of his errors involved unintelligible responses or misidentifications of the object or material; i.e., 80% of his errors were non-numerical). Thus, Alex indeed had a concept of quantity; he was not, however, necessarily counting, as would a human child who understood, for example, the concept of "five" (Fuson, 1988; Mix, Huttenlocher, & Levine, 2002; Pepperberg, 1999); that is, who understood the counting principles: that a stable symbolic list of numerals exists, numerals must be applied to individuals in a set to be enumerated in order, they must be applied in one-to-one correspondence, that the last numeral reached in a count represents the cardinal value of the set, and that each numeral is exactly one more than the previous numeral (Fuson, 1988; Gelman & Gallistel, 1986). Even if he wasn't technically counting, additional tests demonstrated that Alex could quantify even unfamiliar items and those not arranged in any particular (canonical) pattern, such as a square or triangle; he maintained an accuracy of about 75%–80% on novel items in random arrays.

Moreover, he could also quantify subsets within heterogeneous sets; that is, in a mixture of Xs and Ys, he could respond appropriately to "How many X?" "How many Y?" *or* "How many toy?" (70%, first trials; Pepperberg, 1987b). Here, he outperformed some children, who are generally tested on only homogeneous sets (e.g., Starkey & Cooper, 1995) and who, if asked about subsets within a mixed set of toys, usually label the *total* number of items if, like Alex, they have been taught to label homogeneous sets exclusively (see Greeno, Riley, & Gelman, 1984; Siegel, 1982).

Despite these tests, we still could not identify the mechanism(s) Alex might be using to succeed. Notably, our tests ensured that Alex could not use non-numerical cues such as mass, brightness, surface area, odor, object familiarity, or canonical pattern recognition (Pepperberg, 1987b, 1999), because we questioned him on a variety of exemplars of various sizes and of both familiar and novel textures and materials (e.g., metal keys versus bottle corks) often presented by simply tossing them in random arrays on a tray. Such controls did not, however, rule out the possibility that, for the smallest collections, Alex had used a noncounting strategy such as subitizing—a perceptual mechanism that enables humans to quickly quantify sets up to ~4 without counting—or, for larger collections, "clumping" or "chunking"—another form of subitizing (e.g., perception of six as two groups of three; for a review, see von Glasersfeld, 1982). Thus, many other tests would be needed to determine the mechanisms that Alex was indeed using.

Complex Heterogeneous Sets

Some tests to tease apart subitizing/clumping versus counting issues were initially designed for humans by Trick and Pylyshyn (1989, 1994; vanMarle, this volume). In their experiments, subjects had to enumerate a particular set of items embedded within two different types of distractors: (1)

white *or* vertical lines among green horizontals; (2) white vertical lines among green vertical *and* white horizontals. They found subitizing for 1–3 in only the first condition, but counting, even for such small quantities, in the second. When subjects thus must distinguish among various items defined by a collection of competing features (e.g., a conjunction of color *and* shape; see Pepperberg, 1999), subitizing becomes unlikely. Alex could be examined in a comparable manner because he already was being tested on conjunction (e.g., being asked to identify the color of an item that was both triangular and wood in a collection of differently shaped objects of various materials; Pepperberg, 1992). He could thus be asked to label the quantity of a similarly defined subset—for example, the number of green blocks in a set of orange and green balls and blocks. Would his numerical capacities match those of humans?

Alex turned out to be about as accurate as humans (83.3% on 54 trials, Pepperberg, 1994; see Trick & Pylyshyn, 1989), and analyses suggested, but could not yet prove, that he, like humans, was counting. Had he used perceptual strategies similar to those of humans (e.g., subitizing and clumping), rather than counting, he would have made no errors for 1 and 2, few for 3, and more for larger numbers. His errors, however, were random with respect to number of items to be identified (Pepperberg, 1994) and, importantly, his responses were not simply a close approximation to the correct number label (Pepperberg, 1994), which would be expected had he been subitizing or even estimating. In fact, most of Alex's errors seemed unrelated to numerical competence, but rather were in misinterpreting the defining labels and then correctly quantifying the incorrectly targeted subset. Eight of his nine errors were the correct number for an alternative subset (e.g., the number of blue rather than red keys; in those cases, the quantity of the designated set usually differed from that of the labeled set by two or more items). The problem, however, was that there was no way of knowing whether Alex's perceptual capacities might be more sophisticated than those of humans, allowing him to subitize larger quantities; the data, although impressive with respect to exact number, still did not justify claiming that he was definitively counting. A detailed discussion can be found in Pepperberg (1994).

In a subsequent study (Pepperberg & Carey, 2012), we further tested Alex's responses concerning exact number. Here we examined how he might process quantities greater than those he could label; we specifically wanted to see if his label "sih" actually referred to exactly six items, or roughly six; that is, to anything he might perceive as large. We showed him, in individual trials under no time constraints, seven, eight, and nine items, asking "How many X?" There were two trials for each quantity, in random order, interspersed with trials on smaller sets and non-number tasks, to ensure that he could switch between sets and objects he could label and those that (potentially) he could not. He was neither rewarded nor scolded whatever his reply, simply told "OK"; we then went to the next query. In trials for sets greater than six,

Alex usually did not answer initially but remained quietly seated on his perch or asked to return to his cage. Only when we continuously badgered him, asking over and over, did he eventually reply "sih." His actions suggested that he knew his standard number answers would be incorrect and he did not, as when bored with a task (e.g., Pepperberg, 1992; Pepperberg & Gordon, 2005), give strings of wrong answers, request many treats, or turn his back and preen.

ALEX'S MORE ADVANCED NUMERICAL ABILITIES

En route to determining the mechanism—or mechanisms—Alex used to quantify sets, my students and I examined various other numerical capacities. Thus, Alex was tested on comprehension of numerical labels, on his ability to sum small quantities, and on whether he understood the ordinality of his numbers (Beran et al., this volume; Star & Brannon, this volume). The latter task was of particular interest, because, as noted previously, unlike children, he had not been trained in an ordinal manner: He had first learned to label sets of three and four, then five and two, then six and one.

Number Comprehension

Although Alex could label numerical sets, he had never been tested on number label comprehension. In general, researchers who teach nonhumans to use a human communication code must ensure the equivalence of label production and comprehension (e.g., Savage-Rumbaugh, Rumbaugh, & Boysen, 1980; Savage-Rumbaugh et al., 1993), but the issue is particularly important in numerical studies: even a young child who successfully labels the number of items in a small set ("Here's X marbles") might fail when shown a very large quantity and asked "Can you give me X marbles?" That is, the child might not really understand the relationship between the number label and the quantity (Wynn, 1990). If labeling indeed separates animal and human numerical abilities (see above; Pepperberg, 2012b; Pepperberg & Carey, 2012; Watanabe & Huber, 2006), such comprehension–production equivalence would be necessary to demonstrate nonhuman numerical competence (Fuson, 1988).

To test Alex's comprehension abilities, we used a variation of the previous task. Here, we simultaneously presented several sets of different quantities of different items—for example, X red blocks, Y yellow blocks, Z green blocks, or X blue keys, Y blue wood, and Z blue pompons, with X, Y, and Z being different quantities. Alex was then queried, respectively, "What color Z?" or "What matter (material) X?" (Pepperberg & Gordon, 2005). He received no training on this task prior to testing. To succeed, he had to comprehend the auditorily presented numeral label (e.g., X = "four") and use its meaning to direct a search for the cardinal amount specified by that label (e.g., four things), that is, know exactly what a set of "X" items is, even when

FIGURE 3-1   Alex's comprehension task. Trials with blocks were the only ones in which all the objects were exactly the same size; these trials tested whether accuracy improved with same-sized objects.

intermixed with other items representing different numerical sets (Figure 3-1). We again controlled for contour, mass, etc., by using objects of different sizes, within or across trials so that comprehension of the number label was the only way to consistently perform correctly (Pepperberg & Gordon, 2005). To respond correctly, he also had to identify the item or color of the set specified by the numerical label. Some or all this behavior likely occurred as separate steps, each adding to task complexity (Premack, 1983).

Alex's overall score was again impressive (statistically significant 87.9% on 66 trials), with no errors on the first 10 trials (Pepperberg & Gordon, 2005). Interestingly, errors increased with time, suggesting lack of focus or inattention as testing proceeded. He may have been like keas (*Nestor notabilis*; Gajdon, Amann, & Huber, 2011) or large-billed crows (*Corvus macrorhynchos*; Izawa & Watanabe, 2011) that will, after succeeding on various tasks, often later employ other, less successful or simply different methods, possibly from boredom (e.g., to engender more interesting responses from trainers; Pepperberg, 2012c) or maybe to find other possible solutions. In any case, he understood the meaning of his number labels somewhat better than young children (see above, Fuson, 1988; Wynn, 1990, 1992), and, most importantly, he had little difficulty with numbers differing by small amounts, suggesting that his number sense was exact and not approximate. Most of his errors appeared to involve color perception or phonological confusion, not numerical misunderstanding: He sometimes erred in distinguishing orange from red or yellow, a consequence of differences in parrot and human color vision (Bowmaker, Heath, Das, & Hunt, 1994; Bowmaker, Heath, Wilkie, Das, & Hunt, 1996); he also sometimes confused "wool" and "wood," or "truck" and "chalk"; he pronounced the last label a bit like "chuck" (Pepperberg & Gordon, 2005).

Use of "None"

The comprehension study was notable for another reason: Alex's spontaneous transfer of use of "none"—learned as a response to the queries "What's same/ different?" with respect to two objects when no category (color, shape, or material) was the same or different (Pepperberg, 1988)—to the absence of a set of a particular quantity. After responding appropriately for several trials of the standard comprehension task, Alex reacted in a manner quite different from the norm. When, on one particular trial, he was asked "What color three?" to a set of two, three, and six objects, he replied "five"; obviously no such set existed and his response made little sense. The questioner asked twice more, and each time he replied, "Five." Finally, the questioner said, "OK, Alex, tell me, what color 5?" to which he immediately responded "none." The response came as a complete surprise, as he had never been taught the concept of absence of quantity nor to respond to absence of an exemplar. He had, previously, spontaneously transferred use of "none" from the same-different study to "What color bigger?" for two equally sized items in a study on relative size (Pepperberg & Brezinsky, 1991), but that use of "none" still referred to the absence of difference in an attribute. "None," or a zero-like concept, is advanced, abstract, and relies on the violation of an expectation of presence (Bloom, 1970; Hearst, 1984; Pepperberg, 1988). Of additional interest was that Alex not only had provided a correct, novel response, but had also manipulated the trainer into asking the question he apparently wished to answer, which suggested other levels of abstract processing (Pepperberg & Gordon, 2005). Alex also correctly answered additional queries about absent sets of one to six items, showing that his behavior was intentional and meaningful.

A subsequent study (Pepperberg & Carey, 2012) further emphasized Alex's number comprehension and made use of his knowledge of absence. Here, we again tested him with sets larger than he could label: He saw four trays with sets of various numbers of items, including 7 or 8 but omitting 6 (e.g., 3 yellow wool, 4 blue wool, 7 green wool), and was asked "What color six?" to see if he would reply "none" (Pepperberg & Gordon, 2005)—would he require exactly six or accept the set that was roughly six (here, say "green")? These questions tested whether he knew that "six" meant *exactly* six and not approximately six, that is, whether he truly understood that his labels referred to very specific sets. He was also asked about an existing set for two arrays to ensure he did not learn to respond "none." Thus, he had six queries: two probing an existing set (one for a 3-item set, one for 5) and four for which the correct response was "none" if "six" meant exactly six. Alex responded "none" on all four trials involving quantities above six. On trials for colors of sets that were present, he gave the appropriate labels (respectively, "yellow" and "green" to 3- and 5-item sets).

A critical issue for this study was that Alex's initial use of "none" was spontaneous, unlike that of the chimpanzee, Ai, who had to be trained to use the label "zero" (Biro & Matsuzawa, 2001). But our data did not demonstrate whether he really understood the overall *concept* of zero. How similar was his understanding to that of a young child or an adult human? Only additional studies could provide that information.

Addition of Small Quantities

Although I had always wanted to determine if Alex could perform the same kind of small number addition as did chimpanzees (Boysen & Berntson, 1989), I had started to focus on other areas of cognitive processing (e.g., research on optical illusions; Pepperberg, Vicinay, & Cavanagh, 2008) at this time. Thus, studies on addition (Pepperberg 2006a), like those on "zero," were unplanned, and came about as follows. Alex, who routinely interrupted the sessions of a younger parrot, Griffin, with phrases like "Talk clearly" or with an appropriate answer, appeared to sum the clicks over the individual trials that we were using to train Griffin on sequential auditory numbers (training to respond to, e.g., three computer-generated clicks with the vocal label "three"). Given how difficult it would be to demonstrate true summation auditorially, I chose to replicate, as closely as possible, the object-based addition study of Boysen and Berntson (1989) on apes, and to use the experiment to study further Alex's understanding of zero (Pepperberg, 2006a).

I chose the Boysen and Bernston (1989) procedure because it was a formal test of addition—having a subject observe two (or more) separate quantities and provide the *exact* label for their total (Dehaene, 1997)—that is, it required both summation *and* symbolical labeling of the sum by a nonhuman. Most additive and subtractive studies on nonhumans required the subject to choose the larger amount of two sets, not label final quantity (review in Pepperberg, 2006a). Specifically, when the correct response involves choice of relative amount, no information is obtained on whether the subject has ". . .a digital or discrete representation of numbers" (Dehaene, 1997, p. 27; see also Carey, 2009, for a discussion of how such responses can rely on an approximate number system). In contrast to most other addition studies, moreover, I avoided use of only one token type of a standard size (e.g., whole marshmallows), which could allow evaluations to be based on contour and mass, not number (note Mix et al., 2002).

The procedure was as follows: Alex was presented with a tray on which two upside-down cups had been placed (Figure 3-2); prior to presentation, a trainer had hidden items such as randomly shaped nuts, bits of cracker, or different-sized jelly beans under each cup, with items in the same cup less than 1 cm from each other. We occasionally used identical candy hearts to see if accuracy was higher when mass/contour cues were available

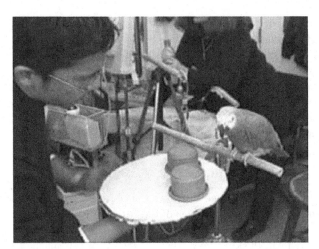

FIGURE 3-2   Alex's addition task.

(Pepperberg, 2006a). After bringing the tray up to Alex's face, the experimenter lifted the cup on his left, showed what was under the cup for 2–3 seconds in initial trials, replaced the cup over the quantity, and then replicated the procedure for the cup on his right. For reasons described later, in trials comprising the last third of the experiment, Alex had ~6–10 seconds to view items under each cup before everything was covered. The experimenter then made eye contact with Alex, who was asked, vocally, and without any training, to respond to "How many total?" He saw collections with all possible addends, totaling to every amount from 1 to 6, plus trials with nothing under both cups to see if he would generalize use of "none" without instruction. No objects other than the cups were visible during questioning. To respond correctly, Alex had to remember the quantity under each cup, perform some combinatorial process, and then produce a label for the total amount. He had no time limit in which to respond, given that his response time generally correlated with his current interest in the items being used in the task, rather than the task itself (Pepperberg, 1988). Appropriate controls for cuing and tests for interobserver agreement were, as usual, in place (Pepperberg, 2006a).

   For sets of countable objects, Alex had a statistically significant accuracy of 85.4% on 48 first-trial responses (Pepperberg, 2006a), and his accuracy did not improve on trials with identical tokens. He had trouble with one set of trials, however. Interestingly, when given only 2–3 seconds, he always erred on the $5+0$ sum, consistently stating "6." Notably, he did not have problems with "fiveness" per se, as he was always accurate on trials involving $4+1$. (Note that with $2+3$ he sometimes labeled the addends, but he did that with $3+1$ as well; the error seemed to be a case of attending only to a single set of objects rather than to a problem with addition; he corrected himself on all those trials when asked a second time but never changed his response to

$5+0$ for the 2–3-second interval.) When given $\sim$6–10 seconds for $5+0$, however, his accuracy went to 100%. Differences in accuracy between the shorter and longer interval trials was significant only on $5+0$ trials (Fisher's exact test, $p=0.01$). Such data suggest that he used a counting strategy for 5: Only when beyond 4 did he, like humans whose subitizing range is 1–4, need time to label the set exactly (i.e., the data suggested his subitizing range matched that of humans; for a detailed discussion, see Pepperberg, 2006a). Overall, his data are comparable to those of children (Mix et al., 2002) and, because he added to six, are more advanced than those published on apes (Boysen & Hallberg, 2000).

In a subsequent study (Pepperberg, 2012a), Alex showed that he could perform with equal accuracy when asked to sum three sets of sequentially presented objects—that is, collections of variously sized objects now hidden under three cups. Here, he had to maintain numerical accuracy under what could be an additional memory load, because the protocol required two updates in memory rather than one. His first trial score was 8/10 correct, 80%, $p<0.001$ (binomial test, chance of either ¼ or 1/6; 1/6 represented a guess of all possible number labels, ¼ represented a guess of using one of the three addends as well as their sum). For all trials, his score was 10/12 correct, or 83.3%. Occasionally, one cup contained no objects, but including only those trials in which all three cups contained items, Alex's first trial score was 4/5 correct, $p=0.015$ (chance of ¼; for chance of 1/6, $p<0.01$); his all-trials score was 5/6 or, again, 83.3%. In this three-cup task, all of the addends were within subitizing range (Boysen & Hallberg, 2000; Pepperberg 2006a); thus, Alex could easily have tracked these without specifically counting. However, he still would have needed to remember the values under each of the three cups, for several seconds for each cup, and update his memory after seeing what was under each cup, even if nothing was present. Again, because he added up to 6, his competence surpassed that of an ape similarly tested (Sheba: Boysen & Hallberg, 2000).

Interestingly, in the two-cup task, Alex did not respond "none" when nothing was under any cup (Pepperberg, 2006a; such trials were not present in the three-cup task). He either looked at the tray and said nothing (five trials) or said "one" (three trials). He never said "two," showing that he understood that the query did not correspond to the number of *cups*. On trials in which he did not respond, his lack of action suggested that he knew his standard number answers ("one" through "six") would be incorrect, and he might not have known what to say. His behavior somewhat resembled that of autistic children (Diane Sherman, personal communication, January 17, 2005), who simply stare at the questioner when asked "How many X?" if nothing exists to count. As for his response of "one," he may, despite never having been trained on ordinality and having learned numbers in random order (see above), have inferred that "none" and "one" represented the lower end of the number spectrum and conflated the two labels. The chimpanzee Ai often confused "none" and "one" (Biro & Matsuzawa, 2001). Alex's inability to use "none" here might have arisen because he was asked to denote the total absence of labeled *objects*; previously, he was responding to the absence of an attribute. Specifically, these data confirmed that Alex's use of "none" was merely *zero-like*: he did not use "none," as he did his number labels, to denote a specific numerosity (Pepperberg, 1987b). In that sense, he was like humans in traditional cultures, or young children, who seem to have to be ~4 years old before achieving full adult-like understanding of the label for zero (e.g., Bialystok & Codd, 2000; Wellman & Miller, 1986).

Ordinality and Equivalence

As noted previously, despite having learned his number labels out of order—quite unlike children—Alex may have deduced something about ordinality, that is, about an exact number line. He had a concept of bigger and smaller (Pepperberg & Brezinsky, 1991) and, without explicit training, may have organized his number labels in that manner. Such behavior would be important for two reasons. First, even for apes that referentially used Arabic symbols, ordinality with respect to those symbols did not emerge but had to be trained (e.g., Biro & Matsuzawa, 2001; Boysen, Berntson, Shreyer, & Quigley, 1993; Matsuzawa, Itakura, & Tomonaga, 1991); if Alex understood ordinality without training, his concepts would be more advanced than those of a nonhuman primate. Second, ordinality is intrinsic to *verbal* counting (e.g., Fuson, 1988; Gelman & Gallistel, 1986). To count, an organism must produce a standard sequence of number tags and know the relationships among and between these tags; for example, that "three" (be it any vocal or physical symbol) not only comes before "four" in the verbal sequence but also represents a quantity exactly one less than "four." An understanding of ordinality, therefore, would help support our possible claims for counting.

Notably, ordinality is not a simple concept. Children acquire ordinal-cardinal abilities in steps. They learn cardinality, slowly, usually over the course of over a year (generally beginning sometime between age 2½ and 3), for very small numbers (<4), and only after having previously acquired a general sense of "more versus less" and while acquiring a meaningless, rote ordinal number series. Only around the time that they acquire an understanding of "fourness" do they connect their knowledge of quantity in the small sets with this number sequence to form one-to-one correspondences that can be extended to larger amounts for both cardinal and ordinal accuracy (e.g., Carey, 2009; see also Mix et al., 2002). Children may give the impression that they have full understanding of cardinality before they actually do, by learning associative rules (i.e., respond correctly using the last number word for a counted set when asked "How many?" but fail on "Give me X"; see above) but cannot consistently act in that manner with ordinality (e.g., Bruce & Threfall, 2004; Teubal & Guberman, 2002).

To test what Alex might know about ordinality and compare his abilities to those of children would require first that he learn to label Arabic numerals so that he could be tested abstractly, that is, in the absence of physical sets of objects. If, after learning English labels for Arabic numerals (production and comprehension) in the absence of the physical quantities to which they refer, Alex could—without any training—use the commonality of these English labels to equate quantities (sets of physical objects) and Arabic numerals, then I could use a task involving these equivalence relations (Pepperberg, 2006c): I could ask him which of two Arabic numerals was bigger or smaller. To ensure that I could repeat the trials enough times to gain statistical significance without Alex learning rote responses to specific pairs, the task would be to identify the *color* of one of a pair of Arabic numbers (e.g., a green 2, a yellow 5, next to each other on a tray; Figure 3-3) that was *numerically* (not physically) bigger or smaller. He already answered "What color/matter bigger/smaller?" for object pairs and responded "none" for same-sized pairs (Pepperberg & Brezinsky, 1991). To succeed on this new task, he would have to use deductions and inferences: deduce that an Arabic symbol has the same numerical value as its *vocal label*, compare *representations* of quantity for which the labels stand, infer rank ordering based on these representations, and then state the result *orally* (Pepperberg, 2006c). Unlike the tasks used in other studies of nonhuman species (e.g., Olthof, Iden, & Roberts, 1997; Olthof & Roberts, 2000), the question would not always be about the larger set, and specific stimuli within pairs would not be associated with reward of the corresponding number of items.

To ensure that Alex really understood not only ordinality but the meaning of the Arabic numerals, he was tested on several related tasks (Pepperberg,

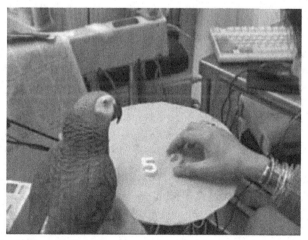

FIGURE 3-3 Alex's ordinality task.

2006c). Trials on identical numerals of different colors but of the same size (e.g., 6:6) tested if Alex would, as expected, reply "none" to the query as to which was bigger or smaller. To determine if he might be tricked into responding based on the physical appearance of the numerals rather than their meanings, he was queried about numerals of the same value but different colors and different sizes (e.g., 2:2). By mixing Arabic symbols and physical items, I could determine whether he really did understand that, for example, one numeral (an Arabic 6) was bigger than five items (or an Arabic 2 was the same as two items) and cleanly separate mass and number.

Alex did indeed succeed on the equivalence task and, as a consequence, demonstrated that, without direct, explicit training, he inferred the ordinality of his number labels (Pepperberg, 2006c). Notably, he had never been trained to recite the labels in order nor to associate any Arabic numeral with any specific set of objects. Nevertheless, for trials on two different Arabic numbers of the same physical size, his first trial score was 63/84, or 75% ($p < .01$, binomial test, chance of ½). If his occasional responses of the Arabic number label rather than the requested color (technically correct, but not with respect to the actual query) were not counted as errors, his score was 74/84, or 88.1% ($p < .001$, binomial test, chance of ½). As in previous studies, errors sometimes involved yellow–orange–red confounds. When numerals were the same value–same size, his accuracy was 10/12, or 83.3%, $p < .01$ (binomial test, chance of 1/3; answers could be one of the two colors or "none"). Importantly, statistical comparisons on his first and final trials for all these sets showed no significant differences in accuracy, suggesting that no training was occurring. For the same value–different size trials, counting as correct either "none" or the color label of the physically targeted number, his accuracy was 12/12, or 100%, $p < .01$ (binomial test, chance of 2/3, a color or "none"). Seven times he gave the correct color of the physically targeted number, five times he said "none," but gave colors most often in earlier trials and "none" most often in later trials, as if he shifted after experience with responses based on symbolic value, even though he had initially been rewarded for responses based on physical size (Pepperberg, 2006c).

Alex's responses to trials that mixed objects and numerals were intriguing. For arrays in which object sets were paired with a single Arabic number representing a quantity larger than or equal to the array (incongruent trials) and in which the single Arabic number represented a quantity less than the array (congruent trials), his accuracy was 16/21, or 76.2%, $p < .01$. However, in five trials in which a *single* object was paired with a *single* Arabic number that represented a larger quantity, Alex consistently replied "none." Only here did the physical set consistently overwhelm symbolic responses.

Overall, Alex did appear to exhibit numerical understanding far closer to that of children than that of other animals. However, he differed from humans and was like other nonhumans in that he had demonstrated no savings in his learning of larger numerals. Once children learn ordinality and the successor function—that each digit in their number line is one more than the previous digit—they no longer need to be taught the values of each individual digit for digits greater than 4 (Carey, 2009). Why was Alex unlike children in this instance? Might the issue be Alex's difficulty in learning to produce the English sounds? In order to produce any given English label, Alex had to learn to coordinate his syrinx, tracheal muscles, glottis, larynx, tongue height and protrusion, beak opening, and even esophagus (Patterson & Pepperberg, 1998); might there be a way to dissociate vocal and conceptual learning to test this possibility?

An Exact Integer System

To test whether such a dissociation existed, colleagues and I devised the following experiment. Initially, I would teach Alex to identify vocally the Arabic numerals 7 and 8 in the absence of their respective quantities, divorcing the time needed to learn the speech patterns from any concept of number. Only after the labels were being produced clearly would I train him to understand that $6 < 7 < 8$, that is, where the new numerals fit on the number line. Training would be based on the way we had tested ordinality: students and I would model "What color (number) bigger/smaller?" He could then be tested as to whether he understood the relationships among 7 and 8 and the quantities represented by his other Arabic labels. If he inferred the new number line in its totality, he could be tested on whether, like children, he could *spontaneously* understand that "seven" represented one more physical object than "six," and that "eight" represented two more than "six" and one more than "seven," by labeling appropriate physical sets on first trials (Pepperberg & Carey, 2012). Nothing in his training at this point would provide specific information about the value of 7 and 8; they could refer to 10 and 20 items, respectively. The question was whether, all other numerals having been taught as either +1 or −1 than those he already knew (that is, after learning "3" and "4," he was taught "5" and "2," then "6" and "1"; Pepperberg, 1987b, 1994), he could use past and present information to induce the cardinal meaning of the labels "seven" and "eight" from their ordinal positions on an implicit count list.

Over the course of the study, Alex did indeed learn to label the novel Arabic numerals, to place them appropriately in his inferred number line, and to label appropriately, *on first trials*, novel sets of seven and eight physical items. Like children, he did *not* have to be taught their cardinal values. Detailed data are presented in the published paper (Pepperberg & Carey, 2012); the conclusion was that Alex, like children, and unlike nonhuman primates tested so far, created a representational structure that allowed him to encode the cardinal value expressed by any numeral in his count list (Carey, 2009).

The Final Study

Once Alex had acquired the numerals through 8, we went back to the addition task to determine if he could, like apes (Boysen & Berntson, 1989), sum the

Arabic numerals that had been hidden under cups (Pepperberg, 2012a). Such a task would demonstrate further knowledge of the representational nature of the numerals. As in the earlier described addition experiment with sets of items, he was sequentially shown two Arabic numerals initially hidden under cups and, in their subsequent absence, was asked to vocally produce a label to indicate their sum. In a separate small set of trials, he was shown the same stimuli in the same manner, but was simultaneously presented with various Arabic numerals of different colors and asked for the color of the numeral representing the sum; colors changed on each trial. The second set of trials ensured that Alex could not learn a particular pattern over time (e.g., "if I see X + Y, I say Z"). Alex's passing precluded completion of this latter task, but had he lived longer, this procedure, with its additional step, would have allowed testing the same sums many more times without training him to produce a specific response, unlike tasks given to other nonhuman subjects (see discussion in Pepperberg, 2012a).

For the Arabic numeral task requiring a numerical response, Alex demonstrated some competence in summing two Arabic numerals, each representing quantities less than or equal to 5, to a total of 1–8. His first trial score was 9/12 (75%), $p = 0.004$ (chance of 1/3; $p = 0.001$ for chance of 1/8). His all trials score was 12/15 (80%). Although the study did not contain enough trials to test all possible sums and combinations of addends or to repeat most queries, Alex was given at least one trial for each sum from 1 to 8. The lack of replication of the various sums over trials, however, emphasizes the first trial nature of the results and shows that no training during assessment could have been involved. Notably, if the numerals had only approximate meanings, Alex's errors would likely have exhibited a range close to the correct response. In contrast, such was the case only once (Pepperberg, 2012a); the other errors were to state "eight" when the sums were five and four. He thus seemed to have some fixation on producing the label "eight," which was his newest. Overall, his data surpassed what would be expected if he were using the kinds of systems employed by most nonhumans or preverbal infants—for example, analog magnitude systems or object files, which cannot represent any positive integer above 4 exactly (see Carey 2009, for a review).

Because of his death, Alex had only three trials on queries requiring a color response; his first trial score, 2/3 (66%), was too low for statistical significance $(p = 0.07)$, but the small number of trials preclude real statistical power. His all-trials score for this set was 3/4 (75%). These data do, however, suggest a capacity for exact number representation: conceivably, his one error, on the first trial, may have represented a misunderstanding of the task. His response, which labeled the numeral representing "two," suggests he might have responded to the number of objects under the cups (i.e., the two numerals) rather than their values, given that no training of any sort had preceded questioning on this novel task, and all previous queries did refer to the number of objects. Note, however, that he did not persist in this response but was correct when asked a second time and responded appropriately on the next two trials. Overall, Alex, like the chimpanzee Sheba, had had no training on summing the Arabic numerals, and, like Sheba, spontaneously transferred from summing items to summing symbols. His data on the color response task (although extremely limited)—a task somewhat like that of Sheba's, in that possible responses were available from which to choose—tended toward significance. In contrast to Sheba, however, he had to indicate the label not just for the sum but for the color of the numeral that represented the correct numerical sum (an additional step), and the total summed quantity on which he was tested could reach 8.

CONCLUSIONS

The preceding data demonstrate the extent to which a nonhuman, nonprimate, nonmammalian subject can form complex, abstract concepts and, specifically, that one particular subject, Alex, understood the cardinal representation of his vocal number labels and their corresponding Arabic numerals. He succeeded at levels that, on occasion, went beyond those of nonhuman primates and approached those of children. And, of course, as noted previously, Alex was not the only avian subject to succeed on numerical tasks.

Corvids are considered "feathered primates" (e.g., Emery & Clayton, 2004); and interestingly, a recent report on hooded crows (Smirnova, 2013) argues for the ability to associate Arabic numbers and object sets in a manner similar to that of Alex, although not in the vocal mode. Thus, we have come a long way from the time when "avian cognition" was considered an oxymoron (Pepperberg, 2011) and exact numerical representations were thought to be an exclusively human accomplishment (Lenneberg, 1971).

ACKNOWLEDGMENTS

Writing of this chapter was supported by donors to *The Alex Foundation*, particularly Anita Keefe, the Sterner Family, and the Marc Haas Foundation. Portions of the chapter were previously published as a tribute to Anthony Wright as Pepperberg (2013).

REFERENCES

Bialystok, E., & Codd, J. (2000). Representing quantity beyond whole numbers: Some, none and part. *Canadian Journal of Experimental Psychology, 54*, 117–128.

Biro, D., & Matsuzawa, T. (2001). Use of numerical symbols by the chimpanzee (*Pan troglodytes*): cardinals, ordinals, and the introduction of zero. *Animal Cognition, 4*, 193–199.

Bloom, L. (1970). *Language development: Form and function in emerging grammars*. Cambridge, MA: MIT Press.

Bogale, B. A., Kamata, N., Mioko, K., & Sugita, S. (2011). Quantity discrimination in jungle crows, *Corvus macrorhynchos. Animal Behaviour, 82*, 635–641.

Bowmaker, J. K., Heath, L. A., Das, D., & Hunt, D. M. (1994). Spectral sensitivity and opsin structure of avian rod and cone visual pigments. *Investigative Ophthalmology & Visual Science*, *35*, 1708.

Bowmaker, J. K., Heath, L. A., Wilkie, S. E., Das, D., & Hunt, D. M. (1996). Middlewave cone and rod visual pigments in birds: Spectral sensitivity and opsin structure. *Investigative Ophthalmology & Visual Science*, *37*, S804.

Boysen, S. T., & Berntson, G. G. (1989). Numerical competence in a chimpanzee (*Pan troglodytes*). *Journal of Comparative Psychology*, *103*, 23–31.

Boysen, S. T., & Hallberg, K. I. (2000). Primate numerical competence: Contributions toward understanding nonhuman cognition. *Cognitive Science*, *24*, 423–443.

Boysen, S. T., Berntson, G. G., Shreyer, T. A., & Quigley, K. S. (1993). Processing of ordinality and transitivity by chimpanzees (*Pan troglodytes*). *Journal of Comparative Psychology*, *107*, 208–215.

Braun, H. (1952). Uber das Unterscheidungsvermögen unbenannter Anzahlen bei Papageien. *Zeitschrift für Tierpsychologie*, *9*, 40–91.

Bruce, B., & Threfall, J. (2004). One, two, three and counting. *Educational Studies in Mathematics*, *55*, 3–26.

Carey, S. (2009). *The origin of concepts*. New York: Oxford University Press.

Davis, H., & Albert, M. (1986). Numerical discrimination by rats using sequential auditory stimuli. *Animal Learning & Behavior*, *14*, 57–59.

Dehaene, S. (1997). *The number sense*. Oxford, UK: Oxford University Press.

Emery, N. J., & Clayton, N. S. (2004). The mentality of crows: Convergent evolution of intelligence in corvids and apes. *Science*, *306*, 1903–1907.

Emmerton, J., & Renner, J. C. (2006). Scalar effects in the visual discrimination of numerosity by pigeons. *Learning & Behavior*, *34*, 176–192.

Frank, M. C., Everett, D. L., Fedorenko, E., & Gibson, E. (2008). Number as a cognitive technology: evidence from Pirahã language and cognition. *Cognition*, *108*, 819–824.

Fuson, K. C. (1988). *Children's counting and concepts of number*. Berlin: Springer-Verlag.

Fuson, K. C., & Hall, J. W. (1983). The acquisition of early number word meanings: A conceptual analysis and review. In H. P. Ginsburg (Ed.), *The development mathematical thinking* (pp. 49–107). New York: Academic Press.

Gajdon, G. K., Amann, L., & Huber, L. (2011). Keas rely on social information in a tool use task but abandon it in favor of overt exploration. *Interaction Studies*, *12*, 304–323.

Gelman, R., & Gallistel, C. R. (1986). *The child's understanding of number* (2nd ed.). Cambridge, MA: Harvard University Press.

Geschwind, N. (1979). Specializations of the human brain. *Scientific American*, *241*, 180–199.

Gordon, P. (2004). Numerical cognition without words: Evidence from Amazonia. *Science*, *306*, 496–499.

Greeno, J. G., Riley, M. S., & Gelman, R. (1984). Conceptual competence and children's counting. *Cognitive Psychology*, *16*, 94–143.

Hearst, E. (1984). Absence as information: Some implications for learning, performance, and representational processes. In H. L. Roitblat, T. G. Bever, & H. S. Terrace (Eds.), *Animal cognition* (pp. 311–332). Hillsdale, NJ: Erlbaum.

Honig, W. K., & Matheson, W. R. (1995). Discrimination of relative numerosity and stimulus mixture by pigeons with comparable tasks. *Journal of Experimental Psychology: Animal Behavior Processes*, *21*, 348–363.

Honig, W. K., & Stewart, K. E. (1989). Discrimination of relative numerosity by pigeons. *Animal Learning & Behavior*, *17*, 134–146.

Izawa, I.-E., & Watanabe, S. (2011). Observational learning in the large-billed crow (*Corvus macrorhynchos*): Effect of demonstrator–observer dominance relationship. *Interaction Studies, 12,* 281–303.

Koehler, O. (1943). 'Zähl'—Versuche an einem Kolkraben und Vergleichsversuche an Menschen. *Zeitschrift für Tierpsychologie, 5,* 575–712.

Koehler, O. (1950). The ability of birds to 'count'. *Bulletin of the Animal Behavioral Society, 9,* 41–45.

Lenneberg, E. (1971). Of language, knowledge, apes, and brains. *Journal of Psycholinguistic Research, 1,* 1–29.

Lögler, P. (1959). Versuche zur Frage des "Zähl" Vermögens an einem Graupapagei und Vergleichsversuche an Menschen. *Zeitschrift für Tierpsychologie, 16,* 179–217.

Mandler, G., & Shebo, B. J. (1982). Subitizing: An analysis of its component processes. *Journal of Experimental Psychology: General, 111,* 1–22.

Matsuzawa, T. (1985). Use of numbers by a chimpanzee. *Nature, 315,* 57–59.

Matsuzawa, T., Itakura, S., & Tomonaga, M. (1991). Use of numbers by a chimpanzee: A further study. In A. Ehara, T. Kimura, O. Tkenaka, & M. Iwamoto (Eds.), *Primatology today* (pp. 317–320). Amsterdam: Elsevie.

Mix, K., Huttenlocher, J., & Levine, S. C. (2002). *Quantitative development in infancy and early childhood.* New York: Oxford University Press.

Olthof, A., & Roberts, W. A. (2000). Summation of symbols by pigeons (*Columba livia*): The importance of number and mass of reward items. *Journal of Comparative Psychology, 114,* 158–166.

Olthof, A., Iden, C. M., & Roberts, W. A. (1997). Judgments of ordinality and summation of number symbols by squirrel monkeys (*Saimiri sciureus*). *Journal of Experimental Psychology: Animal Behavior Processes, 23,* 325–333.

Pastore, N. (1961). Number sense and 'counting' ability in the canary. *Zeitschrift für Tierpsychologie, 18,* 561–573.

Patterson, D. K., & Pepperberg, I. M. (1998). A comparative study of human and grey parrot phonation: Acoustic and articulatory correlates of stop consonants. *Journal of the Acoustical Society of America, 103,* 2197–2213.

Pepperberg, I. M. (1981). Functional vocalizations by an African Grey parrot (*Psittacus erithacus*). *Zeitschrift für Tierpsychologie, 55,* 139–160.

Pepperberg, I. M. (1983). Cognition in the African Grey parrot: Preliminary evidence for auditory/vocal comprehension of the class concept. *Animal Learning & Behavior, 11,* 179–185.

Pepperberg, I. M. (1987a). Acquisition of the same/different concept by an African Grey parrot (*Psittacus erithacus*): Learning with respect to categories of color, shape, and material. *Animal Learning & Behavior, 15,* 423–432.

Pepperberg, I. M. (1987b). Evidence for conceptual quantitative abilities in the African Grey parrot: labeling of cardinal sets. *Ethology, 75,* 37–61.

Pepperberg, I. M. (1988). Acquisition of the concept of absence by an African Grey parrot: learning with respect to questions of same/different. *Journal of the Experimental Analysis of Behavior, 50,* 553–564.

Pepperberg, I. M. (1992). Proficient performance of a conjunctive, recursive task by an African Grey parrot (*Psittacus erithacus*). *Journal of Comparative Psychology, 106,* 295–305.

Pepperberg, I. M. (1994). Evidence for numerical competence in an African Grey parrot (*Psittacus erithacus*). *Journal of Comparative Psychology, 108,* 36–44.

Pepperberg, I. M. (1999). *The Alex studies.* Cambridge, MA: Harvard University Press.

Pepperberg, I. M. (2006a). Grey parrot (*Psittacus erithacus*) numerical abilities: Addition and further experiments on a zero-like concept. *Journal of Comparative Psychology, 120,* 1–11.

Pepperberg, I. M. (2006b). Grey parrot numerical competence: a review. *Animal Cognition, 9,* 377–391.

Pepperberg, I. M. (2006c). Ordinality and inferential abilities of a Grey Parrot (*Psittacus erithacus*). *Journal of Comparative Psychology, 120,* 205–216.

Pepperberg, I. M. (2011). Avian cognition and social interaction: fifty years of advances. *Interaction Studies, 12,* 195–207.

Pepperberg, I. M. (2012a). Further evidence for addition and numerical competence by a Grey parrot (*Psittacus erithacus*). *Animal Cognition, 15,* 711–717.

Pepperberg, I. M. (2012b). Emotional birds—or advanced cognitive processing? In S. J. Watanabe & S. Kuczaj (Eds.), *Emotions of animals and humans: Comparative perspectives* (pp. 49–62). Japan: Springer.

Pepperberg, I. M. (2012c). Symbolic communication in the Grey parrot. In J. Vonk & T. Shackelford (Eds.), *Oxford handbook of comparative evolutionary psychology* (pp. 297–319). New York: Oxford University Press.

Pepperberg, I. M. (2013). Abstract concepts: data from a Grey Parrot. *Behavioural Processes, 93,* 82–90.

Pepperberg, I. M., & Brezinsky, M. V. (1991). Acquisition of a relative class concept by an African Grey parrot (*Psittacus erithacus*): Discriminations based on relative size. *Journal of Comparative Psychology, 105,* 286–294.

Pepperberg, I. M., & Carey, S. (2012). Grey Parrot number acquisition: the inference of cardinal value from ordinal position on the numeral list. *Cognition, 125,* 219–232.

Pepperberg, I. M., & Gordon, J. D. (2005). Number comprehension by a Grey parrot (*Psittacus erithacus*), including a zero-like concept. *Journal of Comparative Psychology, 119,* 197–209.

Pepperberg, I. M., Vicinay, J., & Cavanagh, P. (2008). The Müller-Lyer illusion is processed by a Grey Parrot *(Psittacus erithacus). Perception, 37,* 765–781.

Premack, D. (1983). The codes of man and beast. *Behavioral and Brain Sciences, 6,* 125–176.

Savage-Rumbaugh, E. S., Rumbaugh, D. M., & Boysen, S. (1980). Do apes use language? *American Scientist, 68,* 49–61.

Savage-Rumbaugh, S., Murphy., J., Sevcik, R. A., Brakke, K. E., Williams, S. L., & Rumbaugh, D. M. (1993). Language comprehension in ape and child. *Monographs of the Society for Research in Child Development, 233,* 1–258.

Scharf, D., Hayne, H., & Colombo, M. (2011). Pigeons on par with primates in numerical competence. *Science, 334,* 1664.

Seibt, U. (1982). Zahlbegriff und Zählverhalten bei Tieren: Neue Versuche und Deutungen. *Zeitschrift für Tierpsychologie, 60,* 325–341.

Siegel, L. S. (1982). The development of quantity concepts: perceptual and linguistic factors. In C. J. Brainerd (Ed.), *Children's logical and mathematical cognition* (pp. 123–155). Berlin: Springer-Verlag.

Smirnova, A.A. (2013). Symbolic representation of the numerosities 1–8 by hooded crows (*Corvus cornix L.*). Paper presented at the International Ethological Congress, Newcastle, UK, August.

Smirnova, A. A., Lazareva, O. F., & Zorina, Z. A. (2000). Use of number by crows: investigation by matching and oddity learning. *Journal of the Experimental Analysis of Behavior, 73,* 163–176.

Starkey, P., & Cooper, R. G. (1995). The development of subitizing in young children. *British Journal of Developmental Psychology, 13,* 399–420.

Teubal, E., & Guberman, A. (2002). The development of children's counting ability. *Megamot*, *42*, 83–102.

Todt, D. (1975). Social learning of vocal patterns and modes of their applications in Grey parrots. *Zeitschrift für Tierpsychologie*, *39*, 178–188.

Trick, L., & Pylyshyn, Z. (1989). *Subitizing and the FNST spatial index model*. Ontario, Canada: University of Ontario, COGMEM #44.

Trick, L., & Pylyshyn, Z. (1994). Why are small and large numbers enumerated differently? A limited-capacity preattentive stage in vision. *Psychological Review*, *101*, 80–102.

Ujfalussy, D., Miklosi, A., Bugnyar, T., & Kotrschal, K. (2013). Role of mental representations in quantity judgements by jackdaws (*Corvus monedula*). *Journal of Comparative Psychology*. http://dx.doi.org/10.1037/a0034063.

von Glasersfeld, E. (1982). Subitizing: The role of figural patterns in the development of numerical concepts. *Archives de Psychologie*, *50*, 191–218.

Watanabe, S., & Huber, L. (2006). Animal logics: decision in the absence of human language. *Animal Cognition*, *9*, 235–245.

Wellman, H. M., & Miller, K. F. (1986). Thinking about nothing: Development of concepts of zero. *British Journal of Developmental Psychology*, *4*, 31–42.

Wynn, K. (1990). Children's understanding of counting. *Cognition*, *36*, 155–193.

Wynn, K. (1992). Children's acquisition of the number words and the counting system. *Cognitive Psychology*, *24*, 220–251.

Zorina, Z. A., & Smirnova, A. A. (1996). Quantitative evaluations in gray crows: Generalizations of the relative attribute 'larger set.' *Neuroscience and Behavioral Physiology*, *26*, 357–364.

Chapter 4

Numerical Cognition and Quantitative Abilities in Nonhuman Primates

Michael J. Beran[1], Audrey E. Parrish[2] and Theodore A. Evans[1]
[1]Language Research Center, Georgia State University, Decatur, GA, USA
[2]Language Research Center and Department of Psychology, Georgia State University, Decatur, GA, USA

INTRODUCTION

Does "number" matter to nonhuman animals? For as long as scientists have been conducting experimental investigations of "animal minds," this question has been at the forefront of comparative study (along with a few other questions, such as whether animals have language, or "mindreading" abilities, or can make and use tools). The question itself pertains to whether animals perceive and can manipulate or represent the numerical properties of stimulus sets in the same ways that they can perceive and represent other stimulus dimensions such as color, shape, size, movement, taste, or texture. Number, of course, is more abstract than these other physical attributes of stimuli, and so it was generally reserved as a uniquely human capacity in terms of representational processing of numerical attributes. This changed when comparative psychologists began the long history of trying to assess nonhuman animal numerical cognition. With this long history comes much controversy, namely from concerns that early experimental work suffered from flaws such as poor control over stimulus presentation and worries about unintentional cuing of animals (the "Clever Hans" effect; Pfungst, 1911; also see Beran, 2012a; Candland, 1995).

However, progress continued throughout the 20th century, and this question has, to a large extent, now been answered, using a variety of test paradigms that examine when and how animals respond to numerical stimuli. While these studies suggest that nonhuman animals cannot count in the same way as human children, animals do show many of the hallmark features of numerical cognition that emerge during human development and that support

Mathematical Cognition and Learning, Vol. 1. http://dx.doi.org/10.1016/B978-0-12-420133-0.00004-1

the ultimate human mathematical achievements such as counting, arithmetic, and more advanced mathematical calculations. Whereas it was once considered that animals might use number only as a "last resort" (Davis & Memmott, 1982; Davis & Perusse, 1988), we now know that this is not true and that numerical properties of stimuli are highly salient and relevant in many circumstances. We outline these comparative findings, concentrating on what evidence there is for counting-like abilities in animals and then shift the focus to other quantitative skills that animals (and, particularly, nonhuman primates) show in a variety of contexts. These skills include ordinal learning (e.g., learning to sequence stimuli on the basis of their numerical values), relative quantity discriminations (choosing sets of items on the basis of a more-than or less-than rule), and certain types of quantity illusions or inferential judgments that are made on the basis of quantity information. We also focus closely on the mechanisms that underlie numerical and quantitative abilities shown by animals and directly compare those mechanisms to their counterparts in human numerical cognition.

Although in the present chapter we limit our discussion mostly to tests given to nonhuman primates, it is worth mentioning at least briefly just how broadly the assessment of animal numerical and quantitative competencies has become. Beyond primate species, we now know that relative quantities and even small exact numbers are relevant in the judgments made by all types of animals (see Table 4-1). So, although our focus is on contributions from nonhuman primate research, the competencies shown by primates are, in many cases, also shown by a variety of other animals (for respective reviews of fish, parrots, and fowl, see Agrillo et al., this volume; Pepperberg, this volume; Vallortigara, this volume).

THE QUESTION OF ANIMAL COUNTING

For children, counting is a multifaceted competence that involves learning the list of their culture's number words, mapping these words onto the quantities they represent, and coming to understand that each successive number word represents "one more" than the word right before. These basic skills typically emerge between the ages of 2 and 6 years, but exactly how children acquire them is debated (e.g., Briars & Siegler; 1984; Gelman & Gallistel, 1978). One influential view is that children learning how to count are guided by a set of five implicit principles (e.g., Gelman & Gallistel, 1978; Gelman & Meck, 1983), as shown in Table 4-2.

The comparative approach to the study of counting principles took the form of trying to establish what animals could, or could not, do by way of using these principles in enumerative contexts. In some cases, experimenters tried to match the methods of studies used with children, but in many cases new methods were developed that were more animal friendly but still required numeric labels, cardinal judgments, and ordinal knowledge. The key issue,

TABLE 4.1 Diversity of nonprimate species tested in numerical cognition experiments

Group	Animal	Species	Assessment	Example Citation(s)
Insects	Honeybees	*Apis mellifera*	Landmark counting Navigation	Dacke & Srinivasan (2008) Pahl et al. (2013)
Fish	Mosquitofish Angelfish Guppies	*Gambusia holbrooki* *Pterophyllum scalare* *Poecilia reticulata*	Quantity discrimination	Agrillo et al. (2008, 2011); Dadda et al. (2009) Gómez-Laplaza & Gerlai (2011, 2013) Piffer et al. (2012)
Amphibians	Salamanders	*Plethodon shermani* *Plethodon cinereus*	Quantity discrimination	Krusche et al. (2010) Uller et al. (2003)
Birds	Pigeons Robins Grey parrots	*Columba livia* *Petroica longipes* *Psittacus erithacus*	Quantity discrimination	Emmerton (1998) Garland et al. (2012) Aïn et al. (2009)
	Chicks	*Gallus gallus domesticus*	Counting Ordinality Quantity discrimination	Pepperberg (1987, 1994) Pepperberg (2006a) Rugani et al. (2007, 2008)
Carnivores	Dogs Coyotes Hyenas Bears Lions	*Canis lupus familiaris* *Canis latrans* *Crocuta crocuta* *Ursus americanus* *Panthera leo*	Quantity discrimination	Ward & Smuts (2007) Baker et al. (2011, 2012) Benson-Amram et al. (2010) Vonk & Beran (2012) McComb et al. (1994)
			Auditory enumeration	
Rodents	Rats	*Rattus norvegicus*	Counting	Burns et al. (1995); Capaldi & Miller (1988); Davis & Bradford (1986)
	Voles	*Microtus pennsylvanicus*	Quantity discrimination	Ferkin et al. (2005, 2009)
Marine mammals	Dolphins Whales Sea lions	*Tursiops truncatus* *Delphinapterus leucas* *Otaria flavescens*	Quantity discrimination	Jaakkola et al. (2005); Kilian et al. (2003) Abramson et al. (2013) Abramson et al. (2011)
Ungulates	Horses Elephants	*Equus caballus* *Loxodonta Africana* *Elephas maximus*	Quantity discrimination	Uller & Lewis (2009) Perdue et al. (2012) Irie-Sugimoto et al. (2009)

Note: By no means is this an exhaustive list of studies (or species) that have been conducted to assess numerical cognition and quantitative abilities. Rather, it is representative of the breadth of research in this area beyond the research with primates that is the central focus of this chapter.

TABLE 4-2 The counting principles*

Principle	Definition
One-to-one correspondence	Each item in a counted array must be tagged with a symbolic label once and only once in a counting event.
Stable order	The numeric tags used to represent each item within a counted array must be applied in the same order across counting events so that the labels consistently represent the same specific item number for any count sequence.
Ordinality	The numeric tags, when applied in a stable order, represent each item's sequential place within the array (e.g., first, third, fifth). This principle affords the knowledge for making relational judgments ("more," "less," "bigger," "smaller") and has been one of the most heavily studied principles in the comparative literature.
Cardinality	The final numeric tag applied when one counts a set of items has a special role—namely, that it represents, not only the last item's ordinal position, but also the total number of items in the set. This principle allows one to answer the question "how many?" by responding with the last numeric tag that was applied.
Abstraction	One can count anything that can be individuated, using the principles above—pennies, pistachios, and sneezes are all countable things.

Gelman and Gallistel (1978)

though, was cardinality. Could nonhuman animals use some form of enumerative process to generate a cardinal value for a set of items and just "how high" could animals count. Early evidence was consistent with cardinal-like knowledge, particularly for chimpanzees (*Pan troglodytes:* e.g., Boysen & Berntson, 1989; Matsuzawa, 1985), a Grey parrot (*Psittacus erithacus:* Pepperberg, 1987, 1994), rats (*Rattus norvegicus:* Burns et al., 1995; Capaldi & Miller, 1988; Davis & Bradford, 1986) and pigeons (*Columba livia:* e.g., Roberts & Mitchell, 1994).

For example, Boysen and Berntson (1989) showed that the chimpanzee Sheba could move to different locations in the lab, see varying numbers of items in each location (but not all visible at the same time), and then return to a third location where she labeled the whole array with an Arabic numeral, with high proficiency. This would suggest that Sheba knew the order of the numerals, that she carefully tagged each set of items with a numeral (ordinality; see Boysen, Berntson, Shreyer, & Hannan, 1995), and that she learned that the last tag represented the combined number of items across locations (cardinality). Pepperberg (1987, 1994) also reported that the parrot Alex could survey arrays of items (and even subsets of specific things within larger arrays) and provide a verbal numeral label in response to questions about quantity (e.g., "How many blue key?"), also demonstrating some of the counting principles when faced with a "How many?" task. These are among the most impressive feats of counting-like ability yet demonstrated in nonhuman animals.

However, something else was becoming clear. In other kinds of tests, where animals could learn associations between arbitrary stimuli such as Arabic numerals and specific quantities, they did so with a degree of inexactness that was unlike the improved accuracy and precision of counting in children as they approached 5 or 6 years of age. For example, in our laboratory, we have trained chimpanzees to collect sets of items on a computer screen by using a joystick to move a cursor into contact with those items. In the task, they have to collect a number of items that match the value represented by a presented Arabic numeral (Figure 4-1). This is the equivalent of a "Give me X" task, where the animal must construct the set of items to match a desired target number, and chimpanzees performed above chance levels for numbers up to 7 (Beran, 2004a; Beran & Rumbaugh, 2001; Rumbaugh, Hopkins, Washburn, & Savage-Rumbaugh, 1989; Beran, Rumbaugh, & Savage-Rumbaugh, 1998). But, what is striking is that performance becomes more variable and lower overall as the requested number of items gets larger (Figure 4-2). This is in sharp contrast to children's performance.

Children first learn the relation between specific number words and the quantities they represent but only understand this for a subset of number

FIGURE 4-1 The chimpanzee Lana using a joystick to perform a computerized enumeration task. Lana must move a cursor (the +) into contact with the dots at the bottom of the screen in order to match the target numeral for the trial (in this case, the number 7).

words that they know, but once they understand the concept of cardinal value, this generalizes to all number words they know (Sarnecka & Carey, 2008; Wynn, 1990). The chimpanzees' performance is more similar to that found with children who know the cardinal value of only a subset of number words but do not fully understand cardinality. This suggested that nonhuman animals had a more "approximate" representation of number (Brannon & Roitman, 2003; Gallistel & Gelman, 2000), one that allowed for a range of competencies in responding to numerical tasks, but not in the way one would expect if fully developed counting skills were in place. Coupled with this was the extensive training required in many of these experiments, in some cases years

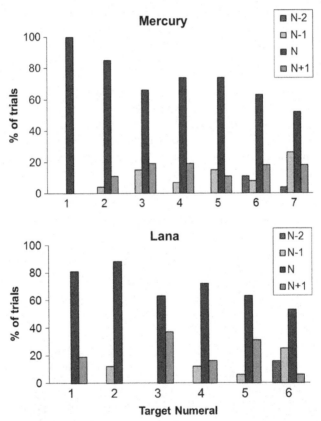

FIGURE 4-2 Construction of quantities to match presented target numerals by the chimpanzees in Beran and Rumbaugh (2001). N represents a quantity equal to the presented target numeral, N+1 represents a quantity that exceeded the presented target numeral by one dot, N–1 represents a quantity that was less than the presented target numeral by one dot, and N–2 represents a quantity that was less than the presented target numeral by two dots. *Reprinted from Beran, M. J., & Rumbaugh, D. M. (2001). "Constructive" enumeration by chimpanzees (Pan troglodytes) on a computerized task.* Animal Cognition, 4, 81–89. *With kind permission from Springer Science and Business Media.*

of training, to establish these impressive but incomplete counting perfor-mances. Although children also take several years to become proficient at counting, typically developing children make continual gains. Once they are armed with a well-learned ordinal number word list and understand the con-cept of cardinality, they do not need explicit experience in mapping each and every numeral or number word to know the quantity it represents. Upon achieving this conceptual insight, they understand that "six" is one more than "five" without explicit demonstrations that six objects are associated with the word "six" (Sarnecka & Carey, 2008; Wynn, 1990). Chimpanzees do not appear to fully arrive at this insight.

What remains unknown is whether these shortcomings are the result of nonhuman species' cognitive or experiential limitations. Although there have been efforts at massive training of numerical competencies in some nonhuman animals such as Sheba or Alex, there has never been an attempt to immerse a nonhuman animal in the daily routines experienced by human children that are full of numerical scripts. The 3-year-old child rarely interacts with his or her environment without being faced with numerical information that requires counting, including when they eat ("*how many* pancakes do you want?"), when they help around the house ("get *four* forks for the table"), and so forth. Thus, a critical question that remains unanswered with regard to the numerical abilities of animals is whether such immersion might pro-mote more human-like counting abilities. The answer may be that it does not, even with the same kinds of environmental stimulation that children receive, and this might be the result of a number of factors ranging from the role of language to differences in neural functioning across species. Yet we know that there are similar neural processes at work in response to quan-titative tasks presented across species (e.g., Nieder & Merten, 2007; Nieder & Miller, 2003, 2004), and we also know that humans (including adults) share with other animals an evolutionarily ancient and foundational system for representing quantities inexactly (see following text). Consequently, we are left with the question of what it takes to spark proficient counting competence—a critical issue as the comparative study of quantitative abilities moves forward.

Next, we outline some of the other capacities that primates exhibit with regard to number and quantity before turning our attention to the question of *how* animals, and specifically primates, represent quantities. These capaci-ties range from seemingly natural judgments between quantities of food items to more complex numerical feats performed within the framework of game-like computer tasks.

RELATIVE QUANTITY JUDGMENTS

A large variety of species, ranging from bees (*Apis mellifera*) to robins (*Petroica longipes*) to elephants (*Loxodonto Africana*), have shown the ability to choose the larger of two or more quantities of food in experimental tests (e.g., Hunt, Low, & Burns, 2008; Krusche et al., 2010; Perdue et al., 2012). In consideration of basic foraging behavior, discrimination and selection of the largest amount of food will maximize intake and increase subsequent survival rates (e.g., Stephens & Krebs, 1986). The great variety and number of animals that have exhibited this capacity support the idea that it provides a clear evolutionary advantage. A large part of this literature consists of nonhuman primate species, and early work in this area was conducted with apes. For example, Dooley and Gill (1977) trained the chimpanzee Lana to choose from two quantities of cereal pieces the quantity that she wanted to receive. Lana often chose the larger quantity of each pair when up to 10 pieces of food were presented in each set. However, her performance declined when both quantities were nearer the upper limit of that range (what is often called a size effect) and when the ratio of the smaller set to the larger set was closer to 1 (e.g., 8 vs. 9 was harder for Lana than 6 vs. 9). Rumbaugh, Savage-Rumbaugh, and Hegel (1987) extended this kind of test to include other chimpanzees and trials where two separate quantities of up to 5 items in each subset and 8 items in total had to be summed (or combined) to determine which pair of quantities was the larger total amount. Again, chimpanzees were successful in performing this summation for all comparison types (as were squirrel monkeys [*Saimiri sciureus sciureus*] in subsequent experiments; Olthof, Iden, & Roberts, 1997; Terrell & Thomas, 1990).

Call (2000) investigated three male orangutans' (*Pongo pygmaeus*) ability to select the larger of two quantities (within the range of 1 to 6) when both quantities could not be viewed at the same time. In that study, the experimenter revealed one quantity at a time by removing a lid from each food container, and the orangutans selected the larger quantity at a level significantly greater than chance. Because the orangutans did not have to see both quantities at once in order to select the larger quantity, Call (2000) concluded that they were using some form of mental representation, rather than a perceptual mechanism. Similarly, Beran (2001) tested two chimpanzees' ability to choose the larger of two sets of one to nine food items when those sets were presented one item at a time. He placed each item (*M&M's* candies) into one of two opaque cups, and the chimpanzees had to track how many candies were in each cup and then select the larger candy quantity. These animals selected the larger quantity on significantly more trials than would be predicted by chance, indicating that they could mentally represent each quantity even though these were never visually accessible in their entirety. The chimpanzees also showed a strong ratio effect, an outcome we discuss in greater detail later in this chapter.

Subsequent experiments with those chimpanzees showed that they could accommodate multiple additions of items to each choice set (Beran, 2004b), over long trial durations in which the additions of items occurred over 20 minutes (Beran & Beran, 2004), and in conditions in which nonvisible sets were compared to a visible set, with the chimpanzees knowing when to take the visible set and knowing when the summed, nonvisible set was larger (Beran, 2004b). These chimpanzees even perform food-based, sequential quantity judgments for auditory stimuli by only being able to hear items that were dropped rather than seeing them being dropped (Beran, 2012b). And, it is not just chimpanzees that perform well in these tests, as the other great apes (gorillas, *Gorilla gorilla gorilla*; bonobos, *Pan paniscus*; orangutans) also show similar abilities (e.g., Anderson et al., 2005; Anderson, Stoinski, Bloomsmith, & Maple, 2007; Hanus & Call, 2007).

Similar work has been conducted with monkey species and has provided consistently positive results. For example, capuchin monkeys (*Cebus apella*) have accurately performed multiple tests requiring choices between different quantities of foods. vanMarle, Aw, McCrink, and Santos (2006) tested whether capuchin monkeys could choose the larger of two quantities of solid food items or non-solid food items. In both types of trials, monkeys were successful, and their performance was negatively correlated with the ratio of the two quantities being compared (i.e., the larger and closer in amount the quantities were, the more difficult they were to compare). In close comparison with some of the chimpanzee work, capuchin monkeys were presented with two sets of food items, identical in food type but differing in number from 1 to 6 (Beran, Evans, Leighty, Harris, & Rice, 2008). Monkeys accurately chose the greater of two quantities when each was presented one at a time (by uncovering and recovering each in succession, as in Call, 2000). Evans, Beran, Harris, and Rice (2009) examined whether capuchin monkeys could perform at the same level when the food sets were presented only one item at a time (as in Beran, 2001). Monkeys succeeded in choosing the more numerous set when one quantity was presented item by item and the alternative set was entirely visible throughout the trial, and some monkeys even accurately compared two sets of food items, each presented item by item at different rates.

Additionally, lemurs (*Eulemur mongoz*) have shown similar sensitivity to quantitative differences in food sets and respond to potential violations of expected numerical outcomes (Lewis, Jaffe, & Brannon, 2005). This result matches those from studies with very young children who also notice violations of expected numerical outcomes (as when, for example, some number of toys move behind an occluder, but when the occluder is removed, a different number of toys are found in that location; e.g., McCrink, this volume; McCrick & Wynn, 2004; Wynn, 1992). Taken together, these studies and others (e.g., Addessi, Crescimbene, & Visalberghi, 2008; Barnard et al., 2013) have demonstrated that the capacity to form and compare representations of food quantities emerged before the divergence of lemurs from the primate lineage.

Investigators have designed other tests that take advantage of these species' ability to form representations of quantity in order to study related phenomena. For example, Beran and colleagues examined how monkeys and apes make quantitative decisions when faced with incomplete information (Beran, Evans, & Harris, 2009; Beran, Owens, Phillips, & Evans, 2012; Beran, Perdue, Parrish, & Evans, 2012). These researchers noticed that, in quantity tests like those described previously, animals sometimes begin to respond to one option or another before encountering both options. This behavior was especially prevalent when the option initially offered represented either the minimum or maximum value seen in a given experiment (e.g., see Evans et al., 2009). In the first study to formally assess this kind of decision making (Beran et al., 2009), four chimpanzees performed sessions in which the first 15 trials were typical quantity judgment trials consisting of two food quantities presented one full set at a time. In the second 15 trials, the experimenter left one of the two quantities covered, and the chimpanzees' choice had to be based on the one set they could see and the range of values they had seen in the session up to that point. Interestingly, the chimpanzees often chose the covered set, and they did so above chance levels, when the visible set represented less than the mean quantity of food items presented in the first half of the session. Moreover, the experimenters varied the range (and mean) of the quantities seen in the first 15 trials across sessions and found that the chimpanzees varied their selection of the visible and covered quantities accordingly. Subsequent studies demonstrated that capuchin monkeys and college undergraduates behaved almost exactly the same way (with the exception being that the nonhuman species slightly undervalued the mean quantity as compared with the human participants; Beran, Owens, Phillips, & Evans, 2012; Beran, Perdue, Parrish, & Evans, 2012). Collectively, these studies suggest that primates are sensitive to probabilities as well as absolute quantities.

In addition to being able to flexibly choose between sets of food quantities, many primate species also make quantity judgments between arrays of artificial items such as dot/stimulus arrays on cards or computer displays, although these kinds of tasks generally require much more training compared with studies in which animals choose between food arrays. Early work with this type of test was performed by Thomas and colleagues. Thomas, Fowlkes, and Vickery (1980) tested whether squirrel monkeys (*Saimiri sciureus*) could choose a card displaying fewer items (different sizes of filled circles) among pairs of cards representing quantities between 2 and 9. Squirrel monkeys were successful in this test, and Thomas and Chase (1980) assessed whether these monkeys could flexibly choose an array from three options that represented the lowest quantity, the highest quantity, or the quantity of intermediate value. Squirrel monkeys continued to succeed in choosing the appropriate quantity in this test. Terrell and Thomas (1990) later trained squirrel monkeys to discriminate between polygons, printed on cards, consisting of different numbers of sides and angles. As in the previous experiments, monkeys were successful in discriminating between these quantitative stimuli, even though they consisted of numerous exemplars of polygons ranging from triangles to octagons. These studies demonstrated that primates not only discriminate on the basis of quantity, but they can also use quantity to direct behavior counter to innate drives (e.g., correctly selecting the lesser quantity) or discriminate quantities of items that do not naturally occur in their environment and that are abstracted individuating characteristics (e.g., number of sides and angles of polygons).

Some of the most interesting tests of numerical cognition have been presented using digital arrays of stimuli on a computer monitor. A well-known computerized numerical study involved training rhesus monkeys (*Macaca mulatta*) to touch pairs of digital arrays representing quantities within the range of one to nine in ascending numerical order (Brannon & Terrace, 1998, 2000). The monkeys were then tested with multiple stimulus categories that assessed the contributions of non-numeric stimulus properties, such as item size and surface area, to quantity discriminations. Brannon and colleagues (Brannon, Cantlon, & Terrace, 2006; Jones, Cantlon, Merritt, & Brannon, 2010) extended this area of study through investigations of the influence of reference points. Specifically, they found that monkeys trained to respond to numerical stimuli in a descending numerical order such as 4-3-2-1 did not generalize this rule to novel values that were larger than the training range (e.g., 5–9), whereas monkeys trained to respond in ascending order had no trouble generalizing to the larger values. Thus, the training stimuli provided the reference points that could create a bias for later discriminations in a predictable fashion. Brannon and colleagues also studied the effects of semantic congruity (Cantlon & Brannon, 2005) and stimulus heterogeneity (e.g., Cantlon & Brannon, 2006a) on monkey quantity judgments, as well as the mechanisms underlying such judgments (e.g., Cantlon & Brannon, 2006b, 2007). This work was followed by a number of replications with other animal species including a baboon (*Papio hamadryas*) and squirrel monkey (Smith, Piel, & Candland, 2003), capuchin monkeys (Judge, Evans, & Vyas, 2005), lemurs (*Lemur catta:* Merritt, MacLean, Crawford, & Brannon, 2011), and pigeons (Scarf, Hayne, & Colombo, 2011).

Other relative quantity tasks have been presented on computer systems to primates. For example, Beran (2007) tested rhesus monkeys with a digital version of the item-by-item quantity presentation method (see Beran, 2001; vanMarle et al., 2006). Monkeys observed different numbers of digital items being dropped into two digital "containers" and then chose the container that "held" more items. They learned to do this despite controls for non-numeric stimulus properties such as presentation duration and the total amount of digital item surface area presented in each set. Beran (2008) also tested capuchin monkeys, rhesus monkeys, and adult humans for the ability to compare arrays of moving digital items. In four experiments, monkeys exhibited the capacity to track, enumerate, and compare sets of moving items, and demonstrated that their judgments were not controlled by non-numeric stimulus properties such as cumulative item area. This test also has been presented to gorillas (Vonk et al., 2014) and to black bears (*Ursus americanus:* Vonk & Beran, 2012), showing that the quantification of moving items is likely a skill possessed by many species.

ORDINALITY JUDGMENTS

Animals also encode the ordinal properties of multiple stimulus sets and make judgments between more than two sets on the basis of quantity/number of items (e.g., Beran, Beran, Harris, & Washburn, 2005; Boysen, Berntson, Shreyer, & Quigley, 1993; Brannon et al., 2006; Brannon & Terrace, 1998, 2000; Pepperberg, 2006b; Washburn & Rumbaugh, 1991). Primates, in particular, have exhibited this capacity with both visible discrete quantities and symbolic representations of number. For example, Washburn and Rumbaugh (1991) trained rhesus monkeys to select among 2, 3, 4, or 5 Arabic numerals within the range of 0 to 9 presented on a computer screen. After selecting one of these numerals, the monkey was presented with a number of food pellets equal to the value of that numeral. These researchers first trained monkeys to select the larger numeral of a pair, and the amount of reward received was closely tied to performance (monkeys were less successful when presented with numeral pairs associated with similar-sized rewards). Critically, Washburn and Rumbaugh withheld several numeral pairings for testing later as probe trials. The monkeys were successful with these novel pairs, despite their lack of experience with them, suggesting that they had learned the ordinal sequence for all of the numerals, rather than learning a matrix of independent two-choice discriminations. Washburn and Rumbaugh then performed another test, this time involving sets of 3, 4, or 5 numerals all being presented at once (e.g., the numerals 2, 4, 6, 7, and 9 might be presented on a trial). The monkeys selected from these larger arrays the numerals in descending order, and their success in doing this provided additional evidence that they were truly applying their knowledge of the ordinal sequence to the full sequence of numerals.

Olthof et al. (1997) also trained squirrel monkeys to order Arabic numerals. However, in this case, the researchers presented the numerals on physical tokens. During training, Olthof and colleagues rewarded monkeys' numeral selections with an equal number of food items to the quantity represented by the selected numerals, although they did not present all pairs of numerals so these could be presented later as novel probe trials during the testing phase (as in Washburn & Rumbaugh, 1991). Then, these researchers allowed monkeys to choose among sets of multiple numerals (i.e., two numerals vs. two numerals, one numeral vs. two numerals, and three numerals vs. three numerals), and they found that the monkeys chose numeral sets representing the largest overall quantity (e.g., $3+3$ over $5+0$) at greater than chance levels. Thus, monkeys were not influenced by the presence of either the largest or smallest individual numeral in any set, but instead integrated the value of the numerals in the each set before choosing a set.

Beran et al. (2008) compared the performance of rhesus monkeys and capuchin monkeys in a computerized ordinality test. The methodology of this study was quite similar to that of Washburn and Rumbaugh (1991), but with two important differences that allowed the researchers to investigate alternative explanations (to ordinality) for monkeys' performance in this type of test. Instead of applying ordinality to the set of numerals, one possibility was that monkeys could learn to assign different hedonic values to each numeral, since each numeral was always rewarded with a different amount of food reward. To examine this hypothesis, Beran and colleagues trained monkeys to select the larger of two numerals, as in the previous studies, but provided only a single food reward for choosing correctly to half of the monkeys of each tested species (the other half of the monkeys were rewarded a number of reward items equal to the value of the numeral, as in previous tests). As a result, all monkeys learned to choose the larger of two numerals in training trials, but only rhesus monkeys that were rewarded with food amounts equal to the value of the numerals succeeded in novel probe trials during the test phase. Thus, those monkeys that were rewarded with just one food item per correct response could discriminate familiar numerals but could not order the larger and smaller numerals in the novel pairings. The results suggest that monkeys can learn to associate numerals with specific quantities, but they learn the ordinal sequence of these numerals only when this association signals differential rewards.

Other "symbolic" stimuli have been used to assess ordinal knowledge in primates. For example, Beran et al. (2005) presented a rhesus monkey and two chimpanzees with different combinations of a set of five colored food containers to test whether they could learn the ordinal relations among the objects based on the food items they contained. The monkey and chimpanzees first learned to choose the more valuable of two presented containers (the one consistently containing more food items). They then transferred that ordinal knowledge to a new test in which they had to choose between containers and entirely visible food quantities. Here, the monkey and chimpanzees demonstrated that they had learned the approximate value of each unique container as well as the relative value of each container compared to the others. Later work replicated these results with capuchin monkeys using similar methods to Beran et al. (2005), as well as by using a method that involved tokens that represented different quantities of food (Evans, Beran, & Addessi, 2010).

Other research teams also have reported success in teaching primates to use tokens to represent quantity, and then to make judgments and decisions about overall set values on the basis of individual token values (e.g., Addessi, Crescimbene, & Visalberghi, 2007; Addessi, Mancinia, Crescimbene, Ariely, & Visalberghi, 2010). These capacities therefore provide evidence of some degree of ordinal knowledge through symbolic representation as would be necessary for simple arithmetic competence, even if animals cannot reach the full representation of ordinality that children show as they begin formal schooling.

QUANTITY ILLUSIONS

A new area of inquiry involves examining whether nonhuman primates fall prey to quantity illusions. These are a subset of visual illusions that create misperceptions of set number or quantity via manipulations of the overall pattern or spatial arrangement of elements in an array. These illusions can disrupt relative quantity judgments by shifting the focus to certain aspects of a given set, resulting in the over- or underestimation of quantity or numerosity. Comparative studies have investigated the performance of both primates and humans on a range of numerical illusions that manipulate stimulus organization, including the regular-random numerosity illusion (RRNI) and the Nested illusion, and other tasks that vary set size perception via manipulation of individual elements in a set or the organization of those elements within an array.

The RRNI was first demonstrated when humans overestimated set size when individual items were arranged in a regular pattern (e.g., in a regular square pattern) and thus appeared more homogenous in comparison to randomly arranged patterns (Ginsburg, 1976, 1978, 1980). Beran (2006) replicated this finding among humans and rhesus monkeys using a relative quantity task in which both species were trained to select the larger of two digital dot arrays. In probe trials, an equal number of dots was displayed in both arrays; however, one array was randomly arranged, whereas the other was arranged in a regular square pattern. For both species, regularly arranged stimuli were judged to be more numerous than the same number of randomly arranged stimuli.

Numerical judgments also are impacted by the visual overlap of stimuli within a set. Chesney and Gelman (2012) presented stimulus arrays to human subjects in which items in an array overlapped such that some elements were wholly contained within other elements. When subjects were asked to compare the set size of multiple circles to an Arabic numeral, their response times were longer and their estimations were lower for sets containing nested stimuli than for those without. These results were conceptually replicated with rhesus monkeys using a numerical bisection task in which subjects classified nested and non-nested arrays as being larger or smaller than a pre-established central value (Beran & Parrish, 2013). Like humans, monkeys underestimated nested arrays more often than non-nested arrays, likely as a result of their failure to individuate nested items.

Another type of quantitative illusion has been observed among chimpanzees in which they misperceived quantities on the basis of container "fullness" (i.e., the proportion of a container occupied by a quantity of food). In this illusion, smaller, but seemingly fuller, quantities were judged to be larger than truly larger, but seemingly less full, quantities (Parrish & Beran, 2014). In this relative quantity judgment task, chimpanzees chose between two amounts of food presented in different-sized containers, a large and small cup. Chimpanzees were highly accurate in choosing the larger food amount when different quantities were presented in same-sized cups or when the smaller cup contained the larger quantity of food. However, when different-sized cups contained the same amount of food or the smaller cup contained the smaller amount of food (but looked relatively fuller), the chimpanzees often showed a bias to select the smaller but fuller cup, even if that meant they got less food. These results demonstrated that the highly accurate system that chimpanzees possess for representing quantities, a system which sometimes rivals or even outperforms that of humans (e.g., Beran, 2001, 2004b, 2012b; Beran & Beran, 2004; Boysen & Berntson, 1995; Dooley & Gill, 1977; Menzel, 1960, 1961; Rumbaugh et al., 1987), can be disrupted by manipulating the context in which quantities are presented.

Beyond overall stimulus organization, presentation style of individual elements contained within sets can also lead to biases in relative quantity judgments. For example, although chimpanzees are excellent at discriminating very small differences in food quantity (e.g., Menzel, 1960, 1961), their performance can be disrupted by too much attention to local features of an array of food such as individual item sizes, and this can lead to errors such as overestimating the total quantity in two choice options when the smaller of the two options contained the individually largest item (Beran, Evans, & Harris, 2008; Boysen, Berntson, & Mukobi, 2001). These studies demonstrate a bias in judgment and decision making surrounding quantity estimation as predicted by the presentation style of the individual elements in a set, which indicates that sometimes judgments regarding quantitative differences cannot be made independently of other perceptual processes that might contribute to predictable biases.

Thus, nonhuman and human primates not only share basic numerical capacities as outlined earlier, but also appear to be subject to some of the same illusory patterns and biases that disrupt their quantitative judgments. Numerosity illusions are specifically influenced by non-numerical cues such as surface area and homogeneity of stimuli that produce these illusory effects. In particular, it seems that numerosity illusions arise when the pattern or arrangement of stimuli is disrupted such that items are no longer spatially continuous (Beran, 2006; Frith & Frith, 1972; Ginsburg, 1976, 1980), visually segregated (e.g., Beran & Parrish, 2013; Chesney & Gelman, 2012), or uniformly equivalent (e.g., Beran, Evans, & Harris, 2008; Boysen et al., 2001; Parrish & Beran, in press). These illusions occur even among adult humans who are mathematically proficient, and therefore, such illusions highlight how some quantitative processes are fallible when quantity or numerical stimulus dimensions come into conflict with other perceptual features of those stimuli.

MECHANISMS FOR REPRESENTING QUANTITY AND NUMBER

Quantity judgments, representation of numerosity, and perhaps even counting-like abilities (to a limited degree) are evident in nonhuman animals.

Additionally, they seem to be susceptible to similar sorts of illusions and biases that disrupt their quantitative systems in a predictable fashion. Thus, the question remains as to whether the mechanisms that support such processes are similar to those seen in humans. One fascinating approach to this question comes from research using neuroimaging techniques (Cantlon, this volume). In humans, the left and right intraparietal cortices have been reported as a critical substrate for processing nonsymbolic and some symbolic quantities (for reviews, see Cohen Kadosh & Walsh, 2009; Cohen Kadosh, Lammertyn, & Izard, 2008; Nieder & Dehaene, 2009), and similar brain regions seem to be active when animals perform their numerical tasks (e.g., Dehaene, Dehaene-Lambertz, & Cohen, 1998; Nieder & Miller, 2004).

Another more prevalent approach has been to look at performance patterns, including errors, to see whether animals look like humans when they make judgments of stimuli, and what characteristics of those judgments might suggest about how numerosity and other quantities are represented. Our abilities to count perfectly and perform complicated mathematical operations appear to be built on a much more basic system for representing number or quantity, and this system is likely one that we and other animals have shared for a very long part of our evolutionary history. This system, sometimes called

---

FIGURE 4-3 *Comparing ratio effects across species. The ratio values represent the small set size divided by the large set size (e.g., 1 versus 10 items is a ratio of .10). Panel A shows the performance of rhesus monkeys Murph and Lou in the task from Beran (2007). From "Rhesus monkeys (*Macaca mulatta*) enumerate large and small sequentially presented sets of items using analog numerical representations," by M. J. Beran, 2007, *Journal of Experimental Psychology: Animal Behavior Processes, 33,* 46. Copyright 2007 by the American Psychological Association. Reprinted with permission. Panel B shows the performance of rhesus monkeys in tracking and enumerating sets of moving stimuli on a computer screen and then choosing the larger number of items. From "Monkeys (*Macaca mulatta* and *Cebus apella*) track, enumerate, and compare multiple sets of moving items," by M. J. Beran, 2008, *Journal of Experimental Psychology: Animal Behavior Processes, 34,* 65. Copyright 2008 by the American Psychological Association. Reprinted with permission. Panel C compares the performance of adult humans who were allowed to count or were prevented from counting to the performance of chimpanzees (reported in Beran, 2001, 2004b) on a sequential enumeration task. From "Nonverbal estimation during numerosity judgements by adult humans," by M. J. Beran, L. B. Taglialatela, T. M. Flemming, F. M. James, and D. A. Washburn, 2006, *Quarterly Journal of Experimental Psychology, 59,* 2073. Copyright 2006 by Taylor & Francis. Panel D shows the performance of adult humans on the task in which moving sets of stimuli had to be enumerated (the task shown for monkeys in Figure 4-3b). Whether number and amount co-varied (congruent condition) or did not (incongruent condition), humans showed ratio effects. From "Monkeys (*Macaca mulatta* and *Cebus apella*) and human adults and children (*Homo sapiens*) enumerate and compare subsets of moving stimuli based on numerosity," by M. J. Beran, S. Decker, A. Schwartz, and N. Schultz, 2011, *Frontiers in Psychology, 2.* Copyright 2011 by Michael J. Beran.

---

*For all panels in the figure, the ratio of set sizes being compared was highly predictive of performance on these tasks and reflects operation of the ANS.

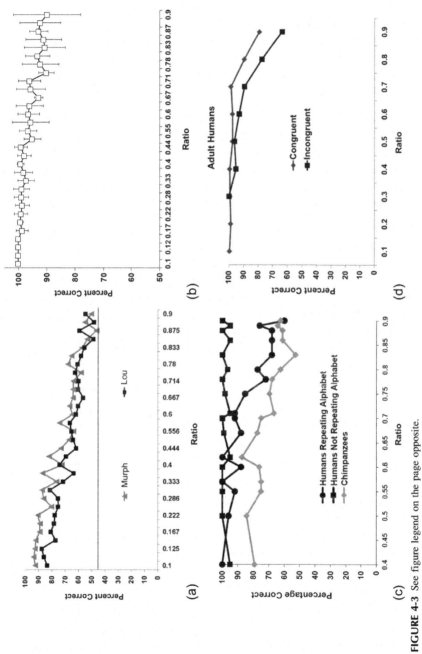

FIGURE 4-3 See figure legend on the page opposite.

the approximate number system (ANS), represents quantities inexactly; small quantities are represented with precision, but as quantity increases, the representations of successive quantities increasingly overlap, resulting in difficulties discriminating quantities that are similar (e.g., 20 vs. 24 items). The system has two signature behavioral effects. The ability to discriminate between quantities varies as a function of ratio; for instance, discriminating quantities with a 2:1 ratio, such as 12 vs. 6 items, is easier than discriminating quantities with a 3:2 ratio, such as 18 vs. 12 items, even when the absolute difference (6 in these examples) across the two comparisons is the same. And, when the difference between sets is held constant, performance is better when comparing small sets rather than large sets (e.g., comparing 3 to 5 is easier than comparing 7 to 9). These two effects, called the *ratio effect* and the *size* (or *magnitude*) *effect* have been found in a large number of species (for reviews, see Brannon & Roitman, 2003; Dehaene, 1997; Gallistel & Gelman, 2000). This is true across a wide range of experimental studies, including those that involve estimating continuous and discrete amounts (see Cantlon, Platt, & Brannon, 2009).

Beran (2007) found strong ratio effects when rhesus monkeys watched differing numbers of items "fall" into containers on a computer screen (Figure 4-3). Beran, Decker, Schwartz, and Schultz (2011) reported the same kind of ratio effect for rhesus monkeys and capuchin monkeys, this time for a test in which monkeys had to enumerate and compare sets of moving dots on a computer screen (Figure 4-3b). This is the pattern that emerges in nearly all studies with nonhuman primates, suggesting that the ANS is a basic mechanism available for representing quantity and numerosity. Moreover, when humans are prevented from counting through the use of other techniques, such as articulatory suppression, they too show greater variability in their estimates of set sizes or increased difficulty in comparing sets of items as a function of increasing set size or ratio (e.g., Cordes, Gelman, Gallistel, & Whalen, 2001; Whalen, Gallistel, & Gelman, 1999). For example, Figure 4-3c shows a comparison of human performance on the same task given to rhesus monkeys in Beran (2007) when humans were prevented from counting or were not. Those prevented from counting looked much more like rhesus monkeys than their counting peers (Beran et al., 2006). And, human adults show ratio effects in other tasks such as when they must compare moving sets of dots on a computer screen, again making them look much like monkeys in terms of their performance (Beran et al., 2011; Figure 4-3d).

Thus, a consensus has emerged that the ANS is likely present in many if not all animals, including humans, and it plays a role in contexts in which exact numerical representations cannot be formed, and in which relational judgments of items occur. However, a second system for numerical representation has been proposed to also account for how animals (and, in some cases, humans) represent number. The *object file* model states that discriminations can be made at high degrees of accuracy for small numerosities but not for large numerosities (vanMarle, this volume). The idea is that there are memory limits for the individuation of items. Typically, it is proposed that there are not more than three or perhaps four files available in short-term memory, and so when one encounters items as part of an enumerative instance, separate items are encoded in terms of separate object files (i.e., stored in those files). Object files then operate as representations of items within the array and can hold information about item identity and features, and the number of filled files also operates as the means of storing the quantity/numerical information about the array as a whole (Feigenson, Carey, & Hauser, 2002; Simon, 1997; Uller, Carey, Huntley-Fenner, & Klatt, 1999). Within the developmental literature, there is support for the existence of this system (e.g., Feigenson & Carey, 2003, 2005; Mou & vanMarle, in press; Xu, 2003), although other reports suggest that the ANS appears to be the dominant system at work (e.g., Cantlon, Safford, & Brannon, 2010; Cordes & Brannon, 2009; Lipton & Spelke, 2003). Thus, comparative data are highly relevant, as they may shed light on the nature of how many (and what kinds of) systems are present to generate numerical and quantitative processes in humans.

We have already noted that nearly all animals tested to date show evidence of the ANS. The question is what comparative evidence exists for the object file model. Some evidence for very good performance with small numbers of items has been reported for fish (e.g., *Pterophyllum scalare:* Gomez-Laplaza & Gerlai, 2011; *Poecilia reticulate:* Piffer et al., 2012), birds (e.g., Garland et al., 2012), beluga whales (*Delphinapterus leucas:* Abramson et al., 2013), and Asian elephants (*Elephas maximus:* Irie-Sugimoto et al., 2009; but see Perdue et al., 2012). The object file model also has drawn support from the data reported by Hauser, Carey, and Hauser (2000). To date, however, that is the only study with primates that has shown set size limits on discrimination performance with food items that are presented in a sequential fashion. In the Hauser et al. (2000) study, monkeys watched experimenters place pieces of apple, one at a time, into two opaque containers, and then the monkeys were allowed to choose one set of items. A wide range of quantity comparisons was given, and those comparisons could involve two small numbers of apple pieces (1 vs. 2, 2 vs. 3, and 3 vs. 4) or they could involve one smaller set and one larger set (3 vs. 5, 4 vs. 5, 4 vs. 6, 4 vs. 8, and 3 vs. 8). Hauser et al. (2000) reported that monkeys were highly proficient in choosing the larger amount with the small-number comparisons, but they did not perform well when a large number was compared to a small one, presumably because this created difficulty for the object file system as it could not represent those large numbers of items in a way that allowed the monkeys to make good choices. This was seen as an important outcome because it suggested that monkeys (like young human children; e.g., vanMarle, 2013) have two distinct systems for representing number that include the object file system and the analog magnitude system and that quantities are not easily compared when one quantity is represented in one system and the other quantity in the other system. This has come to be called the "two core number systems" hypothesis (Feigenson, Dehaene, & Spelke, 2004; Xu, 2003).

However, it is not clear that the comparative literature really supports this notion of two core systems, especially if one focuses on nonhuman primate performance (which seems most prudent, given that they are our closest relatives). For example, recall that Beran (2007) trained two monkeys to make judgments between two sets of sequentially presented arrays of items. The same comparison types were presented in that study as in Hauser et al. (2000) including trials with two small sets and trials with a small set and a large set, and there was no evidence that monkeys had difficulty discriminating between arrays in either of these cases, except where one could show that ratio effects were at work (a signature of the ANS). And, the problematic comparisons that crossed the small-number and large-number divide for the monkeys in Hauser et al. (2000) were easily performed by monkeys in the Beran study (see also Nieder & Miller, 2004). The same outcomes—evidence of a ratio effect and no apparent difficulty in judging small versus large arrays—have been reported in manual tests given to all four great ape species (gorillas, bonobos, chimpanzees, and orangutans; e.g., Beran, 2004b; Hanus & Call, 2007), capuchin monkeys (Evans et al., 2009), and many other species ranging from whales (Abramson et al., 2011) to parrots (Ain et al., 2009) and pigeons (Roberts, 2010).

Feigenson et al. (2004) stated: "Monkeys' restriction to the numerosities 1–4 in situations involving small arrays, coupled with their capacity to create noisy representations of large sets, suggests that monkeys, like humans, have two distinct systems for representing number. The two core number systems therefore offer a strong case of representational continuity across development and across species" (p. 311). The present situation in comparative numerical cognition research is that nonhuman primates do not actually provide much support for the idea of two core systems of number, although some work with other species may provide some support. The object file system likely exists for tracking specific things as they are situated (and even move) through space, as that was the context in which the original model was proposed (Pylyshyn, 1989; Scholl & Pylyshyn, 1999). This system for visual "objectification" then was used to possibly explain why subitizing, or the rapid and highly accurate apprehension of small numbers of items, occurs in humans. Trick and Pylyshyn (1994) suggested that the ability to subitize is a side effect of the way that stages of processing are coordinated in vision, but that is a limited context within the much broader range of circumstances in which enumeration occurs (e.g., with sequentially presented sets, with additions and subtractions of items to sets, and with unlimited processing time rather than rapid judgments).

And, even this model would predict good performance on comparisons of sequentially presented sets with large differences, such as three items versus eight, because of the existence of a set summing mechanism within the model

(Trick & Pylyshyn, 1994). Thus, what seems clear is that there is one dominant system, the ANS, and it appears to have a long evolutionary history
and is brought to bear on a variety of tasks that animals can successfully perform, provided the difference between sets or the absolute size of arrays does
not exceed the sensitivity of that species' ANS. The dominant (and for some
species, perhaps only) form of numerical representation is through use of a
continuous, approximate scaling of numerosity for which the exactness of
the representation increasingly weakens as magnitude increases. One remaining question about the ANS is whether the scaling that occurs in numerical
representation is linear but with increasing variability that accompanies
increasing magnitude, or if it occurs through logarithmic compression of the
mental number line that is used to represent numerosity (e.g., Dehaene,
2003; Roberts, 2006; Siegler & Opfer, 2003).

CONCLUSIONS

Number matters to animals. It is salient, it can be stored and manipulated
(albeit imprecisely), and it can be used to guide decision making in a variety
of contexts. And, there are good reasons why this should be true, and good
reasons why evolution has selected for these abilities. From the perspective
of foraging, sensitivity to the relative differences in the amount of food that
may be obtained from different sources (e.g., patches) is critical, as maximi-
zation of food intake directly impacts survival rates, although there are limits
to the need to discriminate between food patches. For example, choosing a
location with 12 fruits instead of 4 will lead to tangible nutritional benefits,
whereas telling the difference between 12 fruits and 11 does not afford the
same benefit because there is less of a relative difference between the alternatives. The same is true for attending to potential groups of competitors, predators, or prey. Noticing relatively larger differences between those groups
conveys immediate benefits in terms of gaining prey or avoiding danger from
predators or competitors. But, when differences are slight and magnitudes are
large, these benefits fade.

The capacities shown by a variety of nonhuman animals match those seen
in precounting children in some cases and can even match the capacities of
adult humans when those humans are prevented from using additional processes so that they are left with only the ANS that they share with animals.
Thus, there is continuity in the "number sense" that we and other animals
have, a continuity that is reflected in similar behavior and perhaps even in
similar neural processes, and understanding that "number sense" might offer
predictive value in gauging future difficulties or successes in higher level
mathematics for children (e.g., Starr, Libertus, & Brannon, 2013). But continuity of processes does not imply equality of processes, and most humans
by the age of 6 or 7 exceed any capacities that we have yet seen in other animals. Whether higher mathematical learning and proficiency are a result of

rearing and experience, or of brain development, or are an inherent discontinuity between our species and all others remains to be more fully determined. These findings will have important implications for properly understanding the role of symbolic processes and concept use in higher mathematics. Advanced mathematics relies heavily on symbolic representation and certain forms of conceptualization of how mathematical operations can work, and these capacities may rely on a "number sense" that is qualitatively different from the approximate number sense that we share with nonhuman animals. For example, understanding division requires being able to understand that remainders can exist as part of the operation (e.g., one can divide nine pennies equally among four people, but one penny will always be left over). And these kinds of conceptual realizations might be uniquely human (and, even then, only emerge after a certain developmental point), in the same way that other conceptual phenomena have been proposed as uniquely human (e.g., Penn, Holyoak, & Povinelli, 2008; Premack, 2010). But, the most relevant implication of comparative research into numerical cognition is that we share basic cognitive processes with other species that are foundational to whatever higher order mathematics emerge later in development, and a fuller understanding of those foundations offers insight into our own quantitative capabilities, limitations, and potential.

ACKNOWLEDGMENTS

Preparation of the article was supported by grant HD060563 from the National Institutes of Health. For correspondence, contact Michael J. Beran at Language Research Center, Georgia State University, University Plaza, Atlanta, GA 30302. Phone: 404-413-5285; Fax: 404-244-5829; Email to mjberan@yahoo.com.

REFERENCES

Abramson, J. Z., Hernandez-Lloreda, V., Call, J., & Colmenares, F. (2011). Relative quantity judgments in South American sea lions (*Otaria flavescens*). *Animal Cognition, 14*, 695–706.

Abramson, J. Z., Hernández-Lloreda, V., Call, J., & Colmenares, F. (2013). Relative quantity judgments in the beluga whale (*Delphinapterus leucas*) and the bottlenose dolphin (*Tursiops truncatus*). *Behavioural Processes, 96*, 11–19.

Addessi, E., Crescimbene, L., & Visalberghi, E. (2007). Do capuchin monkeys (*Cebus apella*) use tokens as symbols? *Proceedings of the Royal Society of London, 274*, 2579–2585.

Addessi, E., Crescimbene, L., & Visalberghi, E. (2008). Food and token quantity discrimination in capuchin monkeys (*Cebus apella*). *Animal Cognition, 11*, 275–282.

Addessi, E., Mancinia, A., Crescimbene, L., Ariely, D., & Visalberghi, E. (2010). How to spend a token? Trade-offs between food variety and food preference in tufted capuchin monkeys (*Cebus apella*). *Behavioural Processes, 83*, 267–275.

Agrillo, C., Dadda, M., Serena, G., & Bisazza, A. (2008). Do fish count? Spontaneous discrimination of quantity in female mosquitofish. *Animal Cognition, 11*, 495–503.

Agrillo, C., Piffer, L., & Bisazza, A. (2011). Number versus continuous quantity in numerosity judgments by fish. *Cognition, 119*, 281–287.

Ain, S. A., Giret, N., Grand, M., Kreutzer, M., & Bovet, D. (2009). The discrimination of discrete and continuous amounts in African grey parrots (*Psittacus erithacus*). *Animal Cognition, 12,* 145–154.

Anderson, U. S., Stoinski, T. S., Bloomsmith, M. A., & Maple, T. S. (2007). Relative numerousness judgment and summation in young, middle-aged, and old adult orangutans (*Pongo pygmaeus abelii* and *Pongo pygmaeus pygmaeus*). *Journal of Comparative Psychology, 121,* 1–11.

Anderson, U. S., Stoinski, T. S., Bloomsmith, M. A., Marr, M. J., Smith, A. D., & Maple, T. S. (2005). Relative numerousness judgment and summation in young and old Western Lowland gorillas. *Journal of Comparative Psychology, 119,* 285–295.

Baker, J. M., Morath, J., Rodzon, K. S., & Jordan, K. E. (2012). A shared system of representation governing quantity discrimination in canids. *Frontiers in Psychology, 3,* Article 387.

Baker, J. M., Shivik, J., & Jordan, K. E. (2011). Tracking of food quantity by coyotes (*Canis latrans*). *Behavioural Processes, 88,* 72–75.

Barnard, A. M., Hughes, K. D., Gerhardt, R. R., DiVincenti, L. J., Bovee, J. M., & Cantlon, J. F. (2013). Inherently analog quantity representations in olive baboons (*Papio anubis*). *Frontiers in Psychology, 4,* 253.

Benson-Amram, S., Heinen, V. K., Dryer, S. L., & Holekamp, K. E. (2010). Numerical assessment and individual call discrimination by wild spotted hyaenas, *Crocuta crocuta*. *Animal Behaviour, 82,* 743–752.

Beran, M. J. (2001). Summation and numerousness judgments of sequentially presented sets of items by chimpanzees (*Pan troglodytes*). *Journal of Comparative Psychology, 115,* 181–191.

Beran, M. J. (2004a). Long-term retention of the differential values of Arabic numerals by chimpanzees (*Pan troglodytes*). *Animal Cognition, 7,* 86–92.

Beran, M. J. (2004b). Chimpanzees (*Pan troglodytes*) respond to nonvisible sets after one-by-one addition and removal of items. *Journal of Comparative Psychology, 118,* 25–36.

Beran, M. J. (2006). Quantity perception by adult humans (*Homo sapiens*), chimpanzees (*Pan troglodytes*), and rhesus macaques (*Macaca mulatta*) as a function of stimulus organization. *International Journal of Comparative Psychology, 19,* 386–397.

Beran, M. J. (2007). Rhesus monkeys (*Macaca mulatta*) enumerate large and small sequentially presented sets of items using analog numerical representations. *Journal of Experimental Psychology: Animal Behavior Processes, 33,* 55–63.

Beran, M. J. (2008). Capuchin monkeys (*Cebus apella*) succeed in a test of quantity conservation. *Animal Cognition, 11,* 109–116.

Beran, M. J. (2012a). Did you ever hear the one about the horse that could count? *Frontiers in Psychology, 3,* 357.

Beran, M. J. (2012b). Quantity judgments of auditory and visual stimuli by chimpanzees (*Pan troglodytes*). *Journal of Experimental Psychology: Animal Behavior Processes, 38,* 23–29.

Beran, M. J., & Beran, M. M. (2004). Chimpanzees remember the results of one-by-one addition of food items to sets over extended time periods. *Psychological Science, 15,* 94–99.

Beran, M. J., Beran, M. M., Harris, E. H., & Washburn, D. A. (2005). Ordinal judgments and summation of nonvisible sets of food items by two chimpanzees (*Pan troglodytes*) and a rhesus macaque (*Macaca mulatta*). *Journal of Experimental Psychology: Animal Behavior Processes, 31,* 351–362.

Beran, M. J., Decker, S., Schwartz, A., & Schultz, N. (2011). Monkeys (*Macaca mulatta* and *Cebus apella*) and human adults and children (*Homo sapiens*) enumerate and compare subsets of moving stimuli based on numerosity. *Frontiers in Psychology, 2,* Article 61.

Beran, M. J., Evans, T. A., & Harris, E. H. (2008). Perception of food amount by chimpanzees based on the number, size, contour length, and visibility of items. *Animal Behaviour, 75*, 1793–1802.

Beran, M. J., Evans, T. A., & Harris, E. H. (2009). When in doubt, chimpanzees rely on estimates of past reward amounts. *Proceedings of the Royal Society B, 276*, 309–314.

Beran, M. J., Evans, T. A., Leighty, K., Harris, E. H., & Rice, D. (2008). Summation and quantity judgments of sequentially presented sets by capuchin monkeys (*Cebus apella*). *American Journal of Primatology, 70*, 191–194.

Beran, M. J., Owens, K., Phillips, H. A., & Evans, T. A. (2012). Humans and monkeys show similar skill in estimating uncertain outcomes. *Psychonomic Bulletin & Review, 19*, 357–362.

Beran, M. J., & Parrish, A. E. (2013). Visual nesting of stimuli affects rhesus monkeys' (*Macaca mulatta*) quantity judgments in a bisection task. *Attention, Perception, & Psychophysics, 75*, 1243–1251.

Beran, M. J., Perdue, B. M., Parrish, A. E., & Evans, T. A. (2012). Do social conditions affect capuchin monkeys' (*Cebus apella*) choices in a quantity judgment task? *Frontiers in Psychology, 3*, 492.

Beran, M. J., & Rumbaugh, D. M. (2001). "Constructive" enumeration by chimpanzees (*Pan troglodytes*) on a computerized task. *Animal Cognition, 4*, 81–89.

Beran, M. J., Rumbaugh, D. M., & Savage-Rumbaugh, E. S. (1998). Chimpanzee (*Pan troglodytes*) counting in a computerized testing paradigm. *Psychological Record, 48*, 3–20.

Beran, M. J., Taglialatela, L. A., Flemming, T. J., James, F. M., & Washburn, D. A. (2006). Nonverbal estimation during numerosity judgements by adult humans. *Quarterly Journal of Experimental Psychology, 59*, 2065–2082.

Boysen, S. T., & Berntson, G. G. (1989). Numerical competence in a chimpanzee (*Pan troglodytes*). *Journal of Comparative Psychology, 103*, 23–31.

Boysen, S. T., & Berntson, G. G. (1995). Responses to quantity: Perceptual versus cognitive mechanisms in chimpanzees (*Pan troglodytes*). *Journal of Experimental Psychology: Animal Behavior Processes, 21*, 82–86.

Boysen, S. T., Berntson, G. G., & Mukobi, K. L. (2001). Size matters: Impact of item size and quantity on array choice by chimpanzees (*Pan troglodytes*). *Journal of Comparative Psychology, 115*, 106–110.

Boysen, S. T., Berntson, G. G., Shreyer, T. A., & Hannan, M. B. (1995). Indicating acts during counting by a chimpanzee (*Pan troglodytes*). *Journal of Comparative Psychology, 109*, 47–51.

Boysen, S. T., Berntson, G. G., Shreyer, T. A., & Quigley, K. S. (1993). Processing of ordinality and transitivity by chimpanzees (*Pan troglodytes*). *Journal of Comparative Psychology, 107*, 208–215.

Brannon, E. M., Cantlon, J. F., & Terrace, H. S. (2006). The role of reference points in ordinal numerical comparisons by rhesus macaques (*Macaca mulatta*). *Journal of Experimental Psychology: Animal Behavior Processes, 32*, 120–134.

Brannon, E. M., & Roitman, J. D. (2003). Nonverbal representations of time and number in animals and human infants. In W. H. Meck (Ed.), *Functional and neural mechanisms of interval timing* (pp. 143–182). Boca Raton, FL: CRC Press.

Brannon, E. M., & Terrace, H. S. (1998). Ordering of the numerosities 1 to 9 by monkeys. *Science, 282*, 746–749.

Brannon, E. M., & Terrace, H. S. (2000). Representation of the numerosities 1–9 by rhesus macaques (*Macaca mulatta*). *Journal of Experimental Psychology: Animal Behavior Processes, 26*, 31–49.

Briars, D. J., & Siegler, R. S. (1984). A featural analysis of preschoolers' counting knowledge. *Developmental Psychology, 20*, 607–618.

Burns, R. A., Goettl, M. E., & Burt, S. T. (1995). Numerical discriminations with arrhythmic serial presentations. *Psychological Record, 45*, 95–104.

Call, J. (2000). Estimating and operating on discrete quantities in orangutans (*Pongo pygmaeus*). *Journal of Comparative Psychology, 114*, 136–147.

Candland, D. K. (1995). *Feral children and clever animals: Reflections on human nature.* Oxford, MA: Oxford University Press.

Cantlon, J. F., & Brannon, E. M. (2005). Semantic congruity affects numerical judgments similarly in monkeys and humans. *Proceedings of the National Academy of Sciences, 102*, 16507–16511.

Cantlon, J. F., & Brannon, E. M. (2006a). The effect of heterogeneity on numerical ordering in rhesus monkeys. *Infancy, 9*, 173–189.

Cantlon, J. F., & Brannon, E. M. (2006b). Shared system for ordering small and large numbers in monkeys and humans. *Psychological Science, 17*, 401–406.

Cantlon, J. F., & Brannon, E. M. (2007). How much does number matter to a monkey (*Macaca mulatta*)? *Journal of Experimental Psychology: Animal Behavior Processes, 33*, 32–41.

Cantlon, J. F., Platt, M. L., & Brannon, E. M. (2009). Beyond the number domain. *Trends in Cognitive Sciences, 13*, 83–91.

Cantlon, J. F., Safford, K. E., & Brannon, E. M. (2010). Spontaneous analog number representations in 3-year-old children. *Developmental Science, 13*, 289–297.

Capaldi, E. J., & Miller, D. J. (1988). Counting in rats: Its functional significance and the independent cognitive processes that constitute it. *Journal of Experimental Psychology: Animal Behavior Processes, 14*, 3–17.

Chesney, D. L., & Gelman, R. (2012). Visual nesting impacts approximate number system estimation. *Attention, Perception, & Psychophysics, 24*, 1104–1113.

Cohen Kadosh, R., Lammertyn, J., & Izard, V. (2008). Are numbers special? An overview of chronometric, neuroimaging, developmental and comparative studies of magnitude representation. *Progress in Neurobiology, 84*, 132–147.

Cohen Kadosh, R., & Walsh, V. (2009). Numerical representation in the parietal lobes: Abstract or not abstract? *Behavioral and Brain Sciences, 32*, 313–373.

Cordes, S., & Brannon, E. M. (2009). Crossing the divide: Infants discriminate small from large numerosities. *Developmental Psychology, 45*, 1583–1594.

Cordes, S., Gelman, R., Gallistel, C. R., & Whalen, J. (2001). Variability signatures distinguish verbal from nonverbal counting for both large and small numbers. *Psychonomic Bulletin and Review, 8*, 698–707.

Dacke, M., & Srinivasan, M. V. (2008). Evidence for counting in insects. *Animal Cognition, 11*, 683–689.

Dadda, M., Piffer, L., Agrillo, C., & Bisazza, A. (2009). Spontaneous number representation in mosquitofish. *Cognition, 112*, 343–348.

Davis, H., & Bradford, S. A. (1986). Counting behavior by rats in a simulated natural environment. *Ethology, 73*, 265–280.

Davis, H., & Memmott, J. (1982). Counting behavior in animals: A critical evaluation. *Psychological Bulletin, 92*, 547–571.

Davis, H., & Perusse, R. (1988). Numerical competence in animals: Definitional issues, current evidence, and a new research agenda. *Behavioral and Brain Sciences, 11*, 561–615.

Dehaene, S. (1997). *The number sense.* New York: Oxford University Press.

Dehaene, S. (2003). The neural basis of the Weber–Fechner law; A logarithmic mental number line. *Trends in Cognitive Sciences, 7*, 145–147.

Dehaene, S., Dehaene-Lambertz, G., & Cohen, L. (1998). Abstract representations of numbers in the animal and human brain. *Trends in Neurosciences, 21*, 355–361.

Dooley, G. B., & Gill, T. (1977). Acquisition and use of mathematical skills by a linguistic chimpanzee. In D. M. Rumbaugh (Ed.), *Language learning by a chimpanzee: The LANA project* (pp. 247–260). New York: Academic Press.

Emmerton, J. (1998). Numerosity differences and effects of stimulus density on pigeons' discrimination performance. *Animal Learning and Behavior, 26*, 243–256.

Evans, T. A., Beran, M. J., & Addessi, E. (2010). Can nonhuman primates use tokens to represent and sum quantities? *Journal of Comparative Psychology, 124*, 369–380.

Evans, T. A., Beran, M. J., Harris, E. H., & Rice, D. (2009). Quantity judgments of sequentially presented food items by capuchin monkeys (*Cebus apella*). *Animal Cognition, 12*, 97–105.

Feigenson, L., & Carey, S. (2003). Tracking individuals via object files: Evidence from infants' manual search. *Developmental Science, 6*, 568–584.

Feigenson, L., & Carey, S. (2005). On the limits of infants' quantification of small object arrays. *Cognition, 97*, 295–313.

Feigenson, L., Carey, S., & Hauser, M. D. (2002). The representations underlying infants' choice of more: Object files versus analog magnitudes. *Psychological Science, 13*, 150–156.

Feigenson, L., Dehaene, S., & Spelke, E. (2004). Core systems of number. *Trends in Cognitive Sciences, 8*, 307–314.

Ferkin, M. H., Pierce, A. A., & Sealand, R. O. (2009). Gonadal hormones modulate sex differences in judgments of relative numerousness in meadow voles, *Microtus pennsylvanicus*. *Hormones and Behavior, 55*, 76–83.

Ferkin, M. H., Pierce, A. A., Sealand, R. O., & delBarco-Trillo, J. (2005). Meadow voles, *Microtus pennsylvanicus*, can distinguish more over-marks from fewer over-marks. *Animal Cognition, 8*, 182–189.

Frith, C. D., & Frith, U. (1972). The solitaire illusion: An illusion of numerosity. *Perception & Psychophysics, 11*, 409–410.

Gallistel, C. R., & Gelman, R. (2000). Non-verbal numerical cognition: From reals to integers. *Trends in Cognitive Sciences, 4*, 59–65.

Garland, A., Low, J., & Burns, K. C. (2012). Large quantity discrimination by North Island robins (*Petroica longipes*). *Animal Cognition, 15*, 1129–1140.

Gelman, R., & Gallistel, C. R. (1978). *The child's understanding of number*. Cambridge, MA: Harvard University Press.

Gelman, R., & Meck, E. (1983). Preschoolers' counting: Principles before skill. *Cognition, 13*, 343–359.

Ginsburg, N. (1976). Effect of item arrangement on perceived numerosity: Randomness vs regularity. *Perceptual and Motor Skills, 43*, 663–668.

Ginsburg, N. (1978). Perceived numerosity, item arrangement, and expectancy. *American Journal of Psychology, 91*, 267–273.

Ginsburg, N. (1980). The regular-random numerosity illusion: Rectangular patterns. *Journal of General Psychology, 103*, 211–216.

Gomez-Laplaza, L. M., & Gerlai, R. (2011). Spontaneous discrimination of small quantities: Shoaling preferences in angelfish (*Pterophyllum scalare*). *Animal Cognition, 14*, 565–574.

Gomez-Laplaza, L. M., & Gerlai, R. (2013). Quantification abilities in angelfish (*Pterophyllum scalare*): the Influence of continuous variables. *Animal Cognition, 16*, 373–383.

Hanus, D., & Call, J. (2007). Discrete quantity judgments in the great apes (*Pan paniscus, Pan troglodytes, Gorilla gorilla, Pongo pygmaeus*): The effect of presenting whole sets versus item-by-item. *Journal of Comparative Psychology, 121*, 241–249.

Hauser, M., Carey, S., & Hauser, L. (2000). Spontaneous number representation in semi-free-ranging rhesus monkeys. *Proceedings of the National Academy of Sciences, 267*, 829–833.

Hunt, S., Low, J., & Burns, K. C. (2008). Adaptive numerical competency in a food-hoarding songbird. *Proceedings of the Royal Society of London B, 275*, 2373–2379.

Irie-Sugimoto, N., Kobayashi, T., Sato, T., & Hasegawa, T. (2009). Relative quantity judgment by Asian elephants *(Elephas maximus)*. *Animal Cognition, 12*, 193–199.

Jaakkola, K., Fellner, W., Erb, L., Rodriguez, M., & Guarino, E. (2005). Understanding of the concept of numerically "less" by bottlenose dolphins *(Tursiops truncatus)*. *Journal of Comparative Psychology, 119*, 286–303.

Jones, S. M., Cantlon, J. F., Merritt, D. J., & Brannon, E. M. (2010). Context affects the numerical semantic congruity effect in rhesus monkeys *(Macaca mulatta)*. *Behavioural Processes, 83*, 191–196.

Judge, P. G., Evans, T. A., & Vyas, D. K. (2005). Ordinal representation of numeric quantities by brown capuchin monkeys *(Cebus apella)*. *Journal of Experimental Psychology: Animal Behavior Processes, 31*, 79–94.

Kilian, A., Yaman, S., Von Fersen, L., & Gunturkun, O. (2003). A bottlenose dolphin discriminates visual stimuli differing in numerosity. *Learning & Behavior, 31*, 133–142.

Krusche, P., Uller, C., & Dicke, U. (2010). Quantity discrimination in salamanders. *Journal of Experimental Biology, 213*, 1822–1828.

Lewis, K. P., Jaffe, S., & Brannon, E. M. (2005). Analog number representations in mongoose lemurs *(Eulemur mongoz)*: Evidence from a search task. *Animal Cognition, 8*, 247–252.

Lipton, J. S., & Spelke, E. (2003). Origins of number sense: Large-number discrimination in human infants. *Psychological Research, 14*, 396–400.

Matsuzawa, T. (1985). Use of numbers by a chimpanzee. *Nature, 315*, 57–59.

McComb, K., Packer, C., & Pusey, A. (1994). Roaring and numerical assessment in contests between groups of female lions, *Panthera leo*. *Animal Behaviour, 47*, 379–387.

McCrick, K., & Wynn, K. (2004). Large-number addition and subtraction by 9-month-old infants. *Psychological Science, 15*, 776–781.

Menzel, E. W. (1960). Selection of food by size in the chimpanzee, and comparison with human judgments. *Science, 131*, 1527–1528.

Menzel, E. W. (1961). Perception of food size in the chimpanzee. *Journal of Comparative and Physiological Psychology, 54*, 588–591.

Merritt, D. J., MacLean, E. L., Crawford, J. C., & Brannon, E. M. (2011). Numerical rule-learning in ring-tailed Lemurs *(Lemur catta)*. *Frontiers in Psychology, 2*, Article 23.

Mou, Y., & vanMarle, K. (in press). Two core systems of numerical representation in infants. *Developmental Review*.

Nieder, A., & Dehaene, S. (2009). Representation of number in the brain. *Annual Review of Neuroscience, 32*, 185–208.

Nieder, A., & Merten, K. (2007). A labeled-line code for small and large numerosities in the monkey prefrontal cortex. *The Journal of Neuroscience, 27*, 5986–5993.

Nieder, A., & Miller, E. K. (2003). Coding of cognitive magnitude: Compressed scaling of numerical information in the primate prefrontal cortex. *Neuron, 37*, 149–157.

Nieder, A., & Miller, E. K. (2004). Analog numerical representations in rhesus monkeys: Evidence for parallel processing. *Journal of Cognitive Neuroscience, 16*, 889–901.

Olthof, A., Iden, C. M., & Roberts, W. A. (1997). Judgments of ordinality and summation of number symbols by squirrel monkeys *(Saimiri sciureus)*. *Journal of Experimental Psychology: Animal Behavior Processes, 23*, 325–339.

Pahl, M., Si, A., & Zhang, S. (2013). Numerical cognition in bees and other insects. *Frontiers in Psychology, 4*, Article 162.

Parrish, A. E., & Beran, M. J. (2014). Chimpanzees sometimes see fuller as better: Judgments of food quantities based on container size and fullness. *Behavioural Processes, 103*, 184–191.

Penn, D. C., Holyoak, K. J., & Povinelli, D. J. (2008). Darwin's mistake: Explaining the discontinuity between human and nonhuman minds. *Behavioral and Brain Sciences, 31*, 109–178.

Pepperberg, I. M. (1987). Evidence for conceptual quantitative abilities in the African Grey parrot: Labeling of cardinal sets. *Ethology, 75*, 37–61.

Pepperberg, I. M. (1994). Numerical competence in an African Grey parrot (*Psittacus erithacus*). *Journal of Comparative Psychology, 108*, 36–44.

Pepperberg, I. M. (2006a). Grey parrot numerical competence: A review. *Animal Cognition, 9*, 377–391.

Pepperberg, I. M. (2006b). Ordinality and inferential abilities of a Grey parrot (*Psittacus erithacus*). *Journal of Comparative Psychology, 120*, 205–216.

Perdue, B. M., Talbot, C. F., Stone, A. M., & Beran, M. J. (2012). Putting the elephant back in the herd: Elephant relative quantity judgments match those of other species. *Animal Cognition, 15*, 955–961.

Pfungst, O. (1911). *Clever Hans (The horse of Mr. von Osten): A contribution to experimental animal and human psychology (C. L. Rahn, Trans.).* New York: Henry Holt.

Piffer, L., Agrillo, C., & Hyde, D. C. (2012). Small and large number discrimination in guppies. *Animal Cognition, 15*, 215–221.

Premack, D. (2010). Why humans are unique: Three theories. *Perspectives on Psychological Science, 5*, 22–32.

Pylyshyn, Z. W. (1989). The role of location indexes in visual perception: A sketch of the FINST spatial index model. *Cognition, 32*, 65–97.

Roberts, W. A. (2006). Evidence that pigeons represent both time and number on a logarithmic scale. *Behavioural Processes, 72*, 207–214.

Roberts, W. A. (2010). Distance and magnitude effects in sequential number discrimination by pigeons. *Journal of Experimental Psychology: Animal Behavior Processes, 36*, 206–216.

Roberts, W. A., & Mitchell, S. (1994). Can a pigeon simultaneously process temporal and numerical information? *Journal of Experimental Psychology: Animal Behavior Processes, 20*, 66–78.

Rugani, R., Regolin, L., & Vallortigara, G. (2007). Rudimental numerical competence in 5-day-old domestic chicks (*Gallus gallus*): Identification of ordinal position. *Journal of Experimental Psychology: Animal Behavior Processes, 33*, 21–31.

Rugani, R., Regolin, L., & Vallortigara, G. (2008). Discrimination of small numerosities in young chicks. *Journal of Experimental Psychology: Animal Behavior Processes, 34*, 388–399.

Rumbaugh, D. M., Hopkins, W. D., Washburn, D. A., & Savage-Rumbaugh, E. S. (1989). Lana chimpanzee learns to count by "NUMATH": A summary of a videotaped experimental report. *Psychological Record, 39*, 459–470.

Rumbaugh, D. M., Savage-Rumbaugh, E. S., & Hegel, M. T. (1987). Summation in the chimpanzee (*Pan troglodytes*). *Journal of Experimental Psychology: Animal Behavior Processes, 13*, 107–115.

Sarnecka, B. W., & Carey, S. (2008). How counting represents number: What children must learn and when they learn it. *Cognition, 108*, 662–674.

Scarf, D., Hayne, H., & Colombo, M. (2011). Pigeons on par with primates in numerical competence. *Science, 334*, 1664. http://dx.doi.org/10.1126/science.1213357.

Scholl, B. J., & Pylyshyn, Z. W. (1999). Tracking multiple items through occlusion: Clues to visual objecthood. *Cognitive Psychology, 38*, 259–290.

Siegler, R. S., & Opfer, J. E. (2003). The development of numerical estimation: Evidence for multiple representations of numerical quantity. *Psychological Science, 14,* 237–243.

Simon, T. J. (1997). Reconceptualizing the origins of number knowledge: A "non-numerical" account. *Cognitive Development, 12,* 349–372.

Smith, B. R., Piel, A. K., & Candland, D. K. (2003). Numerity of a socially housed hamadryas baboon (*Papio hamadryas*) and a socially housed squirrel monkey (*Saimiri sciureus*). *Journal of Comparative Psychology, 117,* 217–225.

Starr, A., Libertus, M. E., & Brannon, E. M. (2013). Number sense in infancy predicts mathematical abilities in childhood. *Proceedings of the National Academy of Sciences, 110*(45), 18116–18120.

Stephens, D. W., & Krebs, J. R. (1986). *Foraging theory.* Princeton, NJ: Princeton University Press.

Terrell, D. F., & Thomas, R. K. (1990). Number-related discrimination and summation by squirrel monkeys (*Saimiri sciureus sciureus* and *S. boliviensus boliviensus*) on the basis of the number of sides of polygons. *Journal of Comparative Psychology, 104,* 238–247.

Thomas, R. K., & Chase, L. (1980). Relative numerousness judgments by squirrel monkeys. *Bulletin of the Psychonomic Society, 16,* 79–82.

Thomas, R. K., Fowlkes, D., & Vickery, J. D. (1980). Conceptual numerousness judgments by squirrel monkeys. *American Journal of Psychology, 93,* 247–257.

Trick, L. M., & Pylyshyn, Z. W. (1994). Why are small and large numbers enumerated differently? A limited-capacity preattentive stage in vision. *Psychological Review, 101,* 80–102.

Uller, C., Carey, S., Huntley-Fenner, G., & Klatt, L. (1999). What representations might underlie infant numerical knowledge? *Cognitive Development, 14,* 1–36.

Uller, C., Jaeger, R., Guidry, G., & Martin, C. (2003). Salamanders (*Plethodon cinereus*) go for more: Rudiments of number in an amphibian. *Animal Cognition, 6,* 105–112.

Uller, C., & Lewis, J. (2009). Horses (*Equus caballus*) select the greater of two quantities in small numerical contrasts. *Animal Cognition, 12,* 733–738.

vanMarle, K. (2013). Infants use different mechanisms to make small and large number ordinal judgments. *Journal of Experimental Child Psychology, 114,* 102–110.

vanMarle, K., Aw, J., McCrink, K., & Santos, L. A. (2006). How capuchin monkeys (*Cebus apella*) quantify objects and substances. *Journal of Comparative Psychology, 120,* 416–426.

Vonk, J., & Beran, M. J. (2012). Bears "count" too: Quantity estimation and comparison in black bears (*Ursus americanus*). *Animal Behaviour, 84,* 231–238.

Vonk, J., Torgerson-White, L., McGuire, M., Thueme, M., Thomas, J., Beran, & J., M. (2014). Quantity estimation and comparison in Western Lowland gorillas (*Gorilla gorilla gorilla*). *Animal Cognition, 17,* 755–765.

Ward, C., & Smuts, B. B. (2007). Quantity-based judgments in the domestic dog (*Canis lupus familiaris*). *Animal Cognition, 10,* 71–80.

Washburn, D. A., & Rumbaugh, D. M. (1991). Ordinal judgments of numerical symbols by macaques (*Macaca mulatta*). *Psychological Science, 2,* 190–193.

Whalen, J., Gallistel, C. R., & Gelman, R. (1999). Nonverbal counting in humans: The psychophysics of number representation. *Psychological Science, 10,* 130–137.

Wynn, K. (1990). Children's understanding of counting. *Cognition, 36,* 155–193.

Wynn, K. (1992). Addition and subtraction by infants. *Nature, 358,* 749–750.

Xu, F. (2003). Numerosity discrimination in infants: Evidence for two systems of representations. *Cognition, 89,* B15–B25.

Part II

Number and Magnitude in Infants and Young Children

Chapter 5

Evolutionary and Developmental Continuities in Numerical Cognition

Ariel Starr and Elizabeth M. Brannon
Department of Psychology & Neuroscience and Center for Cognitive Neuroscience, Duke University, Durham, NC, USA

INTRODUCTION

The modern world is indebted to mathematics. The ability to represent and manipulate numerical quantities underlies wide-ranging facets of modern life from paying bills to counting calories. All animals can represent number, yet only educated humans create and learn formal mathematics. Where does this exceptional ability come from? Although humans typically think about number symbolically, we also possess a nonsymbolic representation of quantity that is shared by many other animal species. These primitive numerical representations in humans and nonhuman animals are thought to arise from an evolutionarily ancient system termed the approximate number system (ANS).

The ANS represents number in an approximate, noisy fashion using mental magnitudes (Dehaene, 1997; Feigenson, Dehaene, & Spelke, 2004; Gallistel & Gelman, 2000). Critically, it is not dependent on language or formal training. However, as suggested by the name, its representations are limited in that they are imprecise. This imprecision leads to the two key behavioral hallmarks of the ANS: the distance effect (1 vs. 9 items is easier to discriminate than 1 vs. 2 items) and the size effect (1 vs. 2 items is easier to discriminate than 8 vs. 9 items). Accordingly, ANS representations of number follow Weber's law, which states that the discriminability of two numerosities is dependent on their ratio rather than their absolute numerical difference. This means that preverbal infants and nonhuman animals can differentiate sets of 5 from sets of 10 items more easily than they can differentiate 10 items from 15 items, despite the fact that both pairs differ by 5. Furthermore, ANS representations are too noisy to support fine-grained discriminations such as 20 versus 21 items. This ratio-dependence is a hallmark of the ANS, and ratio-dependent performance in

Mathematical Cognition and Learning, Vol. 1. http://dx.doi.org/10.1016/B978-0-12-420133-0.00005-3

numerical tasks is interpreted as engagement of the ANS (see also vanMarle, this volume). The sharpness or acuity of the ANS can be quantified as a Weber fraction, which is an index of the precision of internal numerical representations (Halberda & Odic, this volume). Importantly, the acuity of the ANS varies across individuals.

Educated humans, of course, are not limited to representing numbers approximately. Through the use of numerical symbols and counting routines, we can appreciate that the difference between 20 and 21 is the same as that between 1 and 2. Although the ANS is functional throughout the lifespan, when children learn number words and written symbols, they are learning a system that supports the exact representation of numerical quantity. Numerical symbols therefore endow humans with the unique ability to represent quantities precisely, which moves our numerical representations beyond the ANS and its inherent constraints. Furthermore, these numerical symbols can be manipulated using the many complex mathematical operations that support the modern world.

This chapter focuses on evidence for evolutionary and developmental continuities in numerical cognition across species and the human lifespan and asks how this ancient system that appears to be both phylogenetically universal and developmentally conservative is related to uniquely human symbolic mathematics. The first section reviews the evidence for numerical representations in nonhuman animals. The second section reviews the evidence that the ANS also underlies numerical reasoning in human infants. The third section addresses the malleability of ANS representations and asks whether training and practice can improve the acuity of the ANS. Finally, the last section argues that there is a causal link between the acuity of the ANS and symbolic mathematics performance. The implications of this relationship for mathematics education are also addressed.

CROSS-SPECIES COMPARISONS

Attending to number may enable animals to compare the relative size of a group, keep track of individuals in a group of predators, or decide when one bush is more or less profitable for foraging compared to another bush. Therefore, it should not be surprising that the ability to represent number is widespread across the animal kingdom. Fish, birds, rodents, and primates all possess the capacity to represent and compare numerical quantities (Agrillo et al., this volume; Beran et al., this volume; Geary et al., this volume; Pepperberg, this volume; Vallortigara, this volume). Furthermore, numerical representations in these divergent species exhibit many common features. However, it is not yet known whether the ANS originated in a very ancient ancestor common to most extant species (homology) or whether ANS-like systems evolved independently in different animals species (homoplasy). One possibility is that the ubiquity of the ANS is a homoplasy, especially when comparing birds, mammals, and fish. In this case, the neural substrates of the ANS may differ radically across species. In contrast, if the ubiquity reflects a homology and shared evolutionary history, then we should expect common neural substrates to underlie the ANS as has been proposed for non-human primates and humans.

Whatever the source of the similarities across species, demonstrations of quantitative capacities in animals have varied widely in their stimulus controls, and therefore, it is not always apparent whether the animals in question are relying on number per se or a host of other quantitative variables (e.g., surface area, density, or contour length). However, many studies have carefully controlled for non-numerical stimulus features and have demonstrated that many different animal species attend to numerosity. In one study, Brannon and Terrace (1998) trained rhesus monkeys (*Macaca mulatta*) to touch arrays containing one, two, three, or four elements in ascending order (Figure 5-1a). The monkeys were then tested with pairs of novel numerical

Numerical ordering task

Training   **Test**

(a)

Match-to-sample task

Training   **Test**

(b)

FIGURE 5-1 (a) Schematic of an ordering task used to show that monkeys trained to order the numerical values one through four performed with above-chance accuracy on pairs of novel values. *(After Brannon & Terrace, 1998.)* (b) Schematic of a numerical match-to-sample task where monkeys are trained to match the test stimulus (top center) to the stimulus that is both a numerical and a shape match. At test, monkeys can choose to match either by shape or by number. The results demonstrated monkeys attended to the numerical value of a sample stimulus even when in training there was redundant shape information present in the sample *(After Cantlon & Brannon, 2007b.)*

values outside the training (i.e., pairs of values between five and nine). Monkeys generalized the ordinal rule and successfully touched the novel pairs in ascending order. That finding has since been replicated in a number of primate species using both expanded numerical ranges and various permutations of the task (Cantlon & Brannon, 2005; 2006; Judge, Evans, & Vyas, 2005; Merritt, MacLean, Crawford, & Brannon, 2011). The same training and transfer paradigm has even been used with pigeons revealing parallel ratio-dependent performance and transfer of an ordinal numerical rule (Scarf, Hayne, & Colombo, 2011). These qualitative similarities in numerical representation in species that diverged more than 400 million years ago suggest that either the same solution has been reached multiple times through convergent evolution or alternatively that the ANS can be traced back to a common ancestor of birds, mammals, and possibly even fish (Agrillo, Petrazzini, Tagliapietra, & Bisazza, 2012; Agrillo, Piffer, Bisazza, & Butterworth, 2012).

The ANS is not limited to ordinal comparisons and in fact enables animals to combine collections of objects and identify the resulting sum (McCrink, this volume; vanMarle, this volume). In one study, Cantlon & Brannon (2007a) tested rhesus monkeys with a computerized approximate arithmetic task. Monkeys watched two sets of dots float behind an occluder on a screen and then had to choose between an array that represented the sum of the two arrays or a distractor array that contained a different number of dots. The monkeys succeeded at this task, and their accuracy was modulated by the ratio between the correct array and the distractor array. In other words, they could correctly identify the resulting sum, and their ability to do so was influenced by the discriminability of the two choices. The greater the ratio between the correct array and the distractor, the better the monkeys performed. Numerical representations in animals are also not limited to the visual modality; they can also match and add stimuli across sensory modalities (Baker & Jordan, this volume). For example, rhesus monkeys can choose the visual array that numerically matches the number of sounds in a sequence, and can sum across visually and aurally presented sets (Jordan, MacLean, & Brannon, 2008). Rats (*Rattus norvegicus*) are also capable of summing across the visual and auditory modalities, and spontaneously recognize the total numerosity of a compound stimulus composed of flashes of light and auditory tones (Meck & Church, 1983). These studies demonstrate that the ANS represents number in a format that is independent of the source modality and can support flexible manipulations of quantity.

Many of the preceding studies required thousands of training trials to obtain high levels of performance. There is a clear trade-off between stringent stimulus controls and ecological validity in all studies of animal cognition. The laboratory affords testing with thousands of trials and with a wide array of stimulus controls but often at the expense of ecological validity. An important question, then, is whether the numerical capacities animals demonstrate in the laboratory are tapping numerical skills they make use of in the wild or instead are co-opting other skills that evolved for other purposes. In other words, does the extensive training create numerical abilities that are not used in the wild? This question has been addressed in a few different ways (see also Agrillo et al., this volume). One approach asks whether spontaneous food choices are based on number or other quantitative features (Feigenson, Carey, & Hauser, 2002; Hauser, Carey, & Hauser, 2000). In this paradigm, food items are placed sequentially into two opaque containers, and the animals are allowed to choose one container from which they will receive the food. Results from this task in all four species of great apes (*Pan paniscus, Pan troglodytes, Gorilla gorilla,* and *Pongo pygmaeus*) (Beran, 2004; Hanus & Call, 2007), rhesus monkeys (Flombaum, Junge, & Hauser, 2005; Hauser et al., 2000), prosimian primates (*Lemuridae*) (Jones et al., 2013), and elephants (*Loxodonta africana*) (Perdue, Talbot, Stone, & Beran, 2012) all demonstrate that animals spontaneously track quantity. The studies do not, however, conclusively show that animals are relying on number as opposed to surface area or other continuous variables to make ordinal judgments.

Another approach has been to ask whether animals spontaneously use number when the task could be solved using an alternative stimulus feature. Cantlon and Brannon (2007b) tested rhesus monkeys with a match-to-sample task in which both choices were correct: one stimulus was a numerical match, whereas the other was a non-numerical match (color, shape, or surface area) (Figure 5-1b). For example, if training in the sample array was two squares, the monkey's choices would be two squares (number and shape match) or four circles (number and shape mismatch). In this case the correct rewarded answer matches on both shape and number. After training, monkeys were given probe trials in which one choice matched the sample in shape but not in number and the other matched the sample in number but not in shape. On probe trials monkeys were not reinforced, and any response was allowed. Although monkeys preferred to match based on shape or color rather than number, their choices did exhibit numerical distance effects. This means that they were more likely to make a number match when the shape or color match was very disparate from the sample in number. Furthermore, when surface area was pitted against number rather than shape or color, the monkeys showed a strong preference to match based on number over surface area. Thus, even though the task structure would allow monkeys to completely ignore number and still do well, they exhibited a strong bias to attend to number. This suggests that instead of attending to number only as a last-resort strategy, monkeys spontaneously and in some cases *preferentially* attend to numerical information in their environment. Importantly, one of the monkeys had no prior experience on any numerical judgment task and still showed this pattern of responding.

There is strong evidence for qualitative similarity in numerical discrimination between humans and other animals with ratio dependence being the main characteristic of performance (e.g., Agrillo, Piffer, Bisazza, & Butterworth,

FIGURE 5-2   Photograph of a ring-tailed lemur participating in a numerical discrimination task via a touch-sensitive screen.

2012; Beran, Decker, Schwartz, & Schultz, 2011; Beran, Johnson-Pynn, & Ready, 2008; 2011; Cantlon & Brannon, 2006; 2007a). At the same time, it is important to note that quantitative differences exist, as numerical acuity appears to be better in adult humans compared to other animals. Few studies have directly compared multiple species (Agrillo, Petrazzini, Tagliapietra, & Bisazza, 2012; Hanus & Call, 2007) and to our knowledge, only one study has parametrically tested whether species differ quantitatively in the acuity of the ANS (Jones et al., 2013). Jones and colleagues tested rhesus monkeys and three different lemur species (*Lemur catta, Eulemur mongoz, E. macaco*). On each trial, the monkey or lemur was presented with two arrays of dots and was rewarded for touching the array containing the larger number of dots (see Figure 5-2 for an example of a lemur performing this task). For all species, accuracy increased in proportion to the ratio between the test arrays. In addition to assessing accuracy, Jones and colleagues calculated a Weber fraction (w) for each participant, which is a measure of the acuity of ANS representations (Halberda & Odic, this volume). No species differences were found between monkeys and lemurs or between the three lemur species. This suggests that both qualitative and quantitative similarities exist in the numerical representations of different species of nonhuman primates.

NUMERICAL REPRESENTATIONS IN INFANCY

Before human infants learn language or begin formal schooling, they are already sensitive to numerical information in the world around them. For example, just hours after birth, infants can match numerical information across sensory modalities. As just one example, Izard, Sann, Spelke, Streri, and Gallistel (2009) familiarized newborn infants to auditory sequences of syllables with a fixed number of repetitions. Infants were then presented with visual arrays containing the same number of elements as the auditory sequence or a novel number of elements. The infants preferred to look at the images with the number of elements that matched the auditory sequence. Furthermore, their degree of preference for the matching image was modulated by the ratio of the number of elements in the test images. This suggests that infants come into the world with a sense of number that transcends sensory modalities (see also Baker & Jordan, this issue).

Sensitivity to number increases rapidly throughout the first year of life. Visual habituation and auditory head-turn paradigms (Figure 5-3) are two procedures that rely on infants' preference to attend to novel stimuli. These two paradigms provide convergent evidence that numerical discrimination in

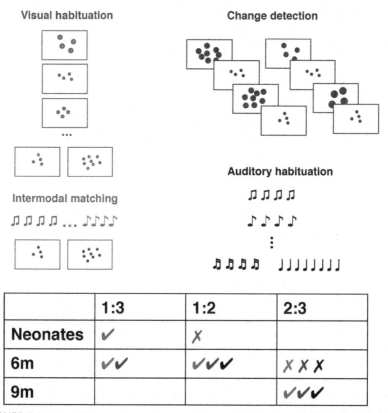

FIGURE 5-3 Four of the most common paradigms used to test numerical discrimination in infancy. The table displays the pattern of successes and failures obtained with each paradigm at different ages and numerical ratios.

infants is ratio-dependent (Brannon, Abbott, & Lutz, 2004; Lipton & Spelke, 2004; Xu & Spelke, 2000; Xu, Spelke, & Goddard, 2005). In both paradigms, 6-month-old infants require a 1:2 ratio (e.g., 10 versus 20 dots) for successful discrimination, whereas by 9 months of age, infants can discriminate a 2:3 ratio (e.g., 10 versus 15 dots). Note that it is the ratio between the numbers of dots that is driving discrimination performance, not the absolute difference. Thus when 6-month-old infants are habituated to arrays with a common number of elements, they look longer at new arrays with a novel number of elements compared to new arrays with the same number of elements, as long as the novel and familiar values differ by a 1:2 ratio (visual habituation). Similarly, if 6-month-old infants are played sequences that have the same number of tones, they orient longer to new sequences with a novel number of elements compared to new sequences with the familiar number of elements (auditory head-turn paradigm).

The numerical change detection paradigm (Figure 5-3) provides additional evidence for ratio-dependent numerical representations in infancy. In this paradigm infants are shown two visual sequences simultaneously on different monitors. One of the sequences changes numerically between two different values, whereas the other remains constant numerically. However, both sequences change in a variety of non-numerical ways (e.g., configuration of the dots, density, surface area). In addition to showing the same developmental trend of increasing sensitivity from 6 to 9 months, infants' preference for the numerically changing stream is graded as a function of numerical ratio, such that infants exhibit greater visual preferences for streams that differ by a greater ratio in accordance with Weber's law (Libertus & Brannon, 2010; Starr, Libertus, & Brannon, 2013a). Furthermore, electrophysiological measures reveal that brain activity is modulated by the ratio between familiar and novel numerosities in visually presented sequences (Hyde & Spelke, 2011; Libertus, Pruitt, Woldorff, & Brannon, 2009). Together, the converging results from different methodologies indicate that the ratio-dependent numerical representations found in nonhuman animals are also operational in human infants.

Parallel to findings with nonhuman animals, the ANS also supports approximate arithmetic calculation in human infants (McCrink, this volume; McCrink & Wynn, 2004; 2009). McCrink and Wynn (2004) showed 9-month-old infants' animations of arithmetic problems with large sets. In the displays, two arrays of shapes sequentially moved behind an occluder (addition) or one array of shapes moved behind an occluder and then a subset of the shapes moved from behind the occluder to offscreen (subtraction). The infants then saw the occluder drop away to reveal either a correct or an impossible outcome (either twice or half as many items). Infants looked significantly longer at the impossible outcomes. This suggests that infants are sensitive to the approximate outcomes of addition and subtraction operations and that they recognized when an inappropriate number of items was present after the transformation. The ANS also enables infants to recognize equivalence between numerical proportions, and this ability is again ratio-dependent. For example, 10-month-old infants are sensitive to differences between arrays in which half of the dots are blue and half are yellow and arrays in which three-fourths of the dots are blue and the other fourth are yellow (McCrink & Wynn, 2007).

Interestingly, there are some situations in which infants' numerical representations appear to be limited by an upper set-size limit rather than ratio (Posid & Cordes, this volume; vanMarle, this volume). In particular, when infants need to discriminate between small numbers of items that have been sequentially occluded, they seem to engage object tracking mechanisms rather than the ANS (Uller, Carey, Huntley-Fenner, & Klatt, 1999). These object-tracking mechanisms, which are also active in adults and at least some nonhuman animals, enable infants to precisely keep track of a small number of objects through space and time (see Feigenson et al., 2004, for a review). Although such mechanisms have likely not evolved for numerical purposes, they do seem to support some quantitative tasks involving up to three items. For example, in manual search and food choice tasks with one to four items, infants appear to be able to track only up to three items simultaneously. This means that infants succeed at discrimination when tested with 1 versus 2, 1 versus 3, or 2 versus 3 items but fail when tested with contrasts of 1 and 4 or 2 and 4 items (Feigenson et al., 2002; Feigenson & Carey, 2003; 2005). Similarly, in visual habituation studies, 6-month-old infants fail to discriminate between arrays with 1 and 2 or 1 and 3 dots (Xu, 2003; Xu et al., 2005).

However, when small numbers of items are presented simultaneously in either moving arrays or flashed very briefly, infants' numerical discrimination once again exhibits the ratio-dependent hallmark of the ANS. When 6-month-old infants are tested with values between 1 and 4 using the habituation paradigm with moving stimuli (Wynn, Bloom, & Chiang, 2002) or with the change detection paradigm (Starr et al., 2013a), discrimination is ratio-dependent (e.g., infants succeed at 2 vs. 4 and fail at 2 vs. 3). A similar dissociation between small and large number discrimination has been observed in some animal studies. In these cases, as with infants, performance appears to be set-size dependent for small numerosities and ratio-dependent for larger numerosities (e.g., Agrillo, Piffer, Bisazza, & Butterworth, 2012; Barner, Wood, Hauser, & Carey, 2008; Hauser et al., 2000), although other studies have found that nonhuman animals exhibit ratio-dependent performance for small and large numerosities alike (Beran, 2004; Beran, Johnson-Pynn, & Ready, 2011; Brannon & Terrace, 1998; Cantlon & Brannon, 2006; Jones et al., 2013).

These seemingly contradictory results with small and large set sizes suggest that while the ANS can operate over the full range of numerosities, there are some contexts in which the ANS is either overruled or not engaged. In these situations, objects may be tracked individually via a limited capacity system (Feigenson et al., 2004). Multiple explanations have been proposed to explain these discrepant findings. One idea that has been supported by electrophysiological evidence is that when stimulus complexity is low and spatial attention can select individual objects, parallel individuation occurs and performance is limited by set size, typically three items for infants and young children. However, when stimulus complexity is high and spatial attention is distributed, the ANS is more likely to be engaged (Hyde et al., 2010). An alternate idea is that early in development (and possibly also in nonhuman animals), the object-tracking mechanism may "trump" the ANS, perhaps due to the relatively poor precision of the ANS early in human development. As the ANS increases in acuity, however, it is more likely to be engaged in a wider variety of situations, which may be why failures to discriminate small and large numbers are not observed in human adults (Mou & vanMarle, 2013). However, this perspective does not fully explain why set-size limits are more likely to occur in some experimental paradigms than others, nor why infants do sometimes exhibit ratio-dependent discrimination for small numerosities.

MALLEABILITY

Between infancy and adulthood, the precision of nonverbal number representations improves dramatically. As shown in Figure 5-3 and reviewed earlier, neonates require a 1:3 difference for numerical discrimination, and this ability improves to a 2:3 difference by 9 months of age. In adulthood, however, only a 9:10 ratio is required on average for numerical discriminations (Halberda & Feigenson, 2008; Piazza et al., 2010). How much of this improvement should be attributed to maturation and how much to experience, culture, and education? One avenue for exploring this question is to study human societies with limited numerical vocabulary and no formal mathematical system.

The Mundurukú are an indigenous group in Brazil whose lexicon contains words only for the numbers one through five. Nevertheless, the Mundurukú possess an intact number sense, and their performance on a nonsymbolic numerical comparison task is qualitatively similar to that of French-speaking controls, although their overall performance is poorer for the more difficult ratios (Pica, 2004). Because of many cultural differences between the French and Mundurukú, the implications of this difference in performance are difficult to interpret. However, there is a great deal of variation in the amount of educational exposure for individuals within the Mundurukú community, which means that the Mundurukú provide an ideal test case for exploring the role of mathematics education on number sense acuity. Recently, Piazza, Pica, Izard, Spelke, and Dehaene (2013) found that Mundurukú subjects with some exposure to formal schooling had significantly better numerical acuity than those without any formal schooling, even after controlling for age. A similar pattern has been found in comparisons of schooled and unschooled adults from Western cultures, with mathematically educated adults exhibiting more precise numerical approximation skills than adults without any formal mathematics education (Nys et al., 2013). Experience with numerical symbols and calculations therefore appears to improve the precision of the ANS.

The anthropological studies just described suggest that culture and education impact ANS acuity. Recent work suggests that the acuity of the ANS can also be improved through brief exposure to explicit training on nonsymbolic tasks. DeWind and Brannon (2012) trained adults on a numerical comparison task for six 1-hour sessions. After the introduction of trial-by-trial feedback, ANS acuity rapidly improved and then remained steady after discontinuing the feedback. The authors propose that improvement in ANS acuity may be due to actual changes in the underlying numerical representations or to enhanced attention to number and ability to inhibit extraneous cues from other dimensions (e.g., surface area or density). However, a third possibility is that improvements in performance are due to the motivational effects of feedback rather than to any refinement in the ANS itself (Lindskog, Winman, & Juslin, 2013).

A return to some research with nonhuman primates may be informative in answering this question. To examine the role of extended training on numerical discrimination in rhesus monkeys, we looked through our archived data to examine performance for three monkeys on a numerical comparison task soon after they learned how to perform task. Since this early training, referred to as Time 1, the monkeys had extensive additional training on some variant of numerical comparison (156, 140, and 55 thousand trials, respectively). We then retested the three monkeys with the same stimulus parameters and same numerical values as that used at Time 1, and we refer to this second testing as Time 2. All three monkeys showed a reduction in Weber fraction from Time 1 to Time 2 (Feinstein: .29 to .2; Mikulski: 1.17 to 1.02, and Schroeder: .66 to .54). These improvements over time suggest that training can produce measurable changes in ANS acuity, although additional work will be needed to ascertain the underlying causes for the changes in the Weber fraction. One explanation is that changes in ANS acuity may reflect increased numerical sensitivity. Within parietal cortex, neurons have been found that are selective to specific numerosities (Nieder & Miller, 2004). Training may therefore serve to actually improve the sensitivity of these neurons. However, it is also possible that training improves the ability to home in on the relevant stimulus features or involves changes in motivation.

CAUSAL RELATION BETWEEN ANS ACUITY AND MATHEMATICS

The ANS is hypothesized to be a conceptual foundation for uniquely human symbolic mathematics skills (Dehaene, 1997; Feigenson, Libertus, & Halberda, 2013;

Piazza, 2010). In support of this proposal, a host of recent studies have found that ANS acuity is correlated with symbolic mathematics skill (Agrillo, Piffer, & Adriano, 2013; Gilmore, McCarthy, & Spelke, 2010; Halberda, Mazzocco, & Feigenson, 2008; Libertus, Feigenson, & Halberda, 2011; 2013a; Libertus, Odic, & Halberda, 2010; Lourenco, Bonny, Fernandez, & Rao, 2012; Mazzocco, Feigenson, & Halberda, 2011; Mundy & Gilmore, 2009; Piazza, Fumarola, Chinello, & Melcher, 2011; Starr, Libertus, & Brannon, 2013b). This relationship has been documented throughout the educational spectrum, from preschoolers who are just beginning formal mathematics education (Libertus et al., 2011; Libertus et al., 2013a; Starr et al., 2013b) all the way through to students taking their college entrance exams (DeWind & Brannon, 2012; Libertus et al., 2010). In addition, children with dyscalculia, a learning disability specific to math, have poorer ANS acuity than their typically developing peers (Piazza et al., 2010).

Other studies, however, have failed to find a relationship between ANS acuity and mathematics ability (Bonny & Lourenco, 2012; Gilmore et al., 2013; Holloway & Ansari, 2009; Nosworthy, Bugden, Archibald, Evans, & Ansari, 2013; Sasanguie, Defever, Maertens, & Reynvoet, 2013; Sasanguie, Göbel, Moll, Smets, & Reynvoet, 2012; Soltész, Szűcs, & Szűcs, 2010; Xenidou-Dervou, De Smedt, van der Schoot, & van Lieshout, 2010). Differences in the methods used to assess both ANS acuity and mathematics ability may account for some of these divergent findings. In addition, because mathematics ability is multifaceted, it is likely that many factors besides the ANS contribute to a child's mathematical proficiency. General cognitive abilities such as working memory, executive functions, and short-term memory are known to impact mathematics performance (e.g., Bull & Scerif, 2001; Espy et al., 2004; Geary, 2004; McLean & Hitch, 1999), in addition to socio-environmental factors such as income level and learning environment (e.g., Jordan, Kaplan, Ramineni, & Locuniak, 2009; Klibanoff, Levine, Huttenlocher, Vasilyeva, & Hedges, 2006). Determining how these factors interact with or even mediate the link between the ANS and formal mathematics is an important avenue for future research.

Even if the positive correlation between ANS acuity and mathematics ability stands, these associations do not exclude the possibility that the causal arrow points in the opposite direction, or that the influence is bidirectional. Does the precision of the ANS influence the acquisition of formal math, or might proficiency with formal mathematics determine the precision of the ANS? A recent study by Starr et al. (2013b) suggests that ANS acuity in infancy, well before children learn number words or acquire any type of formal mathematics education, may explain some of the variance in preschool mathematics ability. In this longitudinal study, numerical sensitivity was measured at six months of age, using the numerical change detection paradigm. Children then returned to the lab three years later and ANS acuity, mathematics achievement, and verbal IQ were measured. The main finding was that infants' numerical acuity at 6 months predicted their mathematics achievement at 3.5 years of age (Figure 5-4). Given that ANS acuity in this study was measured in infancy before any symbolic number skills had been formally taught, this study suggests that the causal arrow points from ANS acuity to symbolic mathematics ability. However, it is still important to remember that while a significant portion of the variance in children's mastery of the symbolic number system was explained by the ANS acuity measure, the effect was not large, and the ANS is clearly not the only determinant of a child's mathematics ability.

We are still far from understanding the mechanisms by which the ANS supports symbolic mathematics skills, but several hypotheses have been proposed. Given that ANS representations can be mentally manipulated in arithmetic operations such as addition and subtraction, the ANS may serve as an intuitive basis for symbolic arithmetic and therefore lay the groundwork for the acquisition of formal arithmetic principles (Gilmore et al., 2010; Park & Brannon, 2013). Gilmore and colleagues found that in kindergarten, typically the start of formal schooling, children's ability to perform arithmetic with nonsymbolic quantities (arrays of dots) correlated with their mastery of symbolic mathematics curriculum. The ANS may also be foundational for anchoring symbolic number knowledge, such that quality of the mapping between number words and their corresponding ANS values is a driving force behind

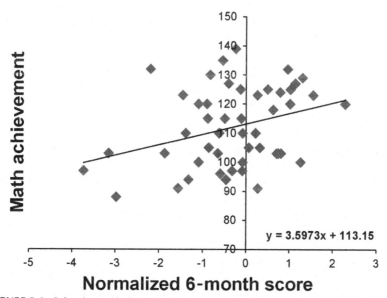

FIGURE 5-4 Infants' numerical change detection scores at 6 months of age (displayed here as standardized z-scores) predict mathematics achievement at 3.5 years of age as measured by the *Test of Early Mathematical Ability (Ginsburg & Baroody, 2003; redrawn from Starr et al., 2013b.)*

the link between ANS acuity and mathematics achievement (Feigenson et al., 2013; Piazza, 2010).

In support of this hypothesis, symbol ordering proficiency has been found to mediate the link between the ANS and symbolic mathematics (Lyons & Beilock, 2011), and ANS acuity is correlated with children's understanding of the count list (Starr et al., 2013b; vanMarle, Chu, Li, & Geary, 2014). An interesting possibility is that ANS acuity in preschool children may correlate more with early-acquired symbolic mathematics concepts (e.g., number word knowledge, arithmetic using fingers or tokens) compared to later-acquired symbolic mathematics concepts (e.g., writing Arabic digits, exact multiplication and division) (Libertus, Feigenson, & Halberda, 2013b), suggesting that the ANS is most critical for acquisition rather than maintenance of symbolic mathematics skills. Yet another proposal is that the ANS serves as an online error-monitoring system during arithmetic, allowing erroneous symbolic answers to be discarded in favor of correct answers (Feigenson et al., 2013; Lourenco et al., 2012).

Although a link between ANS precision and mathematics ability has been documented throughout the lifespan (Halberda, Ly, Wilmer, Naiman, & Germine, 2012), it is also possible that the nature of the relation may change with development and experience. Perhaps in preschool-aged children, the precision of the ANS is critical for learning the meaning of number words, such that children with sharper ANS acuity are better positioned to form mappings between the ANS and the counting words that they are just beginning to understand. Later, in older children and adults, the fact that the ANS supports quantity manipulations may be the critical factor, and it is this manipulation skill, rather than the precision of the underlying representations themselves, that influences later mathematics achievement. Much more work is needed to explore these possibilities and to explore alternative mediating factors that may refute the proposed causal link between and symbolic mathematics proficiency.

Given the inherent limitations in correlational designs, additional methods are critical for understanding the basis of the relationship between the ANS and symbolic math. If the ANS is foundational for formal math, then improvements in ANS acuity should also lead to improvements in mathematics ability. To date, only a few studies have examined this tantalizing hypothesis. In one series of studies, college students were trained on an approximate arithmetic task that involved addition and subtraction problems with arrays of dots (Park & Brannon, 2013, 2014). Training led not just to improvement on the task itself, but also to improvements in symbolic arithmetic and subtraction as assessed by arithmetic tests administered before the first training session and soon after the final training session (Figure 5-5). In contrast, subjects who received training on tasks involving other factors thought to influence mathematics performance (e.g., numerical comparison, working memory, and symbol ordering) showed no improvement in their symbolic arithmetic

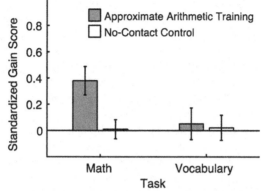

FIGURE 5-5 Standardized gain scores for arithmetic and vocabulary after training. *(Reprinted from Park & Brannon, 2013.)*

performance (Park & Brannon, 2013). In another study, children's symbolic arithmetic performance was enhanced following practice with nonsymbolic arithmetic problems, whereas practice adding and subtracting line length led to no such enhancement (Hyde, Khanum, & Spelke, 2014). Together, these studies suggest that improving nonverbal number sense boosts some aspects of symbolic mathematics ability. These studies thus provide the strongest evidence to date that the ANS supports the acquisition of symbolic mathematics skills.

APPLICATIONS

The possibility that there is a causal relationship between ANS acuity and symbolic mathematics skills leads to the exciting possibility of novel mathematics education strategies. In particular, it suggests that interventions targeting the ANS may be particularly beneficial for young children who have not yet learned to count or begun formal arithmetic instruction. Improving children's nonsymbolic numerical competence may improve the foundation that symbolic mathematics builds upon, thus facilitating the acquisition of symbolic number and arithmetic principles. If formal mathematics draws on the mapping between ANS representations and symbolic number representations, then improving the precision of children's ANS may facilitate the construction of this mapping, leading to an earlier mastery of the count sequence and an advantage for learning to manipulate numerical symbols. In addition, if there is overlap in the cognitive processes between nonsymbolic quantity manipulation and symbolic mathematics, then honing approximate arithmetic may jumpstart formal mathematics education and ultimately improve symbolic arithmetic capabilities.

Recently, there has been a trend toward the development of computerized adaptive software for educational use. Although such software cannot replace trained educators, it has the advantage of making educational interventions accessible to a larger number of students in a broad range of settings (Butterworth, Varma, & Laurillard, 2011). One such program, The Number Race (Wilson et al., 2006a), is an adaptive computer game designed to target number sense in children with the goal that this will transfer to improvements in mathematics achievement. The primary emphasis of The Number Race is numerical comparison. Initially, the game presents nonsymbolic quantities with relatively large ratios. The difficulty of the task is gradually increased by switching to symbolic numerosities, and by decreasing the ratio between the values to be compared. Each time the child makes a choice, corrective feedback is provided that displays the correct quantity both symbolically and nonsymbolically, thus reinforcing the mapping between numeric symbols and the quantities they represent. Although The Number Race is not the only computerized intervention for boosting mathematics achievement (see also Graphogame-Math and Numberbonds; Butterworth et al., 2011), it is the only game explicitly designed to improve nonsymbolic number sense.

Interventions utilizing The Number Race have thus far provided mixed results. In one study of 7- to 9-year-old children with mathematical learning difficulties, playing The Number Race for 2 hours a week for a 5-week period led to improvements in subitizing and numerical comparison speed and sub-traction accuracy, although addition was not affected (Wilson, Revkin, Cohen, Cohen, & Dehaene, 2006b). In a second study, 6- and 7-year-old children played The Number Race daily for 3 weeks. Again, the intervention improved children's number comparison skills, but there was no change in children's arithmetic skill (Räsänen, Salminen, Wilson, Aunio, & Dehaene, 2009).

In a third study, The Number Race game was modified to emphasize either approximate or exact numerical representations (Obersteiner, Reiss, & Ufer, 2013). In the approximate version, children needed to choose the digit that corresponded to the number of dots on the screen. In the exact version, children needed to choose the organized dot pattern that corresponded to a digit on the screen. Note, however, that both the approximate and exact versions of the game required mapping quantities to exact, symbolic representations of number. Seven-year-old children played the approximate version, the exact version, or both versions of the game for ten 30-minute sessions over 4 weeks. Children in all intervention groups exhibited improvements in basic number processing skills, but only the approximate group improved on numerical comparison and approximate arithmetic, whereas only the exact group improved on subitizing. In addition, all three groups exhibited improvement on arithmetic (Obersteiner et al., 2013). Together, these studies demonstrate that interventions that target number sense may be effective in improving some aspects of mathematical achievement, although further fine-tuning of the program may be needed to increase the efficacy of such interventions. Future work should explore whether interventions focusing on approximate arithmetic, as opposed to numerical comparison, are more effective (Hyde et al., 2014; Park & Brannon, 2013).

CONCLUSION

In this chapter we have reviewed evidence for continuity of the approximate number system (ANS) across both phylogeny and ontogeny. The ANS enables humans and nonhuman animals to represent and manipulate numerical quantities without language or formal training. These nonverbal representations may therefore form the foundation for acquisition of symbolic mathematics. Thus far, several studies have found a link between the acuity of the ANS and symbolic mathematics performance, although the matter is not yet resolved. When a relation between acuity of the ANS and symbolic mathematics is found, it appears to be bidirectional: learning numerical symbols leads to improvements in acuity, and children with better ANS acuity may have an advantage for acquiring symbolic arithmetic principles. This relationship gives way to the exciting possibility that educational interventions focused on improving the ANS may be an effective method for improving mathematics education outcomes.

REFERENCES

Agrillo, C., Petrazzini, M. E. M., Tagliapietra, C., & Bisazza, A. (2012). Inter-specific differences in numerical abilities among teleost fish. *Frontiers in Psychology, 3*, 483. http://dx.doi.org/10.3389/fpsyg.2012.00483/abstract.

Agrillo, C., Piffer, L., & Adriano, A. (2013). Individual differences in non-symbolic numerical abilities predict mathematical achievements but contradict ATOM. *Behavioral and Brain Functions, 9*(1), 26. http://dx.doi.org/10.1186/1744-9081-9-26.

Agrillo, C., Piffer, L., Bisazza, A., & Butterworth, B. (2012). Evidence for two numerical systems that are similar in humans and guppies. In S. F. Brosnan (Ed.), *PLoS ONE, 7*(2), e31923. http://dx.doi.org/10.1371/journal.pone.0031923.g004.

Barner, D., Wood, J. N., Hauser, M., & Carey, S. (2008). Evidence for a non-linguistic distinction between singular and plural sets in rhesus monkeys. *Cognition, 107*(2), 603–622. http://dx.doi.org/10.1016/j.cognition.2007.11.010.

Beran, M. J. (2004). Chimpanzees (*Pan troglodytes*) respond to nonvisible sets after one-by-one addition and removal of items. *Journal of Comparative Psychology, 118*(1), 25–36. http://dx.doi.org/10.1037/0735-7036.118.1.25.

Beran, M. J., Decker, S., Schwartz, A., & Schultz, N. (2011). Monkeys (*Macaca mulatta* and *Cebus apella*) and human adults and children (*Homo sapiens*) compare subsets of moving stimuli based on numerosity. *Frontiers in Psychology, 2*, 61. http://dx.doi.org/10.3389/fpsyg.2011.00061.

Beran, M. J., Johnson-Pynn, J. S., & Ready, C. (2008). Quantity representation in children and rhesus monkeys: Linear versus logarithmic scales. *Journal of Experimental Child Psychology, 100*(3), 225–233. http://dx.doi.org/10.1016/j.jecp.2007.10.003.

Beran, M. J., Johnson-Pynn, J. S., & Ready, C. (2011). Comparing children's (*Homo sapiens*) and chimpanzees' (*Pan troglodytes*) quantity judgments of sequentially presented sets of items. *Current Zoology, 57*(4), 419–428.

Bonny, J. W., & Lourenco, S. F. (2012). The approximate number system and its relation to early math achievement: Evidence from the preschool years. *Journal of Experimental Child Psychology, 1–14.* http://dx.doi.org/10.1016/j.jecp.2012.09.015.

Brannon, E. M., Abbott, S., & Lutz, D. (2004). Number bias for the discrimination of large visual sets in infancy. *Cognition, 93*(2), B59–B68.

Brannon, E. M., & Terrace, H. S. (1998). Ordering of the numerosities 1 to 9 by monkeys. *Science, 282*(5389), 746–749. Retrieved from, http://www.jstor.org/stable/2899267.

Bull, R., & Scerif, G. (2001). Executive functioning as a predictor of children's mathematics ability: Inhibition, switching, and working memory. *Developmental Neuropsychology, 19*(3), 273–293.

Butterworth, B., Varma, S., & Laurillard, D. (2011). Dyscalculia: From brain to education. *Science, 332*(6033), 1049–1053. http://dx.doi.org/10.1126/science.1201536.

Cantlon, J. F., & Brannon, E. M. (2005). Semantic congruity affects numerical judgments similarly in monkeys and humans. *Proceedings of the National Academy of Sciences, 102*(45), 16507.

Cantlon, J. F., & Brannon, E. M. (2006). Shared system for ordering small and large numbers in monkeys and humans. *Psychological Science, 17*(5), 401–406.

Cantlon, J. F., & Brannon, E. M. (2007a). Basic math in monkeys and college students. *PLoS Biology, 5*(12), e328. http://dx.doi.org/10.1371/journal.pbio.0050328.sg001.

Cantlon, J. F., & Brannon, E. M. (2007b). How much does number matter to a monkey (*Macaca mulatta*)? *Journal of Experimental Psychology. Animal Behavior Processes, 33*(1), 32–41. http://dx.doi.org/10.1037/0097-7403.33.1.32.

Dehaene, S. (1997). *The number sense: How the mind creates mathematics.* New York, USA: Oxford University.

DeWind, N. K., & Brannon, E. M. (2012). Malleability of the approximate number system: Effects of feedback and training. *Frontiers in Human Neuroscience.* http://dx.doi.org/10.3389/fnhum.2012.00068/abstract.

Espy, K. A., McDiarmid, M. M., Cwik, M. F., Stalets, M. M., Hamby, A., & Senn, T. E. (2004). The contribution of executive functions to emergent mathematic skills in preschool children. *Developmental Neuropsychology, 26*(1), 465–486.

Feigenson, L., & Carey, S. (2003). Tracking individuals via object-files: Evidence from infants' manual search. *Developmental Science, 6*(5), 568–584.

Feigenson, L., & Carey, S. (2005). On the limits of infants' quantification of small object arrays. *Cognition, 97*(3), 295–313.

Feigenson, L., Carey, S., & Hauser, M. (2002). The representations underlying infants' choice of more: Object files versus analog magnitudes. *Psychological Science, 13*(2), 150–156. http://dx.doi.org/10.1111/1467-9280.00427.

Feigenson, L., Dehaene, S., & Spelke, E. S. (2004). Core systems of number. *Trends in Cognitive Sciences, 8*(7), 307–314. http://dx.doi.org/10.1016/j.tics.2004.05.002.

Feigenson, L., Libertus, M. E., & Halberda, J. (2013). Links between the intuitive sense of number and formal mathematics ability. *Child Development Perspectives, 7*(2), 74–79. http://dx.doi.org/10.1111/cdep.12019.

Flombaum, J. I., Junge, J. A., & Hauser, M. D. (2005). Rhesus monkeys (*Macaca mulatta*) spontaneously compute addition operations over large numbers. *Cognition, 97*(3), 315–325. http://dx.doi.org/10.1016/j.cognition.2004.09.004.

Gallistel, C. R., & Gelman, R. (2000). Non-verbal numerical cognition: From reals to integers. *Trends in Cognitive Sciences, 4*(2), 59–65.

Geary, D. C. (2004). Mathematics and learning disabilities. *Journal of Learning Disabilities*, *37*(1), 4–15. http://dx.doi.org/10.1177/00222194040370010201.

Gilmore, C., Attridge, N., Clayton, S., Cragg, L., Johnson, S., Marlow, N., et al. (2013). Individual differences in inhibitory control, not non-verbal number acuity, correlate with mathematics achievement. In C. Chambers (Ed.), *PLoS ONE*, *8*(6), e67374. http://dx.doi.org/10.1371/journal.pone.0067374.g002.

Gilmore, C. K., McCarthy, S. E., & Spelke, E. S. (2010). Non-symbolic arithmetic abilities and achievement in the first year of formal schooling in mathematics. *Cognition*, *115*(3), 394.

Ginsburg, H. P., & Baroody, A. J. (2003). *Test of early mathematical ability* (3rd ed.). Austin, TX: Pro-ed.

Halberda, J., & Feigenson, L. (2008). Developmental change in the acuity of the "number sense": The approximate number system in 3-, 4-, 5-, and 6-year-olds and adults. *Developmental Psychology*, *44*(5), 1457.

Halberda, J., Ly, R., Wilmer, J. B., Naiman, D. Q., & Germine, L. (2012). Number sense across the lifespan as revealed by a massive Internet-based sample. *Proceedings of the National Academy of Sciences*, *109*(28), 11116–11120. http://dx.doi.org/10.1073/pnas.1200196109.

Halberda, J., Mazzocco, M. M. M., & Feigenson, L. (2008). Individual differences in non-verbal number acuity correlate with maths achievement. *Nature*, *455*(7213), 665–668.

Hanus, D., & Call, J. (2007). Discrete quantity judgments in the great apes (*Pan paniscus, Pan troglodytes, Gorilla gorilla, Pongo pygmaeus*): The effect of presenting whole sets versus item-by-item. *Journal of Comparative Psychology*, *121*(3), 241–249. http://dx.doi.org/10.1037/0735-7036.121.3.241.

Hauser, M. D., Carey, S., & Hauser, L. B. (2000). Spontaneous number representation in semi-free-ranging rhesus monkeys. *Proceedings of the Royal Society B: Biological Sciences*, *267*(1445), 829–833.

Holloway, I. D., & Ansari, D. (2009). Mapping numerical magnitudes onto symbols: The numerical distance effect and individual differences in children's mathematics achievement. *Journal of Experimental Child Psychology*, *103*(1), 17–29. http://dx.doi.org/10.1016/j.jecp.2008.04.001.

Hyde, D. C., Khanum, S., & Spelke, E. S. (2014). Brief non-symbolic, approximate number practice enhances subsequent exact symbolic arithmetic in children. *Cognition*, *131*(1), 92–107. http://dx.doi.org/10.1016/j.cognition.2013.12.007.

Hyde, D. C., & Spelke, E. S. (2011). Neural signatures of number processing in human infants: Evidence for two core systems underlying numerical cognition. *Developmental Science*, *4*(2), 360–371. http://dx.doi.org/10.1111/j.1467-7687.2010.00987.x.

Hyde, D. C., Winkler-Rhoades, N., Lee, S. -A., Izard, V., Shapiro, K. A., & Spelke, E. S. (2010). Spatial and numerical abilities without a complete natural language. *Neuropsychologia. 1–13.* http://dx.doi.org/10.1016/j.neuropsychologia.2010.12.017.

Izard, V., Sann, C., Spelke, E. S., Streri, A., & Gallistel, C. R. (2009). Newborn infants perceive abstract numbers. *Proceedings of the National Academy of Sciences*, *106*(25), 10382–10385.

Jones, S. M., Pearson, J., DeWind, N. K., Paulsen, D., Tenekedjieva, A. -M., & Brannon, E. M. (2013). Lemurs and macaques show similar numerical sensitivity. *Animal Cognition*, *17*(3), 503–515. http://dx.doi.org/10.1007/s10071-013-0682-3.

Jordan, N. C., Kaplan, D., Ramineni, C., & Locuniak, M. N. (2009). Early math matters: Kindergarten number competence and later mathematics outcomes. *Developmental Psychology*, *45*(3), 850–867. http://dx.doi.org/10.1037/a0014939.

Jordan, K. E., MacLean, E. L., & Brannon, E. M. (2008). Monkeys match and tally quantities across senses. *Cognition*, *108*(3), 617–625. http://dx.doi.org/10.1016/j.cognition.2008.05.006.

Judge, P. G., Evans, T. A., & Vyas, D. K. (2005). Ordinal representation of numeric quantities by brown capuchin monkeys (*Cebus apella*). *Journal of Experimental Psychology. Animal Behavior Processes, 31*, 79–94.

Klibanoff, R. S., Levine, S. C., Huttenlocher, J., Vasilyeva, M., & Hedges, L. V. (2006). Preschool children's mathematical knowledge: The effect of teacher 'math talk.' *Developmental Psychology, 42*(1), 59–69. http://dx.doi.org/10.1037/0012-1649.42.1.59.

Libertus, M. E., & Brannon, E. M. (2010). Stable individual differences in number discrimination in infancy. *Developmental Science, 13*(6), 900–906.

Libertus, M. E., Feigenson, L., & Halberda, J. (2011). Preschool acuity of the approximate number system correlates with school math ability. *Developmental Science, 14*(6), 1292–1300. http://dx.doi.org/10.1111/j.1467-7687.2011.01080.x.

Libertus, M. E., Feigenson, L., & Halberda, J. (2013a). Is approximate number precision a stable predictor of math ability? *Learning and Individual Differences, 25*, 126–133.

Libertus, M. E., Feigenson, L., & Halberda, J. (2013b). Numerical approximation abilities correlate with and predict informal but not formal mathematics abilities. *Journal of Experimental Child Psychology, 116*(4), 829–838. http://dx.doi.org/10.1016/j.jecp.2013.08.003.

Libertus, M. E., Odic, D., & Halberda, J. (2010). Intuitive sense of number correlates with math scores on college-entrance examination. *Acta Psychologica, 141*(3), 1–7. http://dx.doi.org/10.1016/j.actpsy.2012.09.009.

Libertus, M. E., Pruitt, L. B., Woldorff, M., & Brannon, E. M. (2009). Induced alpha-band oscillations reflect ratio-dependent number discrimination in the infant brain. *Journal of Cognitive Neuroscience, 21*(12), 2398–2406.

Lindskog, M., Winman, A., & Juslin, P. (2013). Are there rapid feedback effects on approximate number system acuity? *Frontiers in Human Neuroscience, 7*, 270. http://dx.doi.org/10.3389/fnhum.2013.00270/abstract.

Lipton, J., & Spelke, E. S. (2004). Discrimination of large and small numerosities by human infants. *Infancy, 5*(3), 271–290.

Lourenco, S. F., Bonny, J. W., Fernandez, E. P., & Rao, S. (2012). Nonsymbolic number and cumulative area representations contribute shared and unique variance to symbolic math competence. *Proceedings of the National Academy of Sciences, 109*(46), 18737–18742. http://dx.doi.org/10.1073/pnas.1207212109/-/DCSupplemental.

Lyons, I. M., & Beilock, S. L. (2011). Numerical ordering ability mediates the relation between number-sense and arithmetic competence. *Cognition. 1–6.* http://dx.doi.org/10.1016/j.cognition.2011.07.009.

Mazzocco, M. M. M., Feigenson, L., & Halberda, J. (2011). Preschoolers' precision of the approximate number system predicts later school mathematics performance. In L. Santos (Ed.), *PLoS ONE, 6*(9), e23749. http://dx.doi.org/10.1371/journal.pone.0023749.t003.

McCrink, K., & Wynn, K. (2004). Large-number addition and subtraction by 9-month-old infants. *Psychological Science, 15*(11), 776–781. http://dx.doi.org/10.1111/j.0956-7976.2004.00755.x.

McCrink, K., & Wynn, K. (2007). Ratio abstraction by 6-month-old infants. *Psychological Science, 18*(8), 740–745. http://dx.doi.org/10.1111/j.1467-9280.2007.01969.x.

McCrink, K., & Wynn, K. (2009). Operational momentum in large-number addition and subtraction by 9-month-olds. *Journal of Experimental Child Psychology, 103*(4), 400–408. http://dx.doi.org/10.1016/j.jecp.2009.01.013.

McLean, J. F., & Hitch, G. J. (1999). Working memory impairments in children with specific arithmetic learning difficulties. *Journal of Experimental Child Psychology, 74*(3), 240–260.

Meck, W. H., & Church, R. M. (1983). A mode control model of counting and timing processes. *Journal of Experimental Psychology. Animal Behavior Processes, 9*(3), 320.

Merritt, D. J., MacLean, E. L., Crawford, J. C., & Brannon, E. M. (2011). Numerical rule-learning in ring-tailed lemurs (*Lemur catta*). *Frontiers in Psychology, 2*, 23. http://dx.doi.org/10.3389/fpsyg.2011.00023/abstract.

Mou, Y., & vanMarle, K. (2013). Two core systems of numerical representation in infants. *Developmental Review. 1–25.* http://dx.doi.org/10.1016/j.dr.2013.11.001.

Mundy, E., & Gilmore, C. K. (2009). Children's mapping between symbolic and nonsymbolic representations of number. *Journal of Experimental Child Psychology, 103*(4), 490–502. http://dx.doi.org/10.1016/j.jecp.2009.02.003.

Nieder, A., & Miller, E. K. (2004). A parieto-frontal network for visual numerical information in the monkey. *Proceedings of the National Academy of Sciences, 101*(19), 7457–7462.

Nosworthy, N., Bugden, S., Archibald, L., Evans, B., & Ansari, D. (2013). A two-minute paper-and-pencil test of symbolic and nonsymbolic numerical magnitude processing explains variability in primary school children's arithmetic competence. In K. Paterson (Ed.), *PLoS ONE, 8*(7), e67918. http://dx.doi.org/10.1371/journal.pone.0067918.t008.

Nys, J., Ventura, P., Fernandes, T., Querido, L., Leybaert, J., & Content, A. (2013). Does math education modify the approximate number system? A comparison of schooled and unschooled adults. *Trends in Neuroscience and Education. 1–10.* http://dx.doi.org/10.1016/j.tine.2013.01.001.

Obersteiner, A., Reiss, K., & Ufer, S. (2013). How training on exact or approximate mental representations of number can enhance first-grade students' basic number processing and arithmetic skills. *Learning and Instruction, 23*(c), 125–135. http://dx.doi.org/10.1016/j.learninstruc.2012.08.004.

Park, J., & Brannon, E. M. (2013). Training the approximate number system improves math proficiency. *Psychological Science, 24*(10), 2013–2019. http://dx.doi.org/10.1177/0956797613482944.

Park, J., & Brannon, E. M. (2014). Improving arithmetic performance with number sense training: An investigation of underlying mechanism. *Cognition, 133*(1), 188–200. http://dx.doi.org/10.1016/j.cognition.2014.06.011.

Perdue, B. M., Talbot, C. F., Stone, A. M., & Beran, M. J. (2012). Putting the elephant back in the herd: Elephant relative quantity judgments match those of other species. *Animal Cognition, 15*(5), 955–961. http://dx.doi.org/10.1007/s10071-012-0521-y.

Piazza, M. (2010). Neurocognitive start-up tools for symbolic number representations. *Trends in Cognitive Sciences, 14*(12), 542–551. http://dx.doi.org/10.1016/j.tics.2010.09.008.

Piazza, M., Facoetti, A., Trussardi, A. N., Berteletti, I., Conte, S., Lucangeli, D., et al. (2010). Developmental trajectory of number acuity reveals a severe impairment in developmental dyscalculia. *Cognition, 116*(1), 33–41. http://dx.doi.org/10.1016/j.cognition.2010.03.012.

Piazza, M., Fumarola, A., Chinello, A., & Melcher, D. (2011). Subitizing reflects visuo-spatial object individuation capacity. *Cognition, 121*(1), 147–153. http://dx.doi.org/10.1016/j.cognition.2011.05.007.

Piazza, M., Pica, P., Izard, V., Spelke, E. S., & Dehaene, S. (2013). Education enhances the acuity of the nonverbal approximate number system. *Psychological Science, 24*(6), 1037–1043. http://dx.doi.org/10.1177/0956797612464057.

Pica, P. (2004). Exact and approximate arithmetic in an Amazonian Indigene group. *Science, 306*(5695), 499–503. http://dx.doi.org/10.1126/science.1102085.

Räsänen, P., Salminen, J., Wilson, A. J., Aunio, P., & Dehaene, S. (2009). Computer-assisted intervention for children with low numeracy skills. *Cognitive Development, 24*(4), 450–472. http://dx.doi.org/10.1016/j.cogdev.2009.09.003.

Sasanguie, D., Defever, E., Maertens, B., & Reynvoet, B. (2013). The approximate number system is not predictive for symbolic number processing in kindergarteners. *The Quarterly Journal of Experimental Psychology, 1–10.* http://dx.doi.org/10.1080/17470218.2013.803581.

Sasanguie, D., Göbel, S. M., Moll, K., Smets, K., & Reynvoet, B. (2012). Approximate number sense, symbolic number processing, or number-space mappings: What underlies mathematics achievement? *Journal of Experimental Child Psychology*, *1–14*. http://dx.doi.org/10.1016/j.jecp.2012.10.012.

Scarf, D., Hayne, H., & Colombo, M. (2011). Pigeons on par with primates in numerical competence. *Science*, *334*(6063), 1664. http://dx.doi.org/10.1126/science.1213357.

Soltész, F., Szűcs, D., & Szűcs, L. (2010). Relationships between magnitude representation, counting and memory in 4- to 7-year-old children: A developmental study. *Behavioral and Brain Functions*, *6*(1), 13.

Starr, A., Libertus, M. E., & Brannon, E. M. (2013a). Infants show ratio-dependent number discrimination regardless of set size. *Infancy*, *18*(6), 927–941. http://dx.doi.org/10.1111/infa.12008.

Starr, A., Libertus, M. E., & Brannon, E. M. (2013b). Number sense in infancy predicts mathematical abilities in childhood. *Proceedings of the National Academy of Sciences*, *110*(45), 18116–18120. http://dx.doi.org/10.1073/pnas.1302751110/-/DCSupplemental/sm01.avi.

Uller, C., Carey, S., Huntley-Fenner, G., & Klatt, L. (1999). What representations might underlie infant numerical knowledge? *Cognitive Development*, *14*(1), 1–36.

vanMarle, K., Chu, F. W., Li, Y., & Geary, D. C. (2014). Acuity of the approximate number system and preschoolers' quantitative development. *Developmental Science*. http://dx.doi.org/10.1111/desc.12143, n/a–n/a.

Wilson, A. J., Dehaene, S., Pinel, P., Revkin, S. K., Cohen, L., & Cohen, D. (2006). Principles underlying the design of "The Number Race," *Behavioral and Brain Functions*, *2*(1), 19. http://dx.doi.org/10.1186/1744-9081-2-19.

Wilson, A. J., Revkin, S. K., Cohen, D., Cohen, L., & Dehaene, S. (2006). An open trial assessment of "The Number Race," an adaptive computer game for remediation of dyscalculia. *Behavioral and Brain Functions*, *2*(1), 20. http://dx.doi.org/10.1186/1744-9081-2-20.

Wynn, K., Bloom, P., & Chiang, W. -C. C. (2002). Enumeration of collective entities by 5-month-old infants. *Cognition*, *83*(3), B55–B62.

Xenidou-Dervou, I., De Smedt, B., van der Schoot, M., & van Lieshout, E. C. D. M. (2010). Individual differences in kindergarten math achievement: The integrative roles of approximation skills and working memory. *Learning and Individual Differences*, *28*, 119–129. http://dx.doi.org/10.1016/j.lindif.2013.09.012.

Xu, F. (2003). Numerosity discrimination in infants: Evidence for two systems of representations. *Cognition*, *89*(1), B15–B25. http://dx.doi.org/10.1016/S0010-0277(03)00050-7.

Xu, F., & Spelke, E. S. (2000). Large number discrimination in 6-month-old infants. *Cognition*, *74*(1), B1–B11.

Xu, F., Spelke, E. S., & Goddard, S. (2005). Number sense in human infants. *Developmental Science*, *8*(1), 88–101. http://dx.doi.org/10.1111/j.1467-7687.2005.00395.x.

Chapter 6

On the Relation between Numerical and Non-Numerical Magnitudes: Evidence for a General Magnitude System

Stella F. Lourenco

Emory University, Atlanta, GA, USA

INTRODUCTION

Human beings living in modern societies have access to a plethora of numerical symbols such as number words and Arabic digits. Numerical symbols allow us to label and quantify experience; indeed, they are critical for quantification because they support the abstraction of numerical magnitude and promote precision. If one were to count a bag of 15 apples, for example, one could refer precisely to 15 apples. There would be *exactly* 15, regardless of their individual sizes, the overall weight of the bag, their ripeness, and so on. But when explicit counting is not possible, how does quantification take place (cf. Gallistel & Gelman, 1992)? Consider having to decide between multiple security lines at the airport, while being pressed for time or already engaged in another task. Counting each person in each line would be inefficient (in the former case) and perhaps even impossible (in the latter case). Yet many of us could still generate good approximations of the number of people in each line. This raises important questions about the nature of non-symbolic representations of number.

Visuospatial displays such as airport lines consist of multiple sources of magnitude information, each of which could potentially inform one's decision about which line might be quickest. One type of information is, as just mentioned, numerical magnitude, or what has come to be known as an approximate sense of number. In the example of airport lines, one might estimate 20, 25, and 35 people in three different lines without counting the individuals in each line. These estimates are inexact, but specifically *numerical*; that is, each person represents a discrete entity and is ultimately enumerable. It has been suggested

Mathematical Cognition and Learning, Vol. 1. http://dx.doi.org/10.1016/B978-0-12-420133-0.00006-5
145

that representations of numerical magnitude engage processing by the *approximate number system* (ANS; Dehaene, 2011; Feigenson, Dehaene, & Spelke, 2004; Halberda, Mazzocco, & Feigenson, 2008), which recruits neural resources from within a region of the parietal cortex called the intraparietal sulcus (IPS; Dehaene, Piazza, Pinel, & Cohen, 2003; for reviews, see Ansari, 2008; Nieder & Dehaene, 2009). Other types of magnitude information are considered *non-numerical*, and at least for descriptive purposes, can be isolated from numerical magnitude. For example, one's estimates of security lines could be based on cues such as the current lengths of the different lines or the movement and rate of the changing lengths of people. In contrast to numerical magnitude, these cues are considered strictly continuous in nature, only ever enumerable by reference to a number system (e.g., Gebuis & Reynvoet, 2012; Mix, Huttenlocher, & Levine, 2002; Leibovich & Henik, 2013).

Human infants (and some young children) do not have access to numerical symbols and are thus never in a position to rely on explicit counting as a strategy for quantification. Developmental psychologists have debated how infants and young children discriminate arrays of objects, or sequences of events, as well as how they perform simple arithmetic operations such as addition and subtraction. On one side of this debate are researchers who hold that these abilities are supported by the ANS—that is, representations of numerical magnitude per se (e.g., Xu & Spelke, 2000; see also, Brannon, Abbott, & Lutz, 2004; Cordes & Brannon, 2008; McCrink & Wynn, 2004). Xu and Spelke, for example, found that when infants were familiarized to a specific numerical value (e.g., 8 dots), they later looked longer at an array that differed in numerosity (i.e., 16 dots) compared with the original numerosity (8 dots), as would be predicted if they had differentiated the arrays (8 vs. 16 dots). Because non-numerical magnitude cues such as cumulative surface area were varied across trials, this finding has been taken as evidence that infants are capable of discriminating visual displays based on numerical magnitude. On the other side, there are those who argue that discrimination and basic calculation build on representations of non-numerical magnitude such as cumulative surface area (in the case of visual displays) and total duration (in the case of visual or auditory sequences; e.g., Clearfield & Mix, 1999; Feigenson, Carey, & Spelke, 2002; Rousselle, Palmers, & Noël, 2004), at least when these cues are reliable across trials. Seminal work by Clearfield and Mix found that when contour length (i.e., the summed perimeter of individual elements in an array) was pitted against number, infants discriminated visual displays on the basis of contour length and not number. This particular study differs from the Xu and Spelke study in that rather than varying non-numerical magnitude information, this information co-varied with number (as is generally the case in the physical environment), with the critical test coming when numerical and non-numerical magnitude cues were pitted against each other. Although this study could not rule out that infants had not encoded numerosity, it certainly demonstrated that when numerical and non-numerical magnitudes were both available, infants weighted the latter more heavily.

A challenge with comparing studies that find either a preference for numerical or non-numerical magnitude is that different systems may be engaged when processing sets of fewer than 4 items or more than 3 items (e.g., Carey, 2009; Feigenson et al., 2004; vanMarle, this volume). One result is that the findings may be inconsistent because the stimuli used in one study engage a different processing system than the stimuli used in another study. Researchers have speculated that smaller values (<4) engage an "object file system" (which draws on attentional resources to individuate items; Leslie, Xu, Tremoulet, & Scholl, 1998; Simon, 1997), whereas larger values (>3) are processed by the ANS. Studies using values greater than 3 typically demonstrate that infants and young children are capable of representing numerical magnitude (e.g., Brannon et al., 2004; Xu & Spelke, 2000; but see Rousselle et al., 2004), but those using smaller values suggest that non-numerical magnitude is represented (e.g., Clearfield & Mix, 1999; Feigenson et al., 2002). This is consistent with claims that object files are not specific to numerical magnitude but, rather, are associated with object-bound properties, including size, shape, and other features (e.g., Carey, 2009).

Parallel disagreements exist within vision science and cognitive psychology where researchers have debated which properties of images are extracted and ultimately represented for an estimate of a specific quantity. Of particular interest to this debate are questions about the content of these representations and the processes that are used to encode the relevant stimulus features. On one view, the content is exclusively numerical, with extant models positing procedures that normalize over non-numerical properties such as the individual sizes of elements and cumulative surface area (e.g., Barth, Kanwisher, & Spelke, 2003; Dehaene & Changeux, 1993; Ross & Burr, 2010; Stoianov & Zorzi, 2012; Verguts & Fias, 2004). On another view, the content is not specifically numerical (e.g., Dakin, Tibber, Greenwood, Kingdom, & Morgan, 2011; Durgin, 2008; Sophian & Chu, 2008; Tokita & Ishiguchi, 2010); in fact, the claim is that non-numerical sources of information such as surface area, element size, density, and/or convex hull (i.e., the area subtended by the set of items within an array as opposed to the cumulative surface area covered by the items) are extracted from visual displays of discrete items (e.g., Allik & Tuulmets, 1991; Gebuis & Reynvoet, 2012). In this case, numerical estimation is a by-product of non-numerical properties.

In the present chapter, I focus on nonverbal representations of number (generally referred to as "numerosity" or "numerical magnitude") and their relation to non-numerical magnitudes such as spatial extent and duration, which, like numerical magnitude, are unambiguous in their *more-versus-less* organization (Stevens, 1957, 1975; see also, Lourenco & Longo, 2011). Infants and young children do not have access to symbolic systems and are thus restricted to processes of magnitude estimation that engage a primitive, nonverbal system of number representation—one that I will argue forms part of a general system of analog magnitude. As described in more detail later, this putative *general magnitude system* is not specific to number, but encompasses the processing of non-numerical magnitudes as well. Based on converging behavioral and neural evidence, I argue for an anatomical functional model of generalized magnitude representation, hosted by fronto-parietal circuitry that is operational early in human development and whose functions show continuity across development.

In addition to highlighting the structural similarity of number and other magnitudes (i.e., their common analog format), this model posits shared neural circuitry for different magnitudes and *partially* overlapping representational content. Later I describe data that point to nonverbal representations of magnitude that support predictions about arbitrary associative mappings from one magnitude to another (see also Baker & Jordan, this volume). These data demonstrate that children's learning based on one type of magnitude (e.g., physical size) is related to their learning about other types of magnitudes (e.g., number). I also describe cognitive interactions such as interference and facilitation across magnitudes (i.e., when judgments related to one magnitude are slower or faster, respectively, because of the presence of another magnitude), and transfer of representational precision from one magnitude to another following training. There is also emerging work showing that, starting early in development, the precision of both numerical and non-numerical magnitude representations predict mathematical competence, including arithmetic and geometry. This particular work not only provides support for a general magnitude system but also suggests an important role for this system in learning relatively formal math concepts and computations.

ANALOG FORMAT FOR NUMBER AND OTHER MAGNITUDES

One way to conceptualize the relation between number and other magnitudes is in terms of their common analog format. In analog format, magnitudes are represented approximately, such that activation for a given value is accompanied by activation for adjacent values (e.g., the presentation of 7 items will activate representations of 7, but also of 6 and 8), albeit to a lesser extent. Moreover, the larger the magnitudes, the greater the "noise" in the underlying representations; that is, the wider the spread of activation of nearby values (Halberda & Odic, this volume). A behavioral consequence of representations that are analog in format is that discrimination follows Weber's law. For example, discriminating 8 versus 4 items, a 2:1 ratio, is easier than discriminating 12 versus 8 items, a 3:2 ratio, even though the absolute difference between the items (4 items) is the same in both cases (e.g., Buckley & Gillman, 1974). Ease of discrimination in human adults is often indexed by decision speed (i.e., faster reaction times) or accuracy. Weber's law is similarly obeyed when discriminating objects that vary in physical size, or inter-object relations varying in distance, or items varying in duration (e.g., Droit-Volet, Clément, & Fayol, 2008; Fias, Lammertyn, Reynvoet, Dupont, & Orban, 2003; Henmon, 1906).

Behavioral studies with infants suggest that analog format is a property of number and other magnitudes from early in human development, with ratio-modulated discrimination documented for different types of magnitude across the first year of life (for development in nonhuman species, see Agrillo et al., this volume; Vallortigara, this volume). Using looking-time paradigms that measure either "dishabituation" (i.e., recovery of attention following a decrease in attention during the "habituation" phase) or "change detection" (i.e., greater attention to stimuli that demonstrate relatively greater variability), 6-month-old infants have been shown to discriminate numerosities by ratios of 2:1 and larger (e.g., 3:1 and 4:1) in visual (Brannon et al., 2004; Libertus & Brannon, 2010; Xu & Spelke, 2000) and auditory (Lipton & Spelke, 2004; vanMarle & Wynn, 2009) modalities. The same threshold of discrimination has been observed at this age for spatial extent (e.g., size of Elmo faces [Brannon, Lutz, & Cordes, 2006] and line lengths [de Hevia & Spelke, 2010]) presented visually, and for duration presented to visual (Lourenco & Longo, 2010) or auditory (Brannon, Suanda, & Libertus, 2007; vanMarle & Wynn, 2006) modalities. In addition to common discrimination thresholds for numerical and non-numerical magnitudes at 6 months of age, infants' sensitivity to numerical values and duration has been found to increase in parallel over the first year of life, and by 9–10 months, infants distinguish a 3:2 ratio for both of these magnitudes (Brannon et al., 2004; vanMarle & Wynn, 2006). Moreover, behavioral evidence of increases in discrimination sensitivity has been documented in preschool and school age children for number, duration, and spatial extent (Bonny & Lourenco, 2014a; Droit-Volet et al., 2008; Holloway & Ansari, 2008; Odic, Libertus, Feigenson, & Halberda, 2013). In summary, the fine-tuning of children's ability to discriminate differences in numerical and non-numerical magnitudes shows similar developmental trends.

Although such findings provide compelling evidence for a common analog format, they are less clear in their support of a general magnitude system, wherein numerical magnitude shares processing resources with non-numerical magnitudes. Common thresholds of discrimination could reflect activation of shared cognitive mechanisms related to, for example, attention or memory, neither of which is specific to the representation of magnitude, numerical or otherwise. Developmental changes in discrimination sensitivity such as parallel increases over ontogenetic time may similarly reflect nonspecific cognitive improvements (e.g., attention or memory), which could serve to constrain performance on different discrimination tasks (Feigenson, 2007; Lourenco & Longo, 2011). Moreover, it should be noted that discrimination sensitivity for a given magnitude (e.g., numerosity or duration) has been found to depend on variables such as the range of stimulus values (e.g., discrimination is better for duration when the values being compared are greater than 500 ms; Burr, Della Rocca, & Morrone, 2013), stimulus format (e.g., discrimination is better for number when the arrays of items being compared are presented across two spatial fields compared with one overlapping field; Bonny & Lourenco, 2014a; see also Gilmore, Attridge, & Inglis, 2011; Price, Palmer, Battista, & Ansari, 2012), and sensory modality (e.g., discrimination is often better for duration in audition compared with vision; Goodfellow, 1934; see also Bratzke, Seifried, & Ulrick, 2012; Grondin & Ulrich, 2011). Such contextual effects *within* magnitude (e.g., contrasts of numbers or contrasts of durations) highlight a major interpretative challenge for using discrimination sensitivity as evidence for (or against) a general magnitude system, especially if differences in discrimination sensitivity *across* magnitude are also documented (Bonny & Lourenco, 2014a; Leibovich & Henik, 2014; Odic et al., 2013).

To summarize, the findings presented thus far provide strong support for a common analog format among numerical and non-numerical magnitudes, but they are ambiguous in their support of a general magnitude system. In other words, children's ability to discriminate among number sets shows the same characteristic features as their ability to discriminate spatial extent and duration. Discriminations are based on the ratio of compared magnitudes, not the absolute difference. However, it is not known whether these similarities are due to a system of general magnitude representation with overlapping content or to completely differentiated representations with similar properties. Evidence of structural similarity (i.e., analog format) across different magnitudes is observed beginning early in development (perhaps even innately [cf. Walsh, 2003]) and shows continuity over the lifespan (Bonny & Lourenco, 2014a; Droit-Volet et al., 2008). Much of this evidence comes from behavioral paradigms, but there is also neural evidence of an analog format, such as ratio-modulated activation when processing different types of magnitudes in posterior parietal cortex (e.g., Fias et al., 2003; Hayashi et al., 2013; Piazza, Izard, Pinel, Le Bihan, & Dehaene, 2004). In the next section, I describe data that go beyond showing common analog format for numerical and non-numerical magnitudes to demonstrating functional connections and shared neural codes for different magnitudes.

A GENERAL MAGNITUDE SYSTEM: THREE TYPES OF SUPPORTING EVIDENCE

Despite general consensus that numerical and non-numerical magnitudes are represented in a common analog format, the question of whether different dimensions of magnitude form part of a general magnitude system with shared processing resources is far more controversial. Do representations of number and other magnitudes share more than an analog format? Accumulating evidence from neuroimaging studies with humans and electrophysiology with nonhuman animals suggests that magnitudes such as numerosity, physical size, and duration are represented in close neural proximity in parietal cortex, particularly in and around the intraparietal sulcus (IPS; for reviews, see Cantlon, Platt, & Brannon, 2009; Cohen Kadosh, Lammertyn, & Izard, 2008; Walsh, 2003). But does this mean that these quantitative dimensions engage the same neural mechanisms or that the representational content is, even to some extent, undifferentiated in the minds of humans and nonhuman animals? Walsh and colleagues (e.g., Bueti & Walsh, 2009; Walsh, 2003) have argued that the coupling of dimensions such as space and time are critical for sensorimotor transformations, that is, for moving through and engaging the environment. On this view, neural codes within parietal cortex support analog computations across a range of magnitude relations such as "bigger–smaller," "nearer–farther," and "faster–slower" that allow for the automatic integration of simultaneously relevant cues in the external environment subserving directed action execution.

As adults, we also readily experience integration for semantic concepts that may not be directly related to moving through and engaging the physical environment. We recognize, for example, the equivalence between "a lot of money" and "a long drive." Whether or not such a correspondence is rooted in a general magnitude system, however, is an open question. Are comparisons made on behalf of conceptual correspondences supported by interactions of analog representations? One possibility is that they are *not*. The correspondences that adults and verbal children make across different magnitudes could be supported by language. Words such as "a lot" and "long" could promote abstraction by highlighting the correspondence in meaning across different magnitudes. Associations in this case would be indirect, mediated by the recoding of distinct magnitude representations into common categorical units because words such as "a lot" and "long," "big," and so on access a similar concept.

In this section, I review evidence suggesting that conceptual correspondences for different magnitudes are already present in human infants and thus not dependent on linguistic recoding mechanisms. In addition, I lay out three types of evidence that provide support for direct, functional connections between analog representations of numerical and non-numerical magnitudes. I begin with cross-magnitude interactions, which have been documented throughout development. I then turn to inter-individual correlations in the precision of different types of magnitude representations (numerical and non-numerical). I then draw on work showing that, following training, increases in representational precision transfer across magnitudes. Taken together, the behavioral data, which I supplement with neural findings, provide strong support for the existence of a general magnitude system.

Cross-Magnitude Interactions

One type of evidence comes in the form of cross-dimensional transfer, with effects based on the congruity of the more-versus-less relations between magnitudes (i.e., increasing values in one magnitude are congruent with increasing values, but incongruent with decreasing values, in another magnitude). Research with prelinguistic children demonstrates transfer across magnitudes as early as infancy. In one of these studies, Lourenco and Longo (2010) found bidirectional interactions for various combinations of numerical and non-numerical magnitudes. In an initial (habituation) phase, 9-month-olds were taught that one magnitude (e.g., physical size) mapped systematically to color and pattern cues. They were then tested on whether the learning of this arbitrary mapping generalized to another magnitude (e.g., numerosity or duration). During habituation, infants might be shown, for example, that larger-sized rectangles were black with white stripes and that smaller-sized rectangles were white with black dots (see Figure 6-1). When subsequently tested on numerosity, trials that maintained the mapping to the color/pattern cues (i.e., congruent test trials) featured a larger numerical array with black/striped rectangles and a smaller numerical array with white/dotted rectangles; trials that violated the mapping (i.e., incongruent test trials) featured a larger numerical array with white/dotted rectangles and a smaller numerical array with black/striped rectangles (see Figure 6-1). The same logic was applied to duration (e.g., congruent test trials: longer-lasting objects as black/striped and shorter-lasting objects as white/dotted; incongruent test trials: longer-lasting objects as white/dotted and shorter-lasting objects as black/striped). Across all combinations, infants looked longer to incongruent than congruent

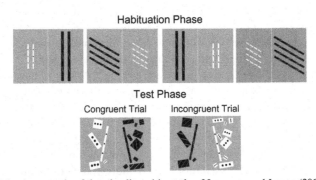

FIGURE 6-1 An example of the stimuli used in study of Lourenco and Longo (2010) On each trial of the habituation phase in this condition (each gray window represents a different trial), infants were shown that larger-sized rectangles were black with white stripes and smaller-sized rectangles were white with black dots (top row). Across these trials, factors such as the orientation and number of rectangles were varied; what remained constant was the mapping between relative magnitude and color/pattern. During the test phase, infants were presented with two types of trials: congruent and incongruent (bottom row). In the congruent trials, the smaller number of rectangles (5 in this example) were white with black dots and the larger number of rectangles (10 in this example) were black with white stripes. In the incongruent trials, the smaller number of rectangles were black with white stripes and the larger number of rectangles were white with black dots. Within trial, non-numerical properties (e.g., cumulative surface area and contour length) for small and large number arrays were held approximately equally.

test trials, providing evidence of transfer across magnitudes, as would be predicted by a general magnitude system.

Focusing on the potential relation between numerosity and spatial extent, de Hevia and Spelke (2010) provided evidence of cross-dimensional transfer among 8-month-olds by visually familiarizing them to sequences of ascending or descending numerical values and then testing them with ascending and descending sequences of line lengths (alternating order). Infants who were familiarized with ascending numerosities looked longer to descending than ascending line lengths during test, whereas those who were familiarized to descending numerosities looked longer to ascending than descending line lengths. In other words, infants generalized the novel ordering across magnitudes (from numerosity to line length). Together with the preceding findings of Lourenco and Longo (2010), these data are inconsistent with claims that interactions between representations of numerical and non-numerical magnitude are linguistically mediated. That is, these studies with infants demonstrate that usage of common magnitude words (e.g., "a lot," "long," "big," and so on) is not solely responsible for similarities in performance across different types of magnitudes. Rather, the data suggest that the more-versus-less content of numerical and non-numerical magnitudes may be functionally integrated, at least to some extent, in the minds of preverbal humans.

It should be noted that both Lourenco and Longo (2010) and de Hevia and Spelke (2010) tested for cross-dimensional transfer within a single sensory modality (vision), leaving open an important question about whether an early developing general magnitude system can accommodate multiple magnitudes and multiple sensory modalities simultaneously. In the case of number, empirical demonstrations of cross-modal matching in infants point to representations of numerical magnitude that integrate across vision and audition. Extending classic work with small numerical values (i.e., 2 and 3; Starkey, Spelke, & Gelman, 1983), Izard and colleagues (Izard, Sann, Spelke, & Streri, 2009) found that newborn infants ($M_{age} = 49$ hours old) matched relatively large numerosities across vision and audition. Specifically, infants who heard 12 syllables (e.g., tu-tu-tu-tu-tu-tu-tu-tu-tu-tu-tu-tu) looked relatively longer to visual displays of 12 objects (i.e., two-dimensional shapes with smiley faces), whereas infants who heard 4 syllables looked relatively longer to displays of 4 objects. The authors also reported that discrimination required a 3-fold difference in visually presented arrays; there was no evidence of cross-modal number matching when visual stimuli involved a 2:1 ratio. These findings are consistent with other data showing that, regardless of sensory modality, nonverbal representations of number are analog in format, and importantly, that an analog format does not preclude the integration of audiovisual stimuli.

Using a similar paradigm, I have asked whether infants integrate audiovisual stimuli across *both* magnitude dimension and sensory modality (Lourenco, 2014a). Given evidence of cross-dimensional transfer for vision (e.g., de Hevia & Spelke, 2010; Lourenco & Longo, 2010) and cross-modal matching for number (e.g., Izard et al., 2009), the clear prediction was that integration should occur for multiple magnitudes and sensory modalities. And indeed, this is what I have observed in my lab when infants are tested under conditions that allow for spontaneous transfer or that involve prescribed pairings. In one experiment, 7-month-olds' looking behaviors were consistent with duration (presented as auditory tones) having interacted with numerosity (presented as visual displays). More specifically, when infants heard a long tone, they looked longer at a larger numerosity compared with a smaller numerosity, but when they heard a short tone, they looked longer at a smaller numerosity compared with a larger numerosity, thus spontaneously matching relative magnitude for numerosity and duration, despite presentation to different sensory modalities (see Figure 6-2). Similar matching was observed for duration and spatial extent, with infants looking longer at a larger object compared with a smaller object when they heard a long tone, but looking longer at a smaller object compared with a larger object when they heard a short tone.

In another experiment, we have found that 9-month-olds showed sensitivity to congruent and incongruent combinations of numerosity, physical size, and duration (when numerosity and size were presented to the visual modality and duration to the auditory modality). Perhaps most striking, though, was that if infants were familiarized to an incongruent pairing between two magnitudes (e.g., long tones preceded small objects and short tones preceded large objects), they generalized this relation to numerosity, subsequently matching long tones to smaller numerosities and short tones to larger numerosities (Lourenco, 2014a). This finding not only provides a further demonstration of the breadth of possible connections made by infants (a total of three magnitudes and two modalities within a given trial), but also points to flexibility in the alignment of magnitude relations. Importantly, however, this flexibility was transient. Although infants showed matching of incongruent pairings on

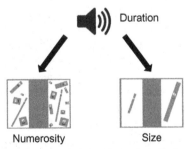

FIGURE 6-2 Layout of the experimental design in the study by Lourenco (2014a). During the familiarization phase, infants heard long and short tones in alternation. During the test phase, one group of infants was presented with visual displays that varied in numerical value (e.g., 5 vs. 10 objects), and another group saw visual displays that varied in physical size. On each test trial, visual displays were preceded by a long or short tone.

the first half of test trials, they reverted to the congruent more-versus-less alignment on the last half of test trials (i.e., matching long tones to large numerosities and short tones to small numerosities).

Taken together, these findings provide support for an early developing general magnitude system that integrates numerical and non-numerical magnitudes across multiple sensory modalities according to a default alignment, which is consistent with other studies showing that congruent cross-magnitude pairings are easier for infants to learn than incongruent pairings (de Hevia & Spelke, 2010; Srinivasan & Carey, 2010) and that congruent pairings (at least in the case of duration and spatial extent) are associated with parietal activation in infants (measured using ERPs; Hyde, Porter, Flom, & Stone, 2013), as has been shown for adults (e.g., Hayashi et al., 2013). That infants are capable of such flexibility, perhaps especially when magnitudes are grounded in separate modalities (Lourenco, 2014a), suggests at least some dissociability between representations of different magnitudes to allow for conditions of reversed alignment. Researchers who advocate for a system of generalized magnitude representation acknowledge some differentiation among magnitudes, with most arguing for a "partially overlapping" system of numerical and' non-numerical magnitudes (e.g., Cappelletti, Freeman, & Cipolotti, 2011; Cohen Kadosh, Lammertyn, & Izard, 2008; Dormal & Pesenti, 2009; Lourenco, Bonny, Fernandez, & Rao, 2012), which these results suggest are present from early in human development (Lourenco & Longo, 2011).

Stroop-like paradigms with adults and older children provide additional demonstrations of cross-magnitude interactions, such that judgments related to one dimension are either interfered with or facilitated by another (irrelevant) dimension. In a classic study, Henik and Tzelgov (1982) found that adults judged the numerical values or physical sizes of Arabic digits faster when number and size were congruent than when they were incongruent. For example, if participants were asked to indicate which number was greater in value, their response times were affected by the physical sizes of the digits (which were irrelevant to the numerical judgment), with facilitation when the size was congruent with number (e.g., 2 7) but interference when it was incongruent (e.g., 2 7). Congruity effects have also been reported between Arabic digits and presentation time (Oliveri et al., 2008). Importantly, these interactions are not specific to symbolic depictions of number; comparable effects have been documented with nonsymbolic displays such as arrays of dots (Dormal & Pesenti, 2007; Dormal, Seron, & Pesenti, 2006; Hurewitz, Gelman, & Schnitzer, 2006; Tokita & Ishiguchi, 2011; Xuan, Zhang, He, & Cheng, 2007). Using such displays, interference and facilitation, at least from spatial extent and duration onto judgments of numerosity, have been observed in children as young as 3 years of age (Gebuis, Cohen Kadosh, de Haan, & Henik, 2009; Lourenco, 2014b; Lourenco, Levine, & Degner, 2014; Piaget, 1965; Rousselle et al., 2004). By the time children reach school age, such effects are generally bidirectional in that numerosity also either interferes with or facilitates judgments of other magnitudes (Rousselle & Noël, 2008; Stavy & Tirosh, 2000). And, critically, at least in adults, congruity effects are modulated by the ratio of the irrelevant dimension, suggesting that the interactions are rooted in analog representations of the different magnitudes.

In summary, the studies described here with looking-time procedures and Stroop-like tasks provide support for a general system of analog magnitude representation that involves at least partial integration of numerical and non-numerical magnitudes. Cross-magnitude interactions that have been documented in infants, children, and adults are consistent with a general magnitude system that is functional from early in development and continues throughout the lifespan. But could such interactions reflect processing by other systems? Previously, I pointed out that language could not mediate the interactions between distinct representations, at least for preverbal infants (for behavioral evidence of cross-magnitude interactions in nonhuman animals, see classic work by Meck & Church, 1983, and more recent work by Merritt, Casasanto, & Brannon, 2010). Here, I acknowledge another possibility—that cross-magnitude interactions between differentiated representations could be mediated by nonlinguistic analogical (comparison) processes (cf. Cantlon et al., 2009; Vicario & Martino, 2010; see also Kotovsky & Gentner, 1996). I return to this issue later.

Correlations in Precision of Numerical and Non-Numerical Representations

A second type of evidence for a general magnitude system comes from an individual differences approach and involves inter-individual correlations in the precision of numerical and non-numerical representations of magnitude. Precision is indexed by the smallest ratio of magnitudes that can be consistently discriminated, with relatively smaller ratios suggesting less noise (more precision) in the underlying representations (e.g., someone who can discriminate 10 vs. 8 items, 5:4 ratio, is more precise than someone who can only make a discrimination of 10 vs. 5 items, 2:1 ratio).

Recent studies demonstrate that discrimination performance related to comparisons of nonsymbolic number displays is positively correlated with performance on tasks of spatial extent (e.g., line lengths or cumulative area) and duration within individual adults (DeWind & Brannon, 2012; Lambrechts, Karolis, Garcia, Obende, & Cappelletti, 2013; Lourenco et al., 2012) as well as within individual children (Lourenco & Bonny, 2014). Positive correlations indicate that more precise representations in one magnitude (e.g., numerosity) are accompanied by greater precision in another magnitude (e.g., spatial extent). In one study, Lourenco et al. (2012) found such a relation in college students when comparing accuracy on two tasks in which participants judged greater numerosity or cumulative area (see also DeWind & Brannon, 2012). As is typical with adults, judgments were speeded and displays were presented briefly to ensure that participants did not engage in counting and could not perform explicit computations; this type of procedure allows for an assessment of the precision of nonverbal magnitude comparisons, analogous to tasks with human infants. In another study, Lambrechts and colleagues found that although comparisons of line length and duration were not correlated in college students, there was a clear relation in older adults (aged 59–74 years). In ongoing work in my lab, we have found similar correlations in children (5- and 6-year-olds) on tasks that involve ordinal judgments of numerosity (i.e., which set has "more") for discrete sets of items or areas for amorphous displays of color (Lourenco & Bonny, 2014; but see Odic et al., 2013).

But could inter-individual correlations reflect cognitive processes not exclusive to a general magnitude system? To address this question, some studies have controlled statistically for factors such as verbal competence and, in the case of children, age. Using this approach, two studies from my lab demonstrate that scores on tasks involving approximations of nonsymbolic number displays and area remain positively correlated even after partialling out verbal IQ and chronological age. The results provide at least some support for inter-individual correlations of numerical and non-numerical magnitude that are not accounted for by other factors (Lourenco & Bonny, 2014; Lourenco et al., 2012). Yet not all potential mediators or alternative explanations are ruled out by partialling out effects such as verbal IQ and chronological age. For example, processes related to executive functioning such as working memory and inhibition as well as common properties of the tasks themselves could contribute to the observed correlations. Another important consideration is how individual differences across magnitudes (e.g., number and spatial extent) compare with individual differences within magnitude (e.g., just number). For example, Gilmore et al. (2011) examined performance across a variety of number tasks (symbolic and nonsymbolic) and found that performance was unrelated across some tasks, despite each engaging representations of numerical magnitude.

To my knowledge, no study examining inter-individual correlations of numerical and non-numerical magnitudes has compared performance across sensory modality (e.g., vision *and* audition). To date, these experiments have focused on comparisons within vision. Given that cognitive interactions have been reported for different combinations of magnitudes, including duration, and across different sensory modalities such as vision and audition, it will be important to determine the extent to which individual differences in representational precision vary by both the type of magnitude dimension and sensory modality. Such data may be especially useful in quantifying the degree of integration across different types of magnitude within a general magnitude system.

Training and Transfer

A third type of evidence for a general magnitude system comes in the form of training studies and involves testing for transfer of representational precision from a "trained" to an "untrained" magnitude. The logic behind this approach is that if numerical and non-numerical magnitudes form part of a general magnitude system with overlapping representations, then training related to one magnitude should transfer to another. However, if different magnitudes share only analog format, then training applied to one magnitude should have no bearing on another. Next, I describe two studies with college students. One is published and found no evidence that training generalized across magnitudes (DeWind & Brannon, 2012). Another study is currently ongoing in my lab, and although the data are preliminary, the effects are consistent with generalization following training, at least under certain conditions (Lourenco, Bonny, Ayzenberg, & Cheung, in progress).

Using a nonsymbolic number comparison task (in which participants judged which of two arrays was larger or smaller in numerosity), DeWind and Brannon (2012) found that the precision of college students' numerical representations improved following multiple sessions of computerized training, with (immediate) feedback on each trial (correct/incorrect). Importantly, the authors also tested for transfer following training with numerical comparisons to comparisons of spatial extent (line lengths). Despite improved performance on the number task, and despite evidence of inter-individual correlations on number and line length tasks (prior to and after training with numerical comparisons), the authors found no evidence that increased numerical precision transferred to line length. That is, representations of line length did not become more precise following training with numerical comparisons. The authors concluded that the lack of transfer (in the face of inter-individual correlations between numerosity and line length) was suggestive of the same comparator mechanism having been engaged (see also Van Opstal, Gevers, De Moor, & Verguts, 2008), rather than a general magnitude system with overlapping representations. On this account, there are distinct representations of numerosity and line length, each of which has access to the same comparator mechanism that supports "more than/less than" judgments because of the structural similarity (i.e., analog format) of magnitudes; inter-individual correlations across the two tasks would reflect the precision of the comparator mechanism, which would be utilized when judging both relative numerosity and line length.

This study motivated recent (ongoing) research in my lab in which we have been focusing on the transfer of representational precision following training with comparisons of either nonsymbolic number stimuli, as in DeWind and Brannon (2012), or non-numerical magnitude (spatial extent). Unlike the DeWind and Brannon study, in our study the non-numerical magnitude of interest was cumulative area rather than the lengths of individual lines. We trained and tested participants with cumulative area for two reasons. One is that it ensured that performance on numerical and non-numerical comparison tasks would be comparable. Comparisons of line lengths are far greater in accuracy than either numerosity or cumulative area, which leaves little room for improvement and makes it difficult to capture transfer related to representational precision. Second, and perhaps most important, line length differs from number in that it is a property of a single element, whereas number applies to a set; cumulative area is thus a better match to numerosity because it is also a property of a set.

Our findings, like those of DeWind and Brannon (2012), demonstrate small, but significant, improvement in representational precision for numerical magnitude. We have also found that the precision of cumulative area representations shows comparable increases following training with area, suggesting malleability of precision for representations of numerical and non-numerical magnitude. On both number and area tasks, corrective feedback following magnitude comparisons resulted in significant improvement in accuracy. But what about transfer of representational precision? As with DeWind and Brannon, we have found that participants trained on numerical comparisons do not seem to generalize greater representational precision to cumulative area. In contrast, when participants receive training with area comparisons, we have found transfer to numerical magnitude; that is, participants who show greater improvement on the area task are more accurate on the number task following training. On the one hand, transfer from area to numerosity provides evidence of representational overlap. On the other hand, it is unclear how a general magnitude system (with overlapping representations of numerical and non-numerical magnitude) might accommodate such an asymmetry; that is, transfer from area to numerosity, but not vice versa.

Data from an additional color-word Stroop task in our study (in which participants judged the font color of a printed word while ignoring the color denoted by the word, which, on some trials, was incongruent with the font; e.g., the word "blue" printed in red) points to a potential effect of the stimuli used on the number task. On any task assessing sensitivity to nonsymbolic number, there must be trials in which the spatial properties such as cumulative area are incongruent with number to ensure that participants are indeed making judgments on the basis of numerical, rather than non-numerical, magnitude. In contrast, our area task involved no incongruent trials; that is, within a given trial, the two arrays being compared were always equal in numerical value, such that there was never any irrelevant information related to a different magnitude (see also Lourenco et al., 2012). This led us to hypothesize that training with numerical comparisons might have served to dissociate numerical and non-numerical magnitudes by promoting inhibition of the irrelevant spatial information. Consistent with this possibility, participants who showed greater improvement in numerical precision also showed less interference on the color-word Stroop task following training with numerical comparisons, suggesting general improvement in inhibitory control. There was no such change on the color-word Stroop task following area training. However, a separate study employing transcranial random noise stimulation (tRNS) of parietal cortex, which temporarily potentiates neural activity, found that improved precision associated with representations of numerical magnitude can transfer to non-numerical magnitudes, namely, spatial extent and duration (Cappelletti et al., in press). Although more research is needed to determine what role tRNS plays in facilitating transfer from number to other magnitudes, this finding provides additional support for representations of numerical and non-numerical magnitude that are functionally overlapping.

Potential Neural Mechanisms

Taken together, the three lines of evidence described in the preceding sections point to a general magnitude system with overlapping representations of number and other magnitudes. Moreover, neuropsychological studies with patients as well as neuroimaging and neural stimulation studies with healthy participants point to a common neural substrate for numerical and non-numerical magnitudes. Patients with deficits in representing magnitude (numerical and/or non-numerical) often evidence parietal damage (for review, see Walsh, 2003), and healthy participants show that parietal activation for different magnitudes is modulated by ratio and congruity effects (e.g., Dormal, Andres, & Pesenti, 2008; Dormal & Pesenti, 2009; Hayashi et al., 2013; Pinel, Piazza, Le Bihan, & Dehaene, 2004). But concerns about the size of brain lesions and the spatial and temporal resolution associated with techniques such as fMRI leave open the possibility that different magnitudes may *not* be processed by the same neural mechanisms. Studies using single cell recording with monkeys, however, are not subject to these criticisms and have found shared coding for different magnitudes at the cellular level (Tudusciuc & Nieder, 2007, 2009). Nieder and colleagues have demonstrated that a proportion of neurons in monkey VIP (ventral intraparietal area) are tuned to both number and line length. Importantly, however, despite this overlap in coding, the preferred numerical values and line lengths do not align in terms of relative magnitude. For example, individual neurons tuned to larger numerosities are not more likely to be tuned to longer line lengths compared with shorter line lengths, at least as reported by Tudusciuc and Nieder (2007); there were no data on this issue in Tudusciuc and Nieder (2009). The lack of alignment with respect to relative magnitude raises obvious questions about how, for example, behavioral congruity effects on magnitude Stroop tasks might be supported by individual neurons where facilitation and interference reflect matching more-versus-less alignment.

One possible explanation is that alignment with respect to relative magnitude may occur, not at the level of an individual neuron, but at the population level, with combinations of neurons contributing dynamically to a coherent (more-versus-less) organization for different dimensions such as numerical magnitude and line length. It should be noted, though, that within magnitude type (e.g., numerosity) there is also a lack of alignment that pertains to sensory modality (e.g., vision and audition). In monkey VIP, neurons tuned to specific cardinal values (e.g., 1, 2, 3, or 4) do not show reliable cross-modal matching, except in the case of 1 (Nieder, 2012). That is, a neuron that responds selectively to numerosity 3 for visual stimuli may respond to auditory sequences of numerosity 2 or 4. And yet, behavioral paradigms show that nonhuman animals (Jordan, Brannon, Logothetis, & Ghazanfar, 2005) and prelinguistic children (Izard et al., 2009; Starkey et al., 1983) match numerosities across sensory modalities, suggesting that alignment, whether *within* or *across* magnitude, may occur at the population level within parietal cortex.

Another possibility is that support for alignment of relative magnitude at the level of individual neurons takes place further downstream, in prefrontal cortex (PFC), rather than the IPS. Although much discussion concerning the neural instantiation of a general magnitude system has tended to focus on the parietal cortex (particularly the IPS), there is clear evidence from humans and nonhuman animals that frontal regions such as the dorsolateral prefrontal cortex and the frontal gyrus play a central role in processing number and other types of magnitudes (Ansari & Dhital, 2006; Coull, Vidal, Nazarian, & Macar, 2004; Fulbright, Manson, Skudlarski, Lacadie, & Gore, 2003; Genovesio, Tsujimoto, & Wise, 2011; Hayashi et al., 2013; Nieder, Freedman, & Miller, 2002; Onoe et al., 2001). Recent findings from the lab of Andreas Nieder suggest that a proportion of neurons in PFC are tuned abstractly to relative magnitude, that is, responding indiscriminately to numerosity and line length when making "greater than/less than" judgments (Eiselt & Nieder, 2013). Not addressed in this work, however, is whether these neurons retain analog format, a critical issue given that the PFC has been implicated in categorization and rule-based decision making (Genovesio, Wise, & Passingham, 2014).

Future studies using single-cell recordings with nonhuman animals would do well to assess ratio-dependent modulation of neuronal responses. Of course, a common criticism of studies with nonhuman animals is that individual monkeys are subjected to massive training prior to testing, which could lead to changes in neuronal tissue, raising concerns about whether experimental effects are due to unique training conditions and thus not representative of typical development. Nevertheless, ratio-modulated responding in neurons tuned to different magnitudes would provide an especially strong test of a general magnitude system with overlapping representations of more-versus-less relations. Consistent with this possibility is that activity in the inferior frontal gyrus (IFGs) during numerosity-duration interactions in human adults on a categorical discrimination task is modulated by the degree of change in the irrelevant dimension, suggesting that magnitudes are not recoded in terms of discrete categories but, rather, retain continuous properties (Hayashi et al., 2013; but see Genovesio et al., 2011).

DO MATHEMATICAL CONCEPTS HAVE A BASIS IN NONSYMBOLIC MAGNITUDES?

As noted at the start of this chapter, infants do not have access to numerical symbols. Moreover, young children take approximately 2 years before they learn to map number words to their numerical (magnitude) meaning, and several more years before they learn to perform formal arithmetic computations (for review, see Carey, 2009) and to understand abstract geometric concepts (for review, Spelke, 2011). Among humans living in modern societies, formal schooling provides children and adolescents with a rich and complex system of mathematics, although even among these cultures there are notable individual differences in math competence based on educational experiences (e.g., Geary, 1994; Jordan & Levine, 2009) and genetic or neuroanatomical factors (e.g., Molko et al., 2003; Rubinsten & Henik, 2009). To what extent do nonsymbolic representations of magnitude, which are present in infancy and used for estimation purposes across cultures and throughout development, interface with a system of formal mathematics? Long-standing theoretical interest in the mental primitives of abstract human cognition (Descartes, 1637/2001; Kant, 1781/2003) sets the stage for fundamental questions about the nonsymbolic basis of mathematics, with claims that formal math builds on core abilities, including analog magnitude (Carey, 2009; Dehaene, 2011; Gallistel & Gelman, 1992; Geary, 1995; Leslie, Gelman, & Gallistel, 2008; McCrink, this volume; Star & Brannon, this volume). It has been suggested that representations of nonsymbolic number, although limited in precision, play a critical role in grounding abstract quantitative concepts (Ansari, 2008; Dehaene, 2009; Gallistel & Gelman, 1992; Libertus, Feigenson, & Halberda, 2011; Verguts & Fias, 2004).

Focusing specifically on numerical magnitude, several investigators have found that the precision of nonsymbolic number comparisons correlates with performance on standardized tests of arithmetic (e.g., Halberda et al., 2008; Lourenco et al., 2012). But there are also instances of others failing to find such a correlation (e.g., Gilmore et al., 2013; Holloway & Ansari, 2009). Nevertheless, positive findings suggest that representations of nonsymbolic number may be important for acquiring symbolic number concepts and for learning to perform precise arithmetic computations. Of course, strong causal interpretations should be made cautiously, especially given that existing studies mostly consist of correlational designs and recent research suggests that formal math instruction affects the precision of numerical magnitude representations (Piazza, Pica, Izard, Spelke, & Dehaene, 2013). There are also important questions about the extent to which the link between nonsymbolic numerical comparisons and mathematical competence is mediated by representations of symbolic number (Holloway & Ansari, 2009; Lyons & Beilock, 2011; vanMarle, Chu, Li, & Geary, in press), whether it can be accounted for by general cognitive functions such as working memory and inhibitory control (Fuhs & McNeil, 2013; Gilmore et al., 2013), and the role of different types of math content (Lourenco et al., 2012). Moreover, and especially relevant to the topic of the present chapter, there are questions about whether the link between nonsymbolic magnitude and formal mathematics is specific to numerical magnitude or whether it also encompasses representations of non-numerical magnitude.

Lourenco et al. (2012) recently examined the extent to which both numerical and non-numerical magnitudes interface with formal mathematics in college students. In separate tasks, participants judged which of two dot arrays was larger in either numerical value or cumulative area. My students and I found that participants' comparisons on each task were both modulated by the ratio of the arrays being compared, and were both positively correlated with performance on tests of advanced arithmetic (specifically, the Calculation subtest of the Woodcock–Johnson III Tests of Achievement; Woodcock, McGrew, Schrank, & Mather, 2001/2007) as well as school-relevant geometry (specifically, the Geometry subtest of KeyMath-3 Diagnostic Assessment; Connolly, 2007), even after accounting for nonmathematical (verbal) intelligence (see also Tibber et al., 2013). Importantly, we also found that the precision of nonsymbolic number representations contributed unique variance to arithmetic competence, predicting performance on problems involving arithmetic above and beyond that contributed by the precision of cumulative area representations. In contrast, representations of cumulative area contributed unique variance to geometric competence, predicting performance on school-relevant geometric concepts and computations above and beyond that contributed by representations of nonsymbolic number. The finding that both numerical and non-numerical magnitudes show connections to formal math competence is suggestive of a general magnitude system that interfaces with learned principles of mathematics. The additional finding that there are specific connections to advanced arithmetic (in the case of numerical magnitude) and school-relevant geometry (in the case of cumulative area) is consistent with a system of generalized magnitude representation that consists of both overlap and differentiation. Indeed, as noted previously, recent neural data point to a system of "partial overlap," with populations of neurons within a frontoparietal network (i.e., IPS and PFC) that either code for a specific type of magnitude or are tuned to multiple magnitudes (e.g., Eiselt & Nieder, 2013; Tudusciuc & Nieder, 2007).

Following from this work are obvious questions about whether the links between nonsymbolic magnitude and abstract concepts related to arithmetic and geometry are present prior to formal math instruction, or whether they only emerge later in development with schooling. One possibility is that representations of analog magnitude, particularly numerical magnitude, play a central role in learning symbolic number concepts and arithmetic computations (vanMarle, this volume). Whether non-numerical magnitude serves a similar function is another open question. Based on the links observed in college students (Lourenco et al., 2012), it would be reasonable to predict that numerical and non-numerical magnitudes play unique roles in learning different types of math. Representations of numerical magnitude, which apply to discrete items and are ultimately enumerable, might be especially well suited to scaffolding arithmetic concepts and computations of symbolic number. In contrast, non-numerical magnitudes such as cumulative area or the sizes of individual elements may be better suited to geometry, where concepts and computations pertain to spatial properties such as shapes, relative location, and multidimensional figures. Alternatively, these unique links could emerge over the course of development following exposure to math instruction in specific domains. Exposure to exact calculations of Arabic numerals and other symbolic notation in the domain of arithmetic could serve to fine-tune connections to representations of numerical magnitude, whereas exposure to geometry where manipulations apply to shapes, locations, and figures might strengthen specific connections to representations of non-numerical magnitude such as spatial extent.

Recent studies with developmental populations are beginning to shed insight on these possibilities. For example, across two studies in my lab, we have found that the precision of children's area representations, assessed via comparisons of individual elements (i.e., irregular 2-D shapes) or the cumulative area of amorphous blobs, positively correlates with the understanding of a broad range of visuospatial geometric concepts that are more or less formal. As described in more detail below, the findings from these studies suggest that, by the time children reach preschool age, non-numerical magnitude representations show links to both arithmetic and geometric concepts, as do representations of nonsymbolic number.

In one study, we assessed 4- to 6-year-olds' understanding of basic geometric concepts such as parallelism, right angles, curvature, and convexity using a multiple-choice visuospatial test (Dehaene, Izard, Pica, & Spelke, 2006; Izard & Spelke, 2009). On each trial of this test, the participant is required to identify the item that violates a particular geometric property (see Figure 6-3). When visuospatial geometric competence was compared to performance on an area discrimination task (where children judged which of two irregular shapes was larger in area), a positive correlation was found.

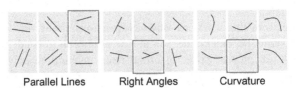

Parallel Lines   Right Angles   Curvature

FIGURE 6-3 Examples of trials from the visuospatial test used in Bonny and Lourenco (2014b) This test was originally developed by Dehaene et al. (2006). Each trial consists of six images, five of which share a given geometric property. In these examples, the one that violates the particular property (i.e., the correct answer) is outlined in red.

Children who performed better on the geometry test were more accurate in their approximations of area, even when controlling for age or verbal competence (Bonny & Lourenco, 2014b). This study extends recent developmental research by pointing to broader links between the precision of analog magnitude representations and knowledge of mathematical concepts. It also provides support for ontogenetic continuity as it relates to the link between non-numerical magnitude and geometry, where tests of magnitude comparisons and geometry are quite different between children and adults. Magnitude comparisons have mostly involved spatial extent of individual forms in the case of children but cumulative area for sets of discrete items in the case of adults. Tests of geometry range from relatively informal visuospatial concepts in the case of children to formally taught principles and computations in the case of adults.

In another currently ongoing study, we are directly comparing representations of numerical and non-numerical magnitude and their potential links to school-relevant math concepts, including those relevant to arithmetic calculations and geometry. We are conducting this study with 5- and 6-year-olds; it is modeled on our previous work with college students in which we found that although the precision of nonsymbolic number and cumulative area representations were both correlated with advanced arithmetic and geometric competence, numerical and non-numerical magnitude each showed specific links to arithmetic and geometry (Lourenco et al., 2012). We have tested 65 children and have found that with them, unlike adults, representations of nonsymbolic number and cumulative area are similarly related to performance on the same math tests; that is, there are no signs of specificity in childhood (Lourenco & Bonny, 2014).

These findings lend support to theories of development that posit a crucial role for representations of numerical magnitude in learning mathematical concepts (e.g., Gallistel & Gelman, 1992). Importantly, they also extend these theories by suggesting that the nonsymbolic basis of mathematical development is not specific to numerical magnitude, perhaps residing instead in a general magnitude system, which includes representations of non-numerical magnitude. Recent work suggests that math education can sharpen the precision of numerical magnitude representations (Piazza et al., 2013). Our findings with adults are further suggestive of math-specific experience shaping the links to numerical and non-numerical magnitudes. Taken together, then, our findings with children and adults suggest that whereas the nonsymbolic basis of mathematical competence may not discriminate between numerical and non-numerical magnitudes, math experiences may fine-tune the connections to representations of analog magnitude.

CONCLUSIONS

There is no question that a critical aspect of quantification is to represent the numerical magnitude of items, either approximately through nonverbal estimation processes or precisely through the use of symbols. Much research from developmental and comparative psychology has pointed to the approximate number system (ANS) as the foundation for representations of numerical magnitude, and studies using neuroimaging techniques point to the intraparietal sulcus (IPS) as the host of these representations, which are analog in format. Although they are nonverbal and imprecise, there are many findings showing that analog number representations support a variety of quantitative judgments, including discrimination, ordinal comparisons, and approximate arithmetic (for reviews, see Carey, 2009; Dehaene, 2009).

In this chapter, I have argued that numerical magnitude shares analog format with non-numerical magnitudes such as spatial extent and duration, a claim that is relatively uncontroversial. As described previously, evidence for such structural similarity comes from classic psychophysical experiments showing that comparisons of number and other magnitudes abide by Weber's law, with easier discrimination for smaller ratios compared to larger ratios (e.g., 5:4 vs. 2:1, respectively). More recent experiments from developmental psychology find that a common analog format for different magnitudes is evident in infancy and throughout the lifespan, with comparable thresholds of sensitivity within the first year of life (e.g., Brannon et al., 2004) and parallel levels of improvement into preschool and school age (e.g., Bonny & Lourenco, 2014a; Droit-Volet et al., 2008). Although some studies find differences in discriminability for numerosity and non-numerical magnitudes such as cumulative area (e.g., Barth, 2008; Rousselle et al., 2004; but see Castelli, Glaser, & Butterworth, 2006; Gebuis et al., 2009), there is evidence of inter-individual correlations in the precision of different magnitudes (e.g., DeWind & Brannon, 2012; Lambrechts et al., 2013; Lourenco & Bonny, 2014; Lourenco et al., 2012), which is consistent with shared processing resources within a general system of analog magnitude.

More than structural similarity, however, the theory of a general magnitude system holds that the processing of numerical and non-numerical magnitudes shares neural circuitry and that representations for different magnitudes are at least partially integrated. It is this type of system that would support an abstract sense of more-versus-less relations across numerical and non-numerical magnitudes. On the one hand, this system would clearly differ from the ANS, which posits that representations of number are distinct from other magnitudes. Proponents of a general magnitude system argue that the cross-magnitude effects observed in behavioral paradigms (described earlier) are due to shared processing within this system, whereas proponents of the ANS appeal to other systems of mediation (e.g., attention) between distinct representations. On the other hand, the general magnitude system also incorporates differentiation, for number (like the ANS) and other magnitudes. There are cases when representations of numerical and non-numerical magnitudes dissociate, as observed in behavioral effects and neural codes. This has led researchers to describe a general magnitude system that consists of both differentiated and overlapping representations (e.g., Cappelletti et al., 2011; Dormal & Pesenti, 2009; Lourenco et al., 2012).

In this chapter, I focused exclusively on "prothetic" dimensions—namely, numerosity, spatial extent, and duration (Stevens, 1957, 1975). Like numerosity, spatial extent such as the physical size of an individual object or the cumulative surface area occupied by multiple objects, and temporal information such as relative duration have a default organization based on intensity, such that increases in values apply to one end of the continuum and decreases to the other end. Cross-magnitude interactions observed in looking-time paradigms with infants as well as magnitude Stroop tasks with children and adults generally reflect the shared organization of numerical and non-numerical magnitudes. Not discussed previously, however, was evidence from behavioral and neural paradigms that points to even broader interactions, that is, cross-dimensional effects that encompass other prothetic dimensions such as weight (Holmes & Lourenco, 2013) and "metathetic" dimensions such as luminance and pitch (Cohen Kadosh, Cohen Kadosh, & Henik, 2008; de Hevia, Vanderslice, & Spelke, 2012; Eitan & Timmers, 2010) for which the organization of more-versus-less may be ambiguous (Stevens, 1957, 1975) and variable across development (Smith & Sera, 1992) and context (Gebuis & van der Smagt, 2011).

Given findings of accompanying spatial orientation for dimensions such as number and pitch, an important question is whether particular dimensions may serve to mediate interactions among others. There is evidence that pitch is represented spatially from top-to-bottom, with high-pitched sounds more likely to be mapped to upper regions of space and low-pitched sounds to lower regions (e.g., Rusconi, Kwan, Giordano, Umilta, & Butterworth, 2006; Walker et al., 2010). This spatial organization has been reported for numerical values as well (e.g., Loetscher, Bockisch, Nicholls, & Brugger, 2010; Schwarz & Keus, 2004; but see Holmes & Lourenco, 2012). In conditions where cognitive interactions apply to numerical magnitude and pitch (Eitan & Timmers, 2010), it may be that a common spatial organization plays a role in supporting the interactions.

In the external environment, cues to numerosity are generally conflated with those that convey information related to other magnitudes. When one is comparing security lines at the airport, for example, the line with more people (number) is also typically the line that takes up more space (longer in length). Given the conflation in physical cues, it is thus perhaps not surprising that representations of numerical and non-numerical magnitude are, to some extent, integrated in the mind and brain. A system of generalized magnitude representation, with such integration, offers a neurocognitive alternative to the well-known ANS, which is specialized for processing numerical magnitude. Many questions remain unanswered, however, about the processes that may lead to integration across magnitudes and about what factors might determine the degree of differentiation and overlap within a general magnitude system (cf. Lambrechts, Walsh, & van Wassenhove, 2013). Future studies interested in further characterizing the nature of magnitude representations as well as their links to formal mathematics are ideally suited to add to this literature, either by contributing details to existing accounts such as a general magnitude system or by providing alternative models that can account for the relations among representations of numerical and non-numerical magnitudes.

ACKNOWLEDGMENTS

This work was supported by a Scholar award from the John Merck Fund to the author.

REFERENCES

Allik, J., & Tuulmets, T. (1991). Occupancy model of perceived numerosity. *Perception & Psychophysics, 49*, 303–314.

Ansari, D. (2008). Effects of development and enculturation on number representation in the brain. *Nature Reviews Neuroscience, 9*, 278–291.

Ansari, D., & Dhital, B. (2006). Age-related changes in the activation of the parietal sulcus during nonsymbolic magnitude processing: An event-related functional magnetic resonance imaging study. *Journal of Cognitive Neuroscience, 18*, 1820–1828.

Barth, H. C. (2008). Judgments of discrete and continuous quantity: An illusory Stroop effect. *Cognition, 109*, 251–266.

Barth, H., Kanwisher, N., & Spelke, E. (2003). The construction of large number representations in adults. *Cognition, 86*, 201–221.

Bonny, J. W., & Lourenco, S. F. (2014a). *Impact of spatial context on cumulative area and non-symbolic number judgments in children and adults.* (Submitted for publication.)

Bonny, J. W., & Lourenco, S. F. (2014b). *Individual differences in children's approximations of area correlate with competence in basic geometry.* (Submitted for publication.)

Brannon, E. M., Abbott, S., & Lutz, D. J. (2004). Number bias for the discrimination of large visual sets in infancy. *Cognition, 93*, B59–B68.

Brannon, E. M., Lutz, D. J., & Cordes, S. (2006). The development of area discrimination and its implications for number representation in infancy. *Developmental Science, 9*(6), F59–F64.

Brannon, E. M., Suanda, S., & Libertus, K. (2007). Temporal discrimination increases in precision over development and parallels the development of numerosity discrimination. *Developmental Science, 10*, 770–777.

Bratzke, D., Seifried, T., & Ulrich, R. (2012). Perceptual learning in temporal discrimination: Asymmetric cross-modal transfer from audition to vision. *Experimental Brain Research, 221*, 205–210.

Buckley, P. B., & Gillman, C. B. (1974). Comparisons of digits and dot patterns. *Journal of Experimental Psychology, 103*, 1131–1136.

Bueti, D., & Walsh, V. (2009). The parietal cortex and the representation of time, space, number and other magnitudes. *Philosophical Transactions of the Royal Society, B: Biological Sciences, 364*, 1831–1840.

Burr, D., Della Rocca, E., & Morrone, M. C. (2013). Contextual effects in interval-duration judgements in vision, audition and touch. *Experimental Brain Research, 230*, 87–98.

Cantlon, J. F., Platt, M. L., & Brannon, E. M. (2009). Beyond the number domain. *Trends in Cognitive Sciences, 13,* 83–91.

Cappelletti, M., Freeman, E. D., & Cipolotti, L. (2011). Numbers and time doubly dissociate. *Neuropsychologia, 49,* 3078–3092.

Cappelletti, M., Gessaroli, E., Hithersay, R., Mitolo, M., Didino, D., Kanai, R., et al. (in press). Transfer of cognitive training across magnitude dimensions achieved with concurrent brain stimulation of the parietal lobe. *The Journal of Neuroscience.*

Carey, S. (2009). *The origin of concepts.* New York: Oxford University Press.

Castelli, F., Glaser, D. E., & Butterworth, B. (2006). Discrete and analogue quantity processing in the parietal lobe: A functional MRI study. *Proceedings of the National Academy of Sciences of the United States of America, 103,* 4693–4698.

Clearfield, M. W., & Mix, K. S. (1999). Number versus contour length in infants' discrimination of small visual sets. *Psychological Science, 10,* 408–411.

Cohen Kadosh, R., Cohen Kadosh, K., & Henik, A. (2008). When brightness counts: The neuronal correlate of numerical–luminance interference. *Cerebral Cortex, 18,* 337–343.

Cohen Kadosh, R., Lammertyn, J., & Izard, V. (2008). Are numbers special? An overview of chronometric, neuroimaging, developmental and comparative studies of magnitude representation. *Progress in Neurobiology, 84,* 132–147.

Connolly, A. J. (2007). *Keymath-3 Diagnostic Assessment: Manual Forms A and B.* Minneapolis, MN: SAGE Publications.

Cordes, S., & Brannon, E. M. (2008). The difficulties of representing continuous extent in infancy: Using number is just easier. *Child Development, 79,* 476–489.

Coull, J. T., Vidal, F., Nazarian, B., & Macar, F. (2004). Functional anatomy of the attentional modulation of time estimation. *Science, 303,* 1506–1508.

Dakin, S. C., Tibber, M. S., Greenwood, J. A., Kingdom, F. A., & Morgan, M. J. (2011). A common visual metric for approximate number and density. *Proceedings of the National Academy of Sciences of the United States of America, 108,* 19552–19557.

Dehaene, S. (2009). Origins of mathematical intuitions. *Annals of the New York Academy of Sciences, 1156,* 232–259.

Dehaene, S. (2011). *The number sense: How the mind creates mathematics.* New York, NY: Oxford University Press.

Dehaene, S., & Changeux, J. (1993). Development of elementary numerical abilities: A neuronal model. *Journal of Cognitive Neuroscience, 5,* 390–407.

Dehaene, S., Izard, V., Pica, P., & Spelke, E. (2006). Core knowledge of geometry in an Amazonian indigene group. *Science, 311,* 381–384.

Dehaene, S., Piazza, M., Pinel, P., & Cohen, L. (2003). Three parietal circuits for number processing. *Cognitive Neuropsychology, 20,* 487–506.

de Hevia, M. D., & Spelke, E. S. (2010). Number–space mapping in human infants. *Psychological Science, 21,* 653–660.

de Hevia, M. D., Vanderslice, M., & Spelke, E. S. (2012). Cross-dimensional mapping of number, length and brightness by preschool children. *PLoS One, 7*(4), e35530.

Descartes, R. (1637/2001). The optics. In P. J. Olscamp (Ed.), *Discourse on method, optics, geometry and meteorology.* Indianapolis, IN: Hackett.

DeWind, N. K., & Brannon, E. M. (2012). Malleability of the approximate number system: Effects of feedback and training. *Frontiers in Human Neuroscience, 6,* 68.

Dormal, V., Andres, M., & Pesenti, M. (2008). Dissociation of numerosity and duration processing in the left intraparietal sulcus: A transcranial magnetic stimulation study. *Cortex, 44,* 462–469.

Dormal, V., & Pesenti, M. (2007). Numerosity-length interference: A Stroop experiment. *Experimental Psychology, 54*, 289–297.

Dormal, V., & Pesenti, M. (2009). Common and specific contributions of the intraparietal sulci to numerosity and length processing. *Human Brain Mapping, 30*, 2466–2476.

Dormal, V., Seron, X., & Pesenti, M. (2006). Numerosity-duration interference: A Stroop experiment. *Acta Psychologica, 121*, 109–124.

Droit-Volet, S., Clément, A., & Fayol, M. (2008). Time, number and length: Similarities and differences in discrimination in adults and children. *The Quarterly Journal of Experimental Psychology, 61*, 1827–1846.

Durgin, F. H. (2008). Texture density adaptation and visual number revisited. *Current Biology, 18*, R855–R856.

Eiselt, A. K., & Nieder, A. (2013). Representation of abstract quantitative rules applied to spatial and numerical magnitudes in primate prefrontal cortex. *The Journal of Neuroscience, 33*, 7526–7534.

Eitan, Z., & Timmers, R. (2010). Beethoven's last piano sonata and those who follow crocodiles: Cross-domain mappings of auditory pitch in a musical context. *Cognition, 114*, 405–422.

Feigenson, L. (2007). The equality of quantity. *Trends in Cognitive Sciences, 11*, 185–187.

Feigenson, L., Carey, S., & Spelke, E. (2002). Infants' discrimination of number vs. continuous extent. *Cognitive Psychology, 44*, 33–66.

Feigenson, L., Dehaene, S., & Spelke, E. (2004). Core systems of number. *Trends in Cognitive Sciences, 8*, 307–314.

Fias, W., Lammertyn, J., Reynvoet, B., Dupont, P., & Orban, G. A. (2003). Parietal representation of symbolic and non-symbolic magnitude. *Journal of Cognitive Neuroscience, 15*, 47–56.

Fuhs, M. W., & McNeil, N. M. (2013). ANS acuity and mathematics ability in preschoolers from low-income homes: Contributions of inhibitory control. *Developmental Science, 16*, 136–148.

Fulbright, R., Manson, C., Skudlarski, P., Lacadie, C. M., & Gore, C. L. J. (2003). Quantity determination and the distance effect with letters, numbers and shapes: A functional MR imaging study of number processing. *American Journal of Neuroradiology, 23*, 197–200.

Gallistel, C., & Gelman, R. (1992). Preverbal and verbal counting and computation. *Cognition, 44*, 43–74.

Geary, D. C. (1994). *Children's mathematical development: Research and practical applications*. Washington, DC: American Psychological Association.

Geary, D. C. (1995). Reflections of evolution and culture in children's cognition: Implications for mathematical development and instruction. *American Psychologist, 50*, 24–37.

Gebuis, T., Cohen Kadosh, R., de Haan, E., & Henik, A. (2009). Automatic quantity processing in 5-year olds and adults. *Cognitive Processing, 10*, 133–142.

Gebuis, T., & Reynvoet, B. (2012). The interplay between nonsymbolic number and its continuous visual properties. *Journal of Experimental Psychology: General, 141*, 642–648.

Gebuis, T., & van der Smagt, M. J. (2011). Incongruence in number–luminance congruency effects. *Attention, Perception, & Psychophysics, 73*, 259–265.

Genovesio, A., Tsujimoto, S., & Wise, S. P. (2011). Prefrontal cortex activity during the discrimination of relative distance. *The Journal of Neuroscience, 31*, 3968–3980.

Genovesio, A., Wise, S. P., & Passingham, R. E. (2014). Prefrontal–parietal function: From foraging to foresight. *Trends in Cognitive Sciences, 18*, 72–81.

Gilmore, C., Attridge, N., Clayton, S., Cragg, L., Johnson, S., Marlow, N., et al. (2013). Individual differences in inhibitory control, not non-verbal number acuity, correlate with mathematics achievement. *PLoS One, 8*(6), e67374.

Gilmore, C., Attridge, N., & Inglis, M. (2011). Measuring the approximate number system. *The Quarterly Journal of Experimental Psychology, 64*, 2099–2109.

Goodfellow, L. D. (1934). An empirical comparison of audition, vision, and touch in the discrimination of short intervals of time. *American Journal of Psychology, 243–258*.

Grondin, S., & Ulrich, R. (2011). Duration discrimination performance: No cross-modal transfer from audition to vision even after massive perceptual learning. In A. Vatakis, A. Esposito, M. Giagkou, F. Cummins, & G. Papadelis (Eds.), *Multidisciplinary aspects of time and time perception* (pp. 92–100). Berlin Heidelberg: Springer.

Halberda, J., Mazzocco, M. M. M., & Feigenson, L. (2008). Individual differences in non-verbal number acuity correlate with maths achievement. *Nature, 455*, 665–668.

Hayashi, M. J., Kanai, R., Tanabe, H. C., Yoshida, Y., Carlson, S., Walsh, V., & Sadato, N. (2013). Interaction of numerosity and time in prefrontal and parietal cortex. *Journal of Neuroscience, 33*, 883–893.

Henik, A., & Tzelgov, J. (1982). Is three greater than five: The relation between physical and semantic size in comparison tasks. *Memory & Cognition, 10*, 389–395.

Henmon, V. A. C. (1906). The time of perception as a measure of differences in sensation. *Archives of Philosophical, Psychological and Science Method, 8*, 5–75.

Holloway, I. D., & Ansari, D. (2008). Domain-specific and domain-general changes in children's development of number comparison. *Developmental Science, 11*, 644–649.

Holloway, I. D., & Ansari, D. (2009). Mapping numerical magnitudes onto symbols: The numerical distance effect and individual differences in children's mathematics achievement. *Journal of Experimental Child Psychology, 103*, 17–29.

Holmes, K. J., & Lourenco, S. F. (2012). Orienting numbers in mental space: Horizontal organization trumps vertical. *Quarterly Journal of Experimental Psychology, 65*, 1044–1051.

Holmes, K. J., & Lourenco, S. F. (2013). When numbers get heavy: Is the mental number line exclusively numerical? *PLoS One, 8*(3), e58381.

Hurewitz, F., Gelman, R., & Schnitzer, B. (2006). Sometimes area counts more than number. *Proceedings of the National Academy of Sciences, 103*, 19599–19604.

Hyde, D. C., Porter, C. L., Flom, R., & Stone, S. A. (2013). Relational congruence facilitates neural mapping of spatial and temporal magnitudes in preverbal infants. *Developmental Cognitive Neuroscience, 6*, 102–112.

Izard, V., Sann, C., Spelke, E. S., & Streri, A. (2009). Newborn infants perceive abstract numbers. *Proceedings of the National Academy of Sciences, 106*, 10382–10385.

Izard, V., & Spelke, E. S. (2009). Development of sensitivity to geometry in visual forms. *Human Evolution, 23*, 213.

Jordan, K. E., Brannon, E. M., Logothetis, N. K., & Ghazanfar, A. A. (2005). Monkeys match the number of voices they hear to the number of faces they see. *Current Biology, 15*, 1034–1038.

Jordan, N. C., & Levine, S. C. (2009). Socioeconomic variation, number competence, and mathematics learning difficulties in young children. *Developmental Disabilities Research Reviews, 15*, 60–68.

Kant, I. (1781/2003). J. M. D. Meiklejohn, Trans. In *Critique of pure reason*. Mineola, NY: Dover.

Kotovsky, L., & Gentner, D. (1996). Comparison and categorization in the development of relational similarity. *Child Development, 67*, 2797–2822.

Lambrechts, A., Karolis, V., Garcia, S., Obende, J., & Cappelletti, M. (2013). Age does not count: Resilience of quantity processing in healthy ageing. *Frontiers in Psychology, 4*, 865.

Lambrechts, A., Walsh, V., & van Wassenhove, V. (2013). Evidence accumulation in the magnitude system. *PLoS One, 8*(12), e82122.

Leibovich, T., & Henik, A. (2014). Comparing performance in discrete and continuous comparison tasks. *Quarterly Journal of Experimental Psychology, 67*, 899–917.

Leibovich, T., & Henik, A. (2014). Magnitude processing in non-symbolic stimuli. *Frontiers in Psychology, 4.*

Leslie, A. M., Gelman, R., & Gallistel, C. R. (2008). The generative basis of natural number concepts. *Trends in Cognitive Sciences, 12*, 213–218.

Leslie, A., Xu, F., Tremoulet, P., & Scholl, B. (1998). Indexing and the object concept: Developing 'what' and 'where' systems. *Trends in Cognitive Sciences, 2*, 10–18.

Libertus, M. E., & Brannon, E. M. (2010). Stable individual differences in number discrimination in infancy. *Developmental Science, 13*, 900–906.

Libertus, M. E., Feigenson, L., & Halberda, J. (2011). Preschool acuity of the approximate number system correlates with school math ability. *Developmental Science, 14*, 1292–1300.

Lipton, J. S., & Spelke, E. S. (2004). Discrimination of large and small numerosities by human infants. *Infancy, 5*, 271–290.

Loetscher, T., Bockisch, C. J., Nicholls, M. E. R., & Brugger, P. (2010). Eye position predicts what number you have in mind. *Current Biology, 20*, R264–R265.

Lourenco, S. F. (2014a). *Cross-dimensional and cross-modal matching of magnitude in infancy* (in progress).

Lourenco, S. F. (2014b). *Numerical abstraction in 3-year-olds: Spatial congruity effects* (in progress).

Lourenco, S. F., Levine, S. C., & Degner, H. (2014). *The influence of temporal cues on preschoolers' numerical judgments* (in progress).

Lourenco, S. F., & Bonny, J. W. (2014). *Representations of numerical and non-numerical magnitude contribute equally to mathematical competence in children* (in progress).

Lourenco, S. F., & Longo, M. R. (2010). General magnitude representation in human infants. *Psychological Science, 21*, 873–881.

Lourenco, S. F., & Longo, M. R. (2011). Origins and development of generalized magnitude representation. In S. Dehaene & E. Brannon (Eds.), *Space, time and number in the brain: Searching for the foundations of mathematical thought* (pp. 225–244). London, UK: Elsevier.

Lourenco, S. F., Bonny, J. W., Fernandez, E. P., & Rao, S. (2012). Nonsymbolic number and cumulative area representations contribute shared and unique variance to symbolic math competence. *Proceedings of the National Academy of Sciences, 109*, 18737–18742.

Lyons, I. M., & Beilock, S. L. (2011). Numerical ordering ability mediates the relation between number-sense and arithmetic competence. *Cognition, 121*, 256–261.

McCrink, K., & Wynn, K. (2004). Large-number addition and subtraction by 9-month-old infants. *Psychological Science, 15*, 776–781.

Meck, W. H., & Church, R. M. (1983). A mode control model of counting and timing processes. *Journal of Experimental Psychology: Animal Behavior Processes, 9*, 320–334.

Merritt, D. J., Casasanto, D., & Brannon, E. M. (2010). Do monkeys think in metaphors? Representations of space and time in monkeys and humans. *Cognition, 117*, 191–202.

Mix, K. S., Huttenlocher, J., & Levine, S. C. (2002). Multiple cues for quantification in infancy: Is number one of them? *Psychological Bulletin, 128*, 278–294.

Molko, N., Cachia, A., Rivière, D., Mangin, J. F., Bruandet, M., Le Bihan, D., et al. (2003). Functional and structural alterations of the intraparietal sulcus in a developmental dyscalculia of genetic origin. *Neuron, 40*, 847–858.

Nieder, A., Freedman, D. J., & Miller, E. K. (2002). Representation of the quantity of visual items in the primate prefrontal cortex. *Science, 297*, 1708–1711.

Nieder, A. (2012). Supramodal numerosity selectivity of neurons in primate prefrontal and posterior parietal cortices. *Proceedings of the National Academy of Sciences, 109*, 11860–11865.

Nieder, A., & Dehaene, S. (2009). Representation of number in the brain. *Annual Review of Neuroscience, 32*, 185–208.

Odic, D., Libertus, M. E., Feigenson, L., & Halberda, J. (2013). Developmental change in the acuity of approximate number and area representations. *Developmental Psychology, 49*, 1103.

Oliveri, M., Vicario, C. M., Salerno, S., Koch, G., Turriziani, P., Mangano, R., et al. (2008). Perceiving numbers alters temporal perception. *Neuroscience Letters, 432*, 308–311.

Onoe, H., et al. (2001). Cortical networks recruited for time perception: A monkey positron emission tomography (PET) study. *NeuroImage, 13*, 37–45.

Piaget, J. (1965). *The child's conception of number*. Oxford, England: W.W. Norton.

Piazza, M., Izard, V., Pinel, P., Le Bihan, D., & Dehaene, S. (2004). Tuning curves for approximate numerosity in the human intraparietal sulcus. *Neuron, 44*, 547–555.

Piazza, M., Pica, P., Izard, V., Spelke, E. S., & Dehaene, S. (2013). Education enhances the acuity of the nonverbal approximate number system. *Psychological Science, 24*, 1037–1043.

Pinel, P., Piazza, M., Le Bihan, D., & Dehaene, S. (2004). Distributed and overlapping cerebral representations of number, size, and luminance during comparative judgments. *Neuron, 41*, 983–993.

Price, G. R., Palmer, D., Battista, C., & Ansari, D. (2012). Nonsymbolic numerical magnitude comparison: Reliability and validity of different task variants and outcome measures, and their relationship to arithmetic achievement in adults. *Acta Psychologica, 140*, 50–57.

Ross, J., & Burr, D. C. (2010). Vision senses number directly. *Journal of Vision, 10*, 1–8.

Rousselle, L., & Noël, M. P. (2008). The development of automatic numerosity processing in preschoolers: Evidence for numerosity–perceptual interference. *Developmental Psychology, 44*, 544.

Rousselle, L., Palmers, E., & Noël, M. P. (2004). Magnitude comparison in preschoolers: What counts? Influence of perceptual variables. *Journal of Experimental Child Psychology, 87*(1), 57–84.

Rubinsten, O., & Henik, A. (2009). Developmental dyscalculia: Heterogeneity might not mean different mechanisms. *Trends in Cognitive Sciences, 13*, 92–99.

Rusconi, E., Kwan, B., Giordano, B. L., Umilta, C., & Butterworth, B. (2006). Spatial representation of pitch height: The SMARC effect. *Cognition, 99*, 113–129.

Schwarz, W., & Keus, I. M. (2004). Moving the eyes along the mental number line: Comparing SNARC effects with saccadic and manual responses. *Perception & Psychophysics, 66*, 651–664.

Simon, T. (1997). Reconceptualizing the origins of number knowledge: A non-numerical account. *Cognitive Development, 12*, 349–372.

Smith, L. B., & Sera, M. D. (1992). A developmental analysis of the polar structure of dimensions. *Cognitive Psychology, 24*, 99–142.

Sophian, C., & Chu, Y. (2008). How do people apprehend large numerosities? *Cognition, 107*, 460–478.

Spelke, E. S. (2011). Natural number and natural geometry. In S. Dehaene & E. Brannon (Eds.), *Space, time and number in the brain: Searching for the foundations of mathematical thought* (pp. 287–317). London, UK: Elsevier.

Srinivasan, M., & Carey, S. (2010). The long and the short of it: On the nature and origin of functional overlap between representations of space and time. *Cognition, 116*, 217–241.

Starkey, P., Spelke, E. S., & Gelman, R. (1983). Detection of intermodal numerical correspondences by human infants. *Science, 222*, 179–181.

Stavy, R., & Tirosh, D. (2000). *How students (mis-)understand science, mathematics: Intuitive rules.* New York, London, UK: Teachers College Press, Columbia University.

Stevens, S. S. (1957). On the psychophysical law. *Psychological Review, 64,* 153–181.

Stevens, S. S. (1975). *Psychophysics: Introduction to its perceptual, neural, and social prospects.* New York: Wiley.

Stoianov, I., & Zorzi, M. (2012). Emergence of a 'visual number sense' in hierarchical generative models. *Nature Neuroscience, 15,* 194–196.

Tibber, M. S., Manasseh, G. S., Clarke, R. C., Gagin, G., Swanbeck, S. N., Butterworth, B., et al. (2013). Sensitivity to numerosity is not a unique visuospatial psychophysical predictor of mathematical ability. *Vision Research, 89,* 1–9.

Tokita, M., & Ishiguchi, A. (2010). How might the discrepancy in the effects of perceptual variables on numerosity judgment be reconciled? *Attention, Perception, & Psychophysics, 72,* 1839–1853.

Tokita, M., & Ishiguchi, A. (2011). Temporal information affects the performance of numerosity discrimination: Behavioral evidence for a shared system for numerosity and temporal processing. *Psychonomic Bulletin & Review, 18,* 550–556.

Tudusciuc, O., & Nieder, A. (2007). Neuronal population coding of continuous and discrete quantity in the primate posterior parietal cortex. *Proceedings of the National Academy of Sciences, 104,* 14513–14518.

Tudusciuc, O., & Nieder, A. (2009). Contributions of primate prefrontal and posterior parietal cortices to length and numerosity representation. *Journal of Neurophysiology, 101,* 2984–2994.

vanMarle, K., & Wynn, K. (2006). Six-month-old infants use analog magnitudes to represent duration. *Developmental Science, 9,* F41–F49.

vanMarle, K., & Wynn, K. (2009). Infants' auditory enumeration: Evidence for analog magnitudes in the small number range. *Cognition, 111,* 302–316.

vanMarle, K., Chu, F., Li, Y., & Geary, D. C. (in press). Acuity of the approximate number system and preschoolers' quantitative development. *Developmental Science.*

Van Opstal, F., Gevers, W., De Moor, W., & Verguts, T. (2008). Dissecting the symbolic distance effect: Comparison and priming distance effects in numerical and nonnumerical orders. *Psychonomic Bulletin & Review, 15,* 419–425.

Vicario, C. M., & Martino, D. (2010). The neurophysiology of magnitude: One example of extraction analogies. *Cognitive Neuroscience, 1,* 144–145.

Verguts, T., & Fias, W. (2004). Representation of number in animals and humans: A neural model. *Journal of Cognitive Neuroscience, 16,* 1493–1504.

Walker, P., Bremmner, J. G., Mason, U., Spring, J., Mattock, K., Slater, A., et al. (2010). Preverbal infants' sensitivity to synaesthetic cross-modality correspondences. *Psychological Science, 21,* 21–25.

Walsh, V. (2003). A theory of magnitude: Common cortical metrics of time, space and quantity. *Trends in Cognitive Sciences, 7,* 483–488.

Woodcock, R. W., McGrew, K. S., Schrank, F. A., & Mather, N. (2001/2007). *Woodcock–Johnson III Normative Update.* Rolling Meadows, IL: Riverside Publishing.

Xu, F., & Spelke, E. S. (2000). Large number discrimination in 6-month-old infants. *Cognition, 74,* B1–B11.

Xuan, B., Zhang, D., He, S., & Chen, X. (2007). Larger stimuli are judged to last longer. *Journal of Vision, 7,* 1–5.

Chapter 7

Foundations of the Formal Number Concept: How Preverbal Mechanisms Contribute to the Development of Cardinal Knowledge

Kristy vanMarle
University of Missouri, Columbia, MO, USA

INTRODUCTION

An understanding of number is one of the most important capacities to develop in early childhood. It provides the foundation for the uniquely human ability to engage in formal mathematics, and is critical to successful functioning in modern society. Children's mathematical skills at school entry predict mathematics achievement throughout schooling (Duncan et al., 2007; Geary, Hoard, Nugent, & Bailey, 2013), and continue to predict economic success and social status into adulthood (Bynner, 1997; Ritchie & Bates, 2013). But what skills predict school entry abilities? Where do number concepts come from, and how do they develop?

The answers to these questions have been the focus of substantial debate for decades. In this chapter, I argue that the foundations for the formal mathematical skills children acquire through instruction depend in part on nonverbal systems that are present at birth (Antell & Keating, 1983; Izard, Sann, Spelke, & Streri, 2009) and exist even in nonhuman animals (Brannon & Merritt, 2011; Brannon & Roitman, 2003). In some ways, the ubiquity of informal, nonverbal quantification abilities throughout development and throughout the animal kingdom is not surprising (Agrillo et al., this volume; Beran et al., this volume; Geary et al., this volume; Pepperberg, this volume; Vallortigara, this volume). Quantities like amount, time, and number are basic aspects of perceptual experience that reflect fundamental attributes of the environment (Dehaene & Brannon, 2011; Gallistel, 2009, 2011). In order

Mathematical Cognition and Learning, Vol. 1. http://dx.doi.org/10.1016/B978-0-12-420133-0.00007-7
175

for any organism to successfully navigate and function in the environment, it must represent behaviorally relevant quantitative information. Contexts such as foraging (e.g., maximizing amount of food per time spent at various locations; Gallistel et al., 2007; Gallistel, Mark, King, & Latham, 2001), determining the number of friends versus foes (ordinal judgments), and keeping track of number of offspring (cardinality), are faced by many species (Gallistel, 1990, 2009). More generally, being able to represent quantitative information routinely and with reasonable accuracy clearly confers an adaptive advantage (Gallistel, 1990, 2012).

Historically, psychologists have been reluctant to attribute numerical knowledge, in particular, to nonhuman animals (Davis & Perusse, 1988). The assumption was that number is abstract, and therefore, it cannot be represented without language. Nonetheless, the past few decades have seen a growing body of evidence suggesting that a wide variety of nonhuman species, as well as preverbal human infants, not only represent quantity but also perform quite sophisticated "computations" over these representations (for reviews, see Gallistel, 1990, and Mou & vanMarle, 2013a). Of course, nonhuman animals never go on to learn formal mathematics or to verbalize the concepts they appear to implicitly understand, whereas human children can and often do learn to read and write numerals and to perform calculations by engaging in the formal routines they acquire through instruction. One reason for this difference between humans and other animals may be that humans are capable (through language or other means) of forming generalizable concepts of cardinality (magnitude), ordinality (greater/less than relations), etc. (Penn, Holyoak, & Povinelli, 2008). The focus here, however, is on the bridge between the evolutionarily ancient, nonverbal systems, and the formal, explicit concepts that children begin to acquire when they learn to count (see also Starr & Brannon, this volume).

In this chapter, I first describe two nonverbal quantity systems and briefly review evidence that they are available to preverbal human infants. I then argue that above and beyond being sensitive to the attribute of numerosity, at least one of the representational systems infants deploy in numerical tasks constitutes a true concept of number, a claim that has been debated (Gallistel & Gelman, 2000; Gelman & Gallistel, 1978; Rips, Bloomfield, & Asmuth, 2008). Finally, I describe recent findings showing that both nonverbal systems play a role in the development of explicit cardinal knowledge and thus directly impact formal mathematics learning in young children. Ultimately, understanding what mechanisms underlie the number concept and whether and how they are related to the acquisition of the earliest formal mathematical skills (e.g., learning to count) will provide a locus to search for individual differences, potentially helping educators identify children at risk for starting school behind, and providing a target for early intervention in at risk children.

Before we delve into the argument, it is useful to first define some of the terminology I use throughout this chapter. In particular, I often describe preverbal infants' performance and abilities using terms and phrases such as "compute," "addition/subtraction," "division," and "engage in arithmetic operations." In using these terms, I am not suggesting that infants can calculate the results of the formal arithmetic problems you might find on a grade school mathematics exam or consciously reason about mathematical operations in the same way educated children and adults do. Infants in their first year (the developmental period that the majority of this chapter focuses on) have not yet developed language; they do not recognize or understand written numerals or number words (spoken or written) and, therefore, are probably not capable of explicitly representing or thinking about numbers at a conscious level.

Instead, it is assumed here that there are algorithms built into the cognitive mechanisms whose function is to combine and evaluate inputs according to the rules of arithmetic (e.g., summing, multiplying, dividing) to produce outputs, which are the results of those processes. The infant likely does not know that it performs computations or how the computations are carried out. Because these processes are occurring at a level below conscious awareness, I refer to them as "implicit," whereas "explicit" processes are those that are available for conscious reflection and verbalization, such as when a child recites the count words as part of the counting routine or writes down an addition problem using conventional written notation, e.g., $4+2=6$. Finally, I use the terms "informal," "nonverbal" or "preverbal," and "compute" to refer to these implicit processes, and the terms "formal," "verbal," and "calculate" to refer to those processes that are explicit and that result from instruction in culturally transmitted knowledge of mathematics (e.g., counting, arithmetic).

TWO CORE MECHANISMS FOR REPRESENTING NUMBER

System 1: Analog Magnitude System (ANS)

After more than three decades of research, it is now clear that humans possess a nonverbal system for representing number and other quantities that is evolutionarily ancient and shared across a wide range of nonhuman animal species (Dehaene & Brannon, 2011; Gallistel, 1990). The analog magnitude system (ANS) is present from birth and continues to support numerical reasoning in adulthood (Cordes, Gelman, Gallistel, & Whalen, 2001; Izard et al., 2009; Whalen, Gallistel, & Gelman, 1999). It represents both discrete and continuous quantities as continuous analog magnitudes, in the same way a line can be used to represent a quantity, with its length being proportional to the number. For example, if a line that is 2 inches long represents the number "2," the line representing "8" would be 8 inches long. Importantly, the magnitude representations are imprecise, and the amount of variability (i.e., error) increases in proportion to the magnitude. This means that the representation for 20 is twice as variable, or "fuzzy," as the representation for 10. The consequence of this *scalar variability* is that our ability to discriminate two numbers depends on their ratio, not their absolute difference (Dehaene & Brannon, 2011; Gallistel, 1990; Halberda & Odic, this volume).

In addition to being imprecise, another important feature of magnitude representations is that they are amodal. Adults can represent the number of dots in a visual array, the number of sounds played in a sequence (Barth et al., 2006), or the number of touches on one's hand (Plaisier, Tiest, & Kappers, 2010). Above and beyond representing number, the ANS also represents continuous spatial and temporal quantities such as surface area and duration (Feigenson, Dehaene, & Spelke, 2004; Walsh, 2003). Representing more than one type of quantity using a single format makes it possible for various quantities to be compared and combined in arithmetical computations, such as generalizing, comparing, and summing quantities across different sensory modalities, as both human and nonhuman animals have been shown to do (Barth, Kanwisher, & Spelke, 2003; Beran, 2012; Meck & Church, 1983), and even across different types of quantities, such as computing rate from representations of time and number (Davison & McCarthy, 1988; Hernstein, 1961). Thus, the ANS represents a variety of quantities from different sensory modalities, and does so using a common mental currency.

There is substantial evidence for the ANS in infants. How do we know this? Because infants cannot speak and tell you, for example, whether two arrays of dots differ in number, the majority of studies exploring these abilities measure infants' visual attention (i.e., their looking time to different displays) to assess their expectations about physical events or to determine whether they can discriminate two stimuli. Most of the studies reviewed here used one of two common paradigms: habituation (Fantz, 1964) or violation-of-expectation (VOE; Baillargeon, Spelke, & Wasserman, 1985). In habituation studies, infants first complete an habituation phase in which they are shown a particular stimulus repeatedly (e.g., 10 dots) over several trials, and their looking time is measured. Other features of the habituation displays may vary (e.g., the spacing of the dots, the size of the dots, the arrangement of the dots), but the number of dots is held constant. Just like adults, infants' attention diminishes with repeated exposure, indicating they are becoming familiar with the stimulus (i.e., 10 dots). Once their looking time declines below a criterion level, they are considered "habituated" and they move to the test phase in which they are presented, on alternating trials, with the familiar (i.e., 10 dots) and a novel (i.e., 20 dots) stimulus. If infants can discriminate between the old and new stimulus, they are expected to recover interest, and therefore look longer, on novel compared to familiar trials.

Studies using the VOE method assume that infants generate expectations about physical events and look longer at events that appear to violate their expectations, just like children and adults do when they see a magician pull a rabbit out of an apparently empty hat! In the typical VOE study, infants are presented with events on a puppet stage involving real objects. The same event usually is shown on every trial. For example, the infant might see an experimenter place two objects on the stage and then raise a screen to occlude them. After a brief pause, the screen is lowered to reveal an "expected" outcome (e.g., two objects), or an "unexpected" outcome (e.g., one object), on alternating trials. Infants' looking time to the outcomes is measured. If infants hold the expectation that objects cannot magically disappear into thin air, then they should look longer at the unexpected outcome compared to the expected outcome. Looking time measures in general, and habituation and VOE procedures specifically, have transformed the study of infant cognition because of their ability to tap infants' implicit knowledge (but see Schöner & Thelen, 2006), and continue to be widely used in infant research.

Using these basic methods, researchers have shown that like nonhuman animals and adults, infants can discriminate both continuous and discrete quantities (vanMarle & Wynn, 2006, 2009). Moreover, their number abilities are not limited to the visual domain, but extend to the auditory and tactile domains, as well (Feron, Gentaz, & Streri, 2006; Lipton & Spelke, 2003; Xu & Spelke, 2000). Importantly, their discrimination of quantities is ratio-dependent. Seminal work by Xu and colleagues (Xu, 2003; Xu & Spelke, 2000; Xu, Spelke, & Goddard, 2005) shows that at 6 months of age, infants can discriminate numbers of visual items (i.e., dots) that differ by a 1:2 ratio, but not a 2:3 ratio. This discrimination function is also seen for both small (e.g., 2 vs. 4) and large (e.g., 8 vs. 16) numbers of sounds (Lipton & Spelke, 2003; vanMarle & Wynn, 2009), as well as for duration (vanMarle & Wynn, 2006), suggesting that like nonhuman animals, infants' ANS uses a common format to represent quantities across different sensory modalities and across different quantitative dimensions. Further, there is a developmental improvement in the precision of infants' number representations that increases substantially over the first year of life and continues to increase throughout childhood (Halberda & Feigenson, 2008). At birth, infants require a 1:3 ratio to reliably discriminate numbers, but succeed at 1:2 ratios by 6 months and 2:3 ratios by 9 months (Izard et al., 2009; Lipton & Spelke, 2003); the same is true for duration, with infants needing at least a 1:2 ratio at 6 months, but succeeding with 2:3 ratios by 9 months (Brannon, Suanda, & Libertus, 2007; vanMarle & Wynn, 2006).

Using a single format allows infants to recognize correspondences, or the equivalency in number, across various sensory modalities, such as matching the number of sounds they hear with the number of objects they see, an ability that is available at birth (Izard et al., 2009). Other studies have shown similar cross-modal matching abilities in 5- to 8-month-olds using a variety of stimuli (Feigenson, 2011; Jordan & Brannon, 2006; Starkey, Spelke, & Gelman, 1983, 1990), including comparisons of tactile to visual stimuli (Feron et al., 2006). Beyond recognizing correspondences, infants are also able to engage in nonverbal arithmetic. Work by McCrink and Wynn (2004) showed that infants are sensitive to computer-animated addition (5 + 5) and subtraction events (10–5), and discriminate between numerically correct (e.g., $5+5=10$, $10-5=5$) and numerically incorrect outcomes (e.g., $5+5=5$, $10-5=10$). For example, when 5 rectangles moved behind a screen, and then 5 more hid behind the same screen, infants looked longer when the screen lowered to reveal only 5 rectangles compared to when 10 were revealed. Building on this finding, McCrink and Wynn (2007) went on to show that 6-month-old infants correctly discriminated ratios that represent a division of two object collections. When familiarized to arrays of blue and yellow dots, where the total number of blue and yellow dots changed across displays (e.g., 10:20, 20:40, 5:10), infants were able to abstract the common ratio (1:2) and generalize this learning to new displays, looking longer at new displays with a novel ratio (15:60), compared to new displays that bore the same, familiarized ratio (15:30). Importantly, infants' arithmetic performance is ratio-dependent and follows the same discrimination function as for simple number discrimination (success with 1:2 ratios and failure with 2:3 ratios), suggesting that the ANS underlies infants' sensitivity to the outcomes of arithmetic events involving collections of objects or sounds (i.e., implicit addition, subtraction, and division; McCrink & Wynn, 2007; McCrink, this volume).

Despite the evidence that infants can enumerate entities from various sensory modalities, some researchers remain skeptical about whether infants in these studies are responding on the basis of number, per se, or to some non-numerical, low-level perceptual property of the displays. For example, there are many studies attempting to show that spatial dimensions that naturally co-vary with number (e.g., surface area, density, perimeter) are more salient to infants than the abstract property of number (Clearfield & Mix, 1999, 2001; Mix, Huttenlocher, & Levine, 2002; cf. Cordes & Brannon, 2008, 2009, 2011). Two ongoing studies in my lab address this debate by testing whether infants' ANS representations are truly abstract. In one study, we explore whether infants can extract ratio information separately from visual and auditory displays, and then detect correspondences across the modalities (Mou & vanMarle, 2013b). To test this, researchers simultaneously presented 6- and 10-month-old infants with two visual displays, each with an array of red and blue dots. The two displays differed only in the ratio of red:blue dots (the total number of dots is held constant, and non-numerical features, such as surface area and density, are controlled). While the displays were visible, an auditory stimulus cycled in the background that contained pitch information matching one of the two visual displays in ratio (i.e., the ratio of high:low pitches was the same as the ratio of red:blue dots). As in previous cross-modal matching studies, infants looked longer at the visual displays that matched the auditory stimulus in ratio, indicating that they not only can extract the ratios from both types of sensory input, but also compare them and recognize when they correspond. Once again, we observe the same ratio-dependent discrimination as found for simple numbers and unimodal ratio discrimination (e.g., success with 1:2 ratios, but not 2:3 ratios, at 6 months, and success with both at 10 months; Mou & vanMarle, 2013b). Because the low-level perceptual properties of the visual displays and auditory sequences share nothing in common, infants' ability to detect correspondences must be based on their sensitivity to the abstract relations of the ratios of pitches and dots, suggesting that ANS representations are highly abstract, even in preverbal infants.

In a separate study, we are exploring whether infants can perform implicit computations involving different quantitative dimensions. In particular, we ask whether 11-month-olds can combine their representations of number and time in order to estimate the number of dots that should appear in a given amount of time, which would suggest sensitivity to rate. The ability to estimate rate has been widely observed among nonhuman animals, as related, for example, to foraging behaviors (see Davison & McCarthy, 1988, for review). To test this in infants, we first familiarize them to computerized displays in which dots appear on the screen one at a time, while a tone plays continuously. The rate is the number of dots per the tone duration. During familiarization, the number of dots (e.g., 4) and the tone duration (e.g., 1250 ms) is held constant. Once familiarized, the infant sees new animated displays in which an occluder rises to hide the center of the screen (where the dots appear). A tone that is either 2x or 4x the familiarized duration (e.g., 2500 ms or 5000 ms, respectively) then plays, depending on test condition. Finally, the occluder lowers to reveal either 8 dots or 16 dots. If infants are sensitive to the rate of the appearance of the dots in the familiarization phase, then in the 2x condition they should expect there to be 8 dots behind the occluder and look longer at the unexpected outcome of 16 dots. Infants in the 4x condition should show the opposite pattern because they should be expecting 16 dots, and should find the 8 dots unexpected. Pilot data support our hypothesis that infants, like animals, can compute rate (Seok & vanMarle, 2014). If these results are replicated in the current study, it will strongly suggest that infants are not only sensitive to the numerical ratio of collections of objects (e.g., "4 blue: 8 yellow"), but also to the relation between number of items and the time needed to present them (i.e., the ratio of "4 dots: 2 seconds"), which would imply again that the ANS is highly abstract.

Together, these studies indicate that from birth, infants can engage in sophisticated quantitative reasoning using the ANS. Like adults and nonhuman animals, they can represent both discrete and continuous quantities, detect correspondences across sensory modalities, and even perform arithmetic computations, including combining representations across different quantitative dimensions.

System 2: Object Tracking System (OTS)

The ANS has undoubtedly received the lion's share of the attention in the infant quantity representation literature, and for good reason. If the animal numerical cognition literature can be used as a guide, there is no need to posit a second system (see also Posid & Cordes, this volume). In animals, there is clear evidence that the ANS represents values throughout the number range, and with the exception of a handful of studies (for review, see Mou & vanMarle, 2013a), there is comparatively little evidence that nonhuman animals routinely use a limited-capacity object tracking system (OTS) to represent and reason about quantities (but see Agrillo and colleagues, this volume). The data for infants is less clear, however, with more support for the OTS in quantitative contexts than seen in the animal literature. And, given the recent hypothesis forwarded by Mou and vanMarle (2013a) that the OTS's role in numerical reasoning is limited primarily to early development, largely disappearing as the ANS gains in precision over the course of the first few years of life, it is important to objectively evaluate the possibility that the OTS may contribute to the development of the number concept.

The object tracking system (OTS) is a recent model that is derived from two well-known theories in the literature on visual attention in adults: the FINST model (Pylyshyn & Storm, 1988) and the object file model (Kahneman, Treisman, & Gibbs, 1992). The OTS consists of a set of indexes that can be used to "point" to objects in the world. Importantly, the indexes are "sticky" and remain fixed on an object as it moves around in space. The indexes can also store featural information about the object (e.g., color, size, orientation, kind), but the location information (i.e., where the index is pointing) is prioritized (Flombaum, Scholl, & Santos, 2009; Scholl, 2001). Functionally, the OTS makes it possible to keep track of the same individual over time, even when that individual goes out of sight for brief periods of time (e.g., when a squirrel runs behind a tree), and even when that individual undergoes radical feature changes (e.g., "it's a bird, it's a plane, it's Superman!"; Kahneman et al., 1992, p. 17).

The most important feature of the OTS is its limited capacity. It can track only as many objects as it has indexes, which in human adults appears to be about four (Pylyshyn & Storm, 1988; Scholl, 2001). However, unlike the ANS, the OTS did not evolve to represent quantity. Indeed, there is no provision in the model for generating representations of continuous spatial quantities (e.g., surface area, volume), although they may be bound to active indexes as features, and it represents discrete quantity (i.e., number) at best only indirectly. For example, you might have three active indexes tracking three dogs running across the street. If all three dogs run behind a house, but only two reappear, you would likely infer that the third was still behind the house. At a mechanistic level, you would make this inference based on the fact that only two of your active indexes were able to reacquire their targets once the dogs reappeared, creating a mismatch between your mental representation (three active indexes pointing at three different dogs) and the actual state of the world (two dogs visible). Importantly, it is only by means of this one-to-one correspondence process that the OTS can be said to represent number at all.

Even so, in contrast to the fuzzy nature of ANS representations, the OTS represents small sets precisely, indexing exactly one or exactly two items, up to its capacity limit.

The evidence for the object tracking system in infants comes primarily from two counterintuitive findings. First, several elegant studies by Feigenson and colleagues (Feigenson, Carey, & Spelke, 2002), as well as work by Xu and colleagues (Xu, 2003; Xu, Spelke, & Goddard, 2005), have shown that 6- and 7-month-old infants often fail to discriminate small numbers of visual items (e.g., 1 vs. 2), even when the same ratio, but for larger sets, would be highly discriminable. Such a finding is inconsistent with ratio-dependent performance and therefore is inconsistent with the ANS model. A second striking finding is based on infants' failure in some cases to represent values beyond three. In particular, Feigenson, Carey, and Hauser (2002) showed that when 10- to 12-month-old infants were allowed to choose between two sequentially hidden sets of food items, they chose the larger amount, but only as long as neither set was larger than 3. They reliably chose 2 crackers over 1, and 3 over 2, but chose randomly when faced with comparisons like 3 vs. 4, 2 vs. 4, 3 vs. 6, and remarkably, even when choosing between 1 vs. 4 crackers, which is a ratio even newborn infants can easily discriminate (Izard et al., 2009). This performance pattern, in which infants succeed with small sets and fail when either set is outside the capacity limit of the OTS, is termed the *set size signature*, and is thought to indicate that the infant used the OTS and not the ANS to represent and compare the sets.

Several other studies also suggest that infants possess and use the OTS to track small sets of visual objects. Wynn's (1992a) classic finding in which 5-month-old infants were reported to be sensitive to simple additions and subtractions of objects from a small set of objects has been explained by appealing to this mechanism. In her experiment, infants tracked visual objects on a puppet stage as they moved around and were briefly occluded. In the "addition" events, infants initially saw one object that was subsequently hidden behind a screen, and then a second object was added behind the screen. When the screen was removed to show the outcome, infants looked longer when only one object was revealed (unexpected, $1+1=1$), compared to when two objects were revealed (expected, $1+1=2$). The "subtraction" condition was the same except two dolls were initially placed and then hidden, and then one was removed from the hidden set before the screen was lowered to reveal one object (expected, $2-1=1$) or two (unexpected, $2-1=2$). Infants in both conditions looked longer at the unexpected outcomes, suggesting they kept track of the objects and detected when one had magically appeared ($2-1=2$) or disappeared ($1+1=1$). Infants' performance in this task is consistent with the OTS. Like adults, infants presumably assigned an index for each object as it was brought out on the stage. When the screen hid the objects, the indexes continued to point to them. And when the screen was finally lowered, the active indexes were put in one-to-one correspondence with the revealed object(s). Mismatches were detected when the wrong number of objects was revealed, leading to longer looking to unexpected outcomes compared to expected outcomes.

Wynn's (1992a) findings have been replicated and extended in various ways that are consistent with the OTS. For example, infants appear to prioritize spatiotemporal information (i.e., an objects' location), looking longer when the wrong number of objects is revealed even if the items changed features behind the screen (Simon, Hespos, & Rochat, 1995), and even when the objects' locations were constantly changing (Koechlin, Dehaene, & Mehler, 1997). Because the objects in Wynn's original study were always placed in the same locations on the stage, infants' longer looking to incorrect outcomes may have been due to attention to whether a given location is empty or filled, rather than attention to the number of objects, per se. Koechlin et al.'s study showed this was not true, and that infants were tracking the individual objects, not the locations on the stage. Thus, infants appear to respond on the basis of number, and not on the basis of missing or extra features, or filled or empty locations. In addition, infants in the original study represented the number of objects precisely. Not only did they look longer when $1 + 1 = 1$ compared to when $1 + 1 = 2$, but they also looked longer when $1 + 1 = 3$ (Wynn, 1992a). This suggests they were not merely expecting there to be *more objects* behind the screen following the addition event but were expecting "exactly two," reflecting the characteristic precision with which the OTS represents small sets.

More recent studies using the same addition/subtraction paradigm suggest interesting limitations to infants' object tracking capacity. For instance, similar to adults (vanMarle & Scholl, 2003), infants' performance breaks down when a tracked object violates cohesion (i.e., falls apart or disintegrates, like a pile of sand when moved; Chiang & Wynn, 2000; Huntley-Fenner, Carey, & Solimando, 2002), indicating that the OTS may only track objects that obey Spelke's object principles (Spelke, 2000) of cohesion (i.e., objects maintain boundedness; Cheries, Mitroff, Wynn, & Scholl, 2008), solidity (i.e., objects cannot pass through each other; Mitroff, Wynn, Scholl, Johnson, & Shuwairi, 2004), and continuity (i.e., objects must travel continuous paths through space and time; Cheries, Mitroff, Wynn, & Scholl, 2009).

Another limitation affects infants' ability to apply one-to-one correspondence when some features of the objects change, and in particular when continuous spatial features (e.g., surface area, perimeter) change while the objects are occluded. In a study by Feigenson, Carey, and Spelke (2002), 7-month-old infants successfully discriminated 1 from 2 objects, but only when the size of the objects did not change. If the objects changed size, infants failed. The same was true in the addition-subtraction paradigm. If one medium-sized object was added to one medium-sized object, and then one large object or two small objects were revealed (equated for total surface area), infants looked equally at the expected and unexpected outcomes. The fact that infants are able to use one-to-one correspondence successfully despite changes in surface features (i.e., color/pattern; Simon et al., 1995), but not changes in spatial dimensions, provides further evidence for the processing preference for spatiotemporal information in the OTS.

In sum, the OTS appears to be an important part of infants' early ability to track objects, and at least indirectly, represents the number of items up to the capacity limit of three items. However, in order to be useful as a foundation for the formal number concept, the OTS needs to do more than just track individuals.

"NUMBER CONCEPT" DEFINED

Before I discuss whether either of the two systems might support a number concept, it is useful to briefly discuss what is meant here by "number concept." Developmentalists have long argued about whether children possess a true concept of number prior to formal schooling, and if so, when it emerges and by what processes. The various accounts reach to both extremes and everything in between, from suggesting that the number concept is innate (Leslie, Gallistel, & Gelman, 2007; Leslie, Gelman, & Gallistel, 2008), to suggesting it is built over the course of the first few years through abstraction and induction of principles available in the two innate core mechanisms (Carey, 2004; Carey & Sarnecka, 2006; LeCorre & Carey, 2007; Spelke, 2011; Spelke & Tsivkin, 2001), to suggesting that it is not until children receive schooling and learn the principles of mathematics as a formal system that they come to have a true concept of number (Rips, Asmuth, & Bloomfield, 2006, 2008; Piaget, 1952).

To my mind, the extreme views in which children have to explicitly understand formal mathematical axioms (e.g., that there is a unique first number, that each number has a unique successor, etc.; Rips et al., 2008) are too stringent. Rips et al., (2008), for example, claim that their view requires only that children be competent of the formal axioms at an implicit, subconscious level. Yet, they go on to assert that it is not enough for children to possess the various axioms themselves. Instead, children must possess a representation of the full set of axioms; that is, they have to know that they know the axioms. Similarly, views suggesting that children do not have a number concept unless they can engage in explicit verbal counting (e.g., Carey, 2004; LeCorre & Carey, 2007) are too stringent because there may be an age at which children possess a number concept and all the necessary principles, but still be unable to act on it, a typical competence-performance dilemma. Indeed, as Gelman and colleagues (Gelman, 1972, 1993; Leslie et al., 2008) have pointed out repeatedly, the principles within a conceptual domain should not have to be represented explicitly, either symbolically or linguistically, particularly for an evolved domain. In Gelman's view, it is enough for the principles to be structurally inherent or implicit in the developmental mechanism, guiding reasoning and behavior, but unavailable to conscious reflection or verbalization. This is the view I adopt here, precisely because it is rigorous and requires the organism to possess all the necessary axiomatic principles (those that define the natural numbers), but also because it is not overly strict. Because the principles may be realized implicitly, and their existence is observed through behaviors guided by these implicit processes, Gelman's view does not *a priori* rule out species for whom, or ages at which, explicit reproduction of the principles is impossible or improbable. It leaves open as an empirical question the possibility that nonverbal animals and preverbal human infants may possess the number concept, but not as explicitly as evaluated in a school setting. As will become clear in the following sections, this claim is central to the present chapter.

I therefore assume Gallistel and Gelman's (2000) view (see also Cordes, Williams, & Meck, 2007) that a nonverbal mechanism can instantiate the number concept inasmuch as the principles that define the domain are part of the structure and process of the mechanism; the child need not be able to reflect on them or consciously implement them in order for the concept to be in place. In the case of number, this means that the mechanism(s) must adhere to the basic counting (arithmetic) principles, including the stable order and abstraction principles. That is, the representations must meet all of the following criteria: (1) *Abstract*. The representations cannot be limited to specific sensory modalities or operate over a restricted set of entities. (2) *Cardinality*. Cardinality is an abstract property of sets that represents the total number of objects in a set, and allows sets to enter into arithmetic computations (i.e., addition, subtraction, multiplication, division). Number representations must denote the cardinality of the sets they represent. (3) *Ordinality*. Numbers form an ordered system; representations of number must be similarly ordered according to their cardinalities to allow judgments of "larger than" and "smaller than." (4) *Arithmetic*. The representations must be able to enter into arithmetic operations. In particular, the representations must be subject to addition, subtraction, multiplication, and division. Meeting all four of these criteria will be taken to indicate that the system in question is capable of supporting a true number concept.

DOES THE ANS AND/OR OTS MEASURE UP?

Abstract Representations

The first requirement for a number concept is that the representations are abstract. The OTS fails to meet this criterion in every sense of the word—in terms of being modality-general, in the sense of representing more than one type of quantity, and in the sense of abstracting away from individual items, which is the very essence of what it means for numbers to be abstract. The OTS is not abstract in the sense of being "amodal" because it is, by definition, a mechanism of *visual* attention. It is true that auditory information can be bound to object indexes in both adults and infants (Jordan, Clark, & Mitroff, 2010; Kobayashi, Hiraki, & Hasegawa, 2005; Kobayashi, Hiraki, Mugitani, & Hasegawa, 2004). However, the OTS does not track auditory individuals, but rather, uses auditory information as an additional cue in the service of its primary function, which is to track visual objects. Nor can the OTS be said to be abstract in the sense of representing more than one type of quantity. Even if we grant that it represents number indirectly (which has been challenged by Gallistel, 2007, and Leslie et al., 2007), it does not itself generate representations of continuous spatial quantities (surface area, volume, etc.) or duration, making it moot to even ask whether it can compute an abstract quantity such as rate. And, perhaps even more important, is that because the primary evolved function of the OTS to maintain the identity of individuals over time, the system is inherently non-numerical. As Gallistel (2007) so clearly stated: "The essence of numerical meanings is their abstraction from the particular. It is that abstraction that enables us to judge that the number of chairs is sufficient for the number of people present" (p. 445).

Clearly then, the OTS falls far short of meeting the criterion of abstraction. The ANS, on the other hand, meets the criterion in all respects. Infants represent numbers of visual items (Xu & Spelke, 2000), auditory items (Lipton & Spelke, 2003; vanMarle & Wynn, 2009), and also actions (i.e., puppet jumps; Wood & Spelke, 2005; Wynn, 1996), and already at birth, they can match the number of stimuli they hear with the number they see or feel (Feron et al., 2006; Izard et al., 2009). The ANS represents both discrete and continuous quantities, including duration (Feigenson et al., 2004), and I described preliminary evidence that infants are able to combine their representations of time and number allowing them to compute a rate (Seok & vanMarle, 2014). Together, such findings indicate a high level of abstraction, and underscore the fundamental utility of having a single representational currency. The ANS also captures the very essence of what it means for a number to be abstract. After enumerating a set of individuals, the resulting magnitude carries information only about the total number in the set, collapsing over the individuals that make up the set. Thus, the ANS stands as the paragon of abstraction, fulfilling the first criterion of the number concept, while the OTS does not.

Cardinality

A critical part of the number concept is the notion of cardinality, which in layman's terms might be thought of as "number sense." Cardinality is a property of sets and a cardinal representation refers to the total number of items in a set. To illustrate the nature of cardinality, consider that on some models (e.g., Gallistel & Gelman, 1992; Meck & Church, 1983), analog magnitude representations of number are formed through an iterative process analogous to filling a beaker with water, one cupful for each item counted. After all the items have been counted, the height of water in the beaker will be proportional to the number of items counted; that is, the height is a representation of the cardinality of the set (Gallistel & Gelman, 2000). In fact, as continuous representations (e.g., height of water, length of a line), analog magnitudes carry information only about cardinality, and not any of the enumerated individuals. This is in stark contrast to the OTS, whose most central function is to maintain distinct representations of individuals. In fact, Gallistel (2007) has argued that the OTS cannot play any role in the acquisition of counting in children precisely because it does not have separate symbols to indicate how many indexes are active at any given time. Thus, it can detect a mismatch between a set of indexes and a set of objects, but it cannot represent how many objects there are in a set, making it unclear how it could ever imbue the number words with cardinal meaning, as some have suggested (e.g., Carey, 2004; LeCorre & Carey, 2007).

Ordinality

Because numbers form an ordered series, it is possible to judge whether one number is larger or smaller than another by knowing where they are located within the series. Children come to learn this explicitly only after they can recite the count list (Geary, 1994). But long before they have explicit knowledge of ordinality, they appear to understand ordinality implicitly (Brannon & Van de Walle, 2001; Bullock & Gelman, 1977; Huntley-Fenner & Cannon, 2000; Rousselle, Palmers, & Noel, 2004; Siegel, 1974; Sophian & Adams, 1987; Strauss & Curtis, 1981). It is well established that infants are sensitive to ordinal relations. For example, in their first year, they can discriminate ascending and descending sequences for both number of objects and other quantitative stimulus dimensions, such as object size (Brannon, 2002; Macchi-Cassia, Picozzi, Girelli, & de Hevia, 2012; Picozzi, de Hevia, Girelli, & Macchi-Cassia, 2010; Suanda, Tompson, & Brannon, 2008). It has been assumed that infants' ordinal skills rely on the ANS, but no published studies have specifically tested whether their performance is ratio-dependent. In addition, as reviewed earlier, there are data to suggest that the OTS may support ordinal judgments, at least for small sets of visual objects (Feigenson, Carey, & Hauser, 2002). Recent work in my lab, however, shows that infants are not limited by set size when making ordinal judgments, as suggested by Feigenson, Carey, and Hauser (2002), When both sets are large, infants succeed and their performance is clearly ratio-dependent (vanMarle, 2013; vanMarle, Mou, & Seok, 2014; vanMarle & Wynn, 2011). The apparent set size signature instead may reflect the incommensurability of OTS and ANS representations. Mou and vanMarle (2013a) proposed such an idea, suggesting that the precision offered by the OTS allows it to trump the ANS for small visual sets early in life (roughly until about 2 years; vanMarle, Seok, &

Mou, 2014). Once the ANS becomes precise enough to reliably discriminate sets within the small number range, infants may abandon the OTS for making judgments that involve cardinality and ordinality. Further research is needed to test this hypothesis. An alternative possibility, not mutually exclusive to the first, is that the OTS may have attentional priority in infancy, perhaps being activated earlier than the ANS for small sets (Hyde & Spelke, 2011).

Arithmetic

As long argued by Gelman and colleagues (Gallistel & Gelman, 1992; 2000; Gelman & Gallistel, 1978; Leslie et al., 2007), in order for a representational system to constitute a number concept, the symbols it generates must be subject to arithmetic manipulation (see also McCrink, this volume). That is to say, they must count as inputs to computational processes that are isomorphic to addition, subtraction, multiplication, and division. In other words, the ANS supports nonverbal addition and subtraction (McCrink & Wynn, 2004), as well as division (ratio, McCrink & Wynn, 2004; Mou & vanMarle, 2013b; rate, Seok & vanMarle, 2014). Nonverbal multiplication has also been shown in 5- to 7-year-old children before they received formal schooling in mathematics (Barth, Baron, Spelke, & Carey, 2009; McCrink & Spelke, 2010). Researchers have also claimed that the OTS supports addition and subtraction as in, for example, Wynn's (1992a) task described earlier (Feigenson, Carey, & Spelke, 2002; Scholl & Leslie, 1999; Simon, 1997; Wynn & Chiang, 1998). While the kinds of events in this task may be conceived as "addition/subtraction" events, they are not true arithmetic. True arithmetic requires operating over cardinalities. Although the process of one-to-one correspondence may very well drive longer looking to displays that mismatch the number of active indexes, or searching for objects in a box when three went in but only two came out, that does not make it, in and of itself, arithmetic (Gallistel, 2007).

So, where does this leave us? The ANS meets all four criteria and therefore, by the present definition, instantiates a true concept of number. It is abstract, in every relevant sense of the word; it generates representations that are cardinalities, albeit imprecisely; the cardinal representations it generates are inherently orderable and support ordinal judgments; and its magnitudes are subject to arithmetic manipulation. The OTS, on the other hand, meets none of the criteria. The mechanism is situated within the visual modality and represents other information (auditory, surface area, volume, color, etc.) only as features of visual objects. In addition, the OTS has no means for representing the cardinality of the sets it indexes and thus cannot order sets on the basis of their cardinality. It also does not support true arithmetic, at most, only detecting mismatches between objects in the world and active indexes via a one-to-one correspondence operation. It appears, therefore, that the ANS has everything to offer, and the OTS nothing (except precision, perhaps), in the way of a preverbal number concept. But does it support the development of the formal number concept?

FOUNDATIONS OF THE FORMAL NUMBER CONCEPT

Children are not born knowing formal mathematics. Explicit knowledge of the verbal number system, arithmetic, algebra, geometry, and so on requires years of formal instruction and indoctrination into the relevant symbolic systems, theorems, and computational rules. The first step in this process begins before preschool when children start to learn number words and how to count. One important empirical question is whether the preverbal magnitude systems contribute anything to the acquisition of this knowledge (Starr & Brannon, this volume). As mentioned earlier, Rips and colleagues (Rips, Asmuth, & Bloomfield, 2006, 2008; Rips, Bloomfield, & Asmuth, 2008) argue that they do not. Other researchers argue that the preverbal systems have a very direct influence by providing the foundation on which the formal knowledge is built. There is substantial debate, however, about which system plays the lead role. Gelman and colleagues (Gallistel & Gelman, 1992, 2000; Gelman and Gallistel, 1978) suggest the ANS alone provides the foundation, whereas Carey and colleagues (Carey, 2004; LeCorre & Carey, 2007) claim that the OTS is the primary mechanism. And finally, Spelke and Tsivkin (2001; also Spelke, 2011) take a middle-of-the-road approach, suggesting that the two systems are complementary and together provide children the representational bases for learning to count.

Learning to count does not happen overnight. In fact, although children often can recite the verbal count list (i.e., "one, two, three, four, . . . ten") as early as age 2, the number words do not initially carry any numerical meaning (Wynn, 1992b). When children are this age, reciting the count list is much like reciting the alphabet; it is just another ordered list that has been committed to memory. In time, children come to realize that number words are special and used in specific contexts (e.g., counting), after which the number words become imbued with meaning (i.e., cardinality), and children slowly, over the course of approximately 1½ to 2 years, explicitly come to understand the rules of counting (i.e., the counting principles, Gelman & Gallistel, 1978) and how to apply them (LeCorre & Carey, 2007; Wynn, 1992b). This slow learning progression is thought to be observable through their performance in the *GiveN* task (Wynn, 1992b). In this task, children are asked to give a puppet a specific number of items, e.g., "exactly one cookie." Children start on set size 1 and move forward in the count list following a correct response. If they answer incorrectly (e.g., giving "three" cookies when asked for "two"), they move back to the previous set size. This continues until they successfully reach set size 6, or respond incorrectly on 2/3 attempts at a given set size. The highest set size for which they respond correctly at least 2/3 times is considered their "knower status." For example, a child who can give exactly one, two, and three items, but fails on set size four, would be considered a "three-knower."

Performance on the GiveN task is characteristically stage-like (Wynn, 1992b). First, children learn the meaning of "one" but seem to think all the other number words mean some indeterminate number that is >1. Thus, they can give the puppet "exactly one" but typically give a handful of cookies when asked to give any other set size. After about 6 months, children progress to being "two-knowers" and can give exactly one and two cookies, but give a large and random number for all other set sizes. Several months later, they become "three-knowers." And then suddenly, they seem to understand counting. Once they can give "exactly four," they can typically also give "exactly five," "six," and any other number up to the limit of their verbal count list and are categorized as "CP-knowers" (i.e., cardinal principle knowers). This sudden shift is considered by some to be a genuine instance of conceptual change (Carey, 2004; Spelke & Tsivkin, 2001), with children moving from their limited, rudimentary, preverbal number capacities to a genuine, explicit (verbal) knowledge of counting, which equips them with a qualitatively different and more powerful means of conceptualizing number.

Although many studies have explored the nature of children's early counting abilities, those arguing for the ANS-only or OTS-only models have depended primarily on indirect data (e.g., Carey, 2004; Gallistel & Gelman, 1992; LeCorre & Carey, 2007; Wynn, 1992b), and almost exclusively on children's performance in a single task—the *GiveN* task. Because these studies did not include independent measures of ANS and OTS performance, they are limited to basing their conclusions on whether children's performance in GiveN is consistent or inconsistent with either model. Recent work conducted by David Geary and myself (Chu, vanMarle, & Geary, 2013; vanMarle, Chu, Li, & Geary, 2014) moves beyond previous studies by exploring children's performance on a range of quantitative tasks, including independent measures of the ANS and OTS, while controlling for a host of other general cognitive capacities (e.g., IQ, executive function, preliteracy aptitude). In addition, we followed children longitudinally, which allowed us to examine the relations between the ANS, the OTS, and cardinal knowledge over time.

Children in our sample completed each of three tasks (ANS, OTS, and GiveN) twice, at the beginning and near the end of their first year of preschool (roughly 4 years of age), allowing us to examine concurrent and prospective relations between the two systems and children's cardinal knowledge. Our findings support a "dual-mechanism" view. At the beginning of preschool, both the ANS and OTS were related to children's cardinal knowledge as measured by the GiveN task; however, by the end of the year, only the ANS remained a significant predictor of cardinal knowledge. Importantly, the likelihood that children transitioned from "non-CP-knower" (i.e., one-, two-, or three-knower) at the first time point, to "CP-knower" (i.e., four-, five-, or six-knower) at the second time point, was predicted by their ANS, but not OTS, performance at the beginning of the year. In addition, the relations to cardinality at the first time point were stronger for the ANS than the OTS, suggesting that although both systems contributed to children's learning the meanings of the first few number words, the ANS played a more substantial role (vanMarle, Chu, Mou, & Geary, 2014).

Our findings are clearly inconsistent with both the ANS-only (Gallistel & Gelman, 1992, 2005; Gelman & Gallistel, 1978) and OTS-only (Carey, 2004; LeCorre & Carey, 2007) models. If one mechanism was solely responsible for imbuing the count words with meaning, then it alone should be related to cardinal knowledge, especially at the first time point. Instead, our data support a "dual-mechanism" view, akin to that proposed by Spelke and Tsivkin (2001; Spelke, 2011) in which both systems are believed to play a role. Our view, however, termed the *Merge* model, differs from Spelke's in at least two important respects. First, because language (i.e., the verbal count list) links the ANS and OTS, Spelke's model predicts that both systems should continue to support cardinal knowledge over time, perhaps even into adulthood (e.g., Dehaene, Spelke, Pinel, Stanescu, & Tsivkin, 1999). Our data, however, suggest that only the ANS remains related to cardinal knowledge, with the OTS falling away relatively early, by the end of the first year of preschool.

Second, the Merge model (vanMarle, Chu, Mou, & Geary, 2014) posits that the transition to CP-knower status occurs not as the result of the ANS and OTS being combined through language, but rather as the result of the OTS fading into the background. It is our contention that rather than being complementary and cooperative, the OTS's role early on is antagonistic, as it is in infancy (Feigenson, Carey, & Hauser, 2002; vanMarle, 2013; Xu, 2003; Xu, Spelke, & Goddard, 2005; for review, see Mou & vanMarle, 2013a). Our model explains the slow, piecemeal learning of the meanings of the number words as a consequence of this antagonism. Once the capacity limit of the OTS is breached, the OTS no longer competes with the ANS to represent the sets, and the cardinal meaning available through the ANS becomes more easily accessible, making it possible for the mapping mechanism to easily link the words with their meanings (vanMarle, Chu, Mou, & Geary, 2014). Part of the motivation for the Merge model comes from the fact that the ANS, but not the OTS, embodies the necessary attributes of a true concept of number. Given this, it is perhaps not surprising that the ANS is important for cardinality at both time points, while the OTS plays a role only at the beginning. Because the OTS really has just one advantage over the ANS—its precision—it plays an accordingly limited role in the building of the verbal number concept.

CONCLUSIONS

There is a wealth of data suggesting that infants possess two nonverbal systems that are capable of representing number. However, only the ANS possesses the necessary qualities to be considered a system that instantiates a true concept of number. The OTS falls short in almost every respect. One question that has been debated for decades is which of these systems provides the foundation for children's first lesson in formal mathematics, which involves developing an explicit understanding of the meaning of number words (their cardinal value), the rules of counting, and how to apply them. When children begin to learn the meanings of the count words, our findings suggest that both mechanisms take part in the process. However, unlike dual-mechanism views in which the two systems are complementary partners aiding in the development of cardinal knowledge, the Merge model suggests that the OTS is antagonistic to the ANS. This may turn what should be a relatively straightforward task of mapping number words onto the cardinal representations within the ANS into a lengthy and difficult process. The Merge model has yet to be directly tested. However, if it or a similar model were to be true, it would suggest that children could benefit from learning contexts that make it more difficult for the OTS to compete. For example, counting exercises that utilize two-dimensional images on cards rather than three-dimensional objects (i.e., manipulatives), or counting large as opposed to small sets, may help children by increasing the probability of engaging the ANS rather than the OTS. Further research will no doubt shed light on whether the Merge model proves useful as a theory, as well as on whether there are ways to enhance children's ANS abilities to make it a stronger competitor in the early years.

REFERENCES

Antell, S. E., & Keating, D. P. (1983). Perception on numerical invariance in neonates. *Child Development, 54*, 695–701.

Baillargeon, R., Spelke, E. S., & Wasserman, S. (1985). Object permanence in five-month-old infants. *Cognition, 20*, 191–208.

Barth, H., Baron, A., Spelke, E., & Carey, S. (2009). Children's multiplicative transformations of discrete and continuous quantities. *Journal of Experimental Child Psychology, 103*, 441–454.

Barth, H., Kanwisher, N., & Spelke, E. S. (2003). The construction of large number representations in adults. *Cognition, 86*, 201–221.

Barth, H., La Mont, K., Lipton, J., Dehaene, S., Kanwisher, N., & Spelke, E. (2006). Nonsymbolic arithmetic in adults and young children. *Cognition, 98*, 199–222.

Beran, M. J. (2012). Quantity judgments of auditory and visual stimuli by chimpanzees (*Pan troglodytes*). *Journal of Experimental Psychology. Animal Behavior Processes, 38*, 23–29.

Brannon, E. M. (2002). The development of ordinal numerical knowledge in infancy. *Cognition, 83*, 223–240.

Brannon, E. M., & Merritt, D. (2011). Evolutionary foundations of the approximate number system. In S. Dehaene & E. M. Brannon (Eds.), *Space, time, and number in the brain: Searching for the foundations of mathematical thought* (pp. 207–224). New York: Elsevier.

Brannon, E. M., & Roitman, J. (2003). Nonverbal representations of time and number in non-human animals and human infants. In W. Meck (Ed.), *Functional and neural mechanisms of interval timing* (pp. 143–182). New York: CRC Press.

Brannon, E. M., Suanda, U., & Libertus, K. (2007). Temporal discrimination increases in precision over development and parallels the development of numerosity discrimination. *Developmental Science, 10,* 770–777.

Brannon, E. M., & Van de Walle, G. (2001). Ordinal numerical knowledge in young children. *Cognitive Psychology, 43,* 53–81.

Bullock, M., & Gelman, R. (1977). Numerical reasoning in young children: The ordering principle. *Child Development, 48,* 427–434.

Bynner, J. (1997). Basic skills in adolescents' occupational preparation. *Career Development Quarterly, 45,* 305–321.

Carey, S. (2004). Bootstrapping and the origins of concepts. *Daedalus, 59–68.*

Carey, S., & Sarnecka, B. W. (2006). The development of human conceptual representations. In Y. Munakata & M. Johnson (Eds.), *Processes of change in brain and cognitive development: Attention and performance XXI* (pp. 473–496). Oxford, UK: Oxford University Press.

Cheries, E. W., Mitroff, S. R., Wynn, K., & Scholl, B. J. (2008). Cohesion as a principle of object persistence in infancy. *Developmental Science, 11*(3), 427–432.

Cheries, E. W., Mitroff, S. R., Wynn, K., & Scholl, B. J. (2009). Do the same principles constrain persisting object representations in infant cognition and adult perception?: The cases of continuity and cohesion. In B. Hood & L. Santos (Eds.), *The origins of object knowledge* (pp. 107–134). Oxford, UK: Oxford University Press.

Chiang, W.-C., & Wynn, K. (2000). Infants' representation and tracking of multiple objects. *Cognition, 77,* 169–195.

Chu, F., vanMarle, K., & Geary, D. C. (2013). Quantitative deficits of preschool children at risk for mathematical learning disability. *Frontiers in Developmental Psychology, 4,* 195. http://dx.doi.org/10.3389/fpsyg.2013.00195.

Clearfield, M. W., & Mix, K. S. (1999). Number versus contour length in infants' discrimination of small visual sets. *Psychological Science, 10,* 408–411.

Clearfield, M. W., & Mix, K. S. (2001). Infant use continuous quantity—not number—to discriminate small visual sets. *Journal of Cognition and Development, 2,* 243–260.

Cordes, S., & Brannon, E. M. (2008). Discrimination of continuous quantities in 6-month-old infants: Using number is just easier. *Child Development, 79*(2), 476–489.

Cordes, S., & Brannon, E. M. (2009). The relative salience of discrete and continuous quantities in infants. *Developmental Science, 12*(3), 453–463.

Cordes, S., & Brannon, E. M. (2011). Attending to one of many: When infants are surprisingly poor at discriminating an item's size. *Frontiers in Psychology, 2*(65), 1–8. http://dx.doi.org/10.3389/fpsyg.2011.00065.

Cordes, S., Gelman, R., Gallistel, C. R., & Whalen, J. (2001). Variability signatures distinguish verbal from nonverbal counting for both large and small numbers. *Psychonomic Bulletin & Review, 8,* 698–707.

Cordes, S., Williams, C. L., & Meck, W. H. (2007). Common representations of abstract quantities. *Current Directions in Psychological Science, 16*(3), 156–161.

Davis, H., & Perusse, R. (1988). Numerical competence in animals: Definitional issues, current evidence, and a new research agenda. *Behavioral and Brain Sciences, 11*(4), 561–579.

Davison, M., & McCarthy, D. (1988). *The matching law: A research review.* Hillsdale, NJ: Lawrence Erlbaum Associates, Inc.

Dehaene, S., & Brannon, E. M. (2011). *Space, time, and number in the brain: Searching for the foundations of mathematical thought.* Oxford: Elsevier.

Dehaene, S., Spelke, E. S., Pinel, P., Stanescu, R., & Tsivkin, S. (1999). Sources of mathematical thinking: Behavioral and brain-imaging evidence. *Science, 284,* 970–974.

Duncan, G. J., Dowsett, C. J., Claessens, A., Magnuson, K., Huston, A. C., Klebanov, P., et al. (2007). School readiness and later achievement. *Developmental Psychology*, *43*, 1428–1446.

Fantz, R. L. (1964). Visual experience in infants: Decreased attention to familiar patterns relative to novel ones. *Science*, *146*, 668–670.

Feigenson, L. (2011). Predicting sights from sounds: 6-month-old infants' intermodal numerical abilities. *Journal of Experimental Child Psychology*, *110*(3), 347–361.

Feigenson, L., Carey, S., & Hauser, M. (2002a). The representations underlying infants' choice of more: Object-files versus analog magnitudes. *Psychological Science*, *13*, 150–156.

Feigenson, L., Carey, S., & Spelke, E. S. (2002b). Infants' discrimination of number vs. continuous extent. *Cognitive Psychology*, *44*, 33–66.

Feigenson, L., Dehaene, S., & Spelke, E. S. (2004). Core systems of number. *Trends in Cognitive Sciences*, *8*, 307–314.

Feron, J., Gentaz, E., & Streri, A. (2006). Evidence of amodal representation of small numbers across visuo-tactile modalities in 5-month-old infants. *Cognitive Development*, *21*, 81–92.

Flombaum, J. I., Scholl, B. J., & Santos, L. R. (2009). Spatiotemporal priority as a fundamental principle of object persistence. In B. Hood & L. Santos (Eds.), *The Origins of Object Knowledge* (pp. 135–164). New York: Oxford University Press.

Gallistel, C. R. (1990). *The organization of learning*. Cambridge, MA: MIT Press.

Gallistel, C. R. (2007). Commentary on Le Corre & Carey. *Cognition*, *105*, 439–445.

Gallistel, C. R. (2009). The foundational abstractions. In M. Piattelli-Palmirini, J. Uriagereka, & P. Salaburu (Eds.), *Of minds and language: A dialogue with Noam Chomsky in the Basque Country* (pp. 58–73). New York: Oxford University Press.

Gallistel, C. R. (2011). Mental magnitudes. In S. Dehaene & E. M. Brannon (Eds.), *Space, time, and number in the brain: Searching for the foundations of mathematical thought* (pp. 3–12). New York: Elsevier.

Gallistel, C. R. (2012). On rationalism and optimality: Responses to the Miller and Nevin commentaries. *Behavioural Processes*, *90*, 87–88.

Gallistel, C. R., & Gelman, R. (1992). Preverbal and verbal counting and computation. *Cognition*, *44*, 43–74.

Gallistel, C. R., & Gelman, R. (2000). Non-verbal numerical cognition: From reals to integers. *Trends in Cognitive Sciences*, *4*, 59–65.

Gallistel, C. R., & Gelman, R. (2005). Mathematical cognition. In K. Holyoak & R. Morrison (Eds.), *The Cambridge handbook of thinking and reasoning* (pp. 559–588). New York: Cambridge University Press.

Gallistel, C. R., King, A. P., Gottlieb, D., Balci, F., Papachristos, E. B., Szalecki, M., et al. (2007). Is matching innate? *Journal of the Experimental Analysis of Behavior*, *7*(2), 161–199.

Gallistel, C. R., Mark, T. A., King, A. P., & Latham, P. E. (2001). The rat approximates an ideal detector of changes in rates of reward: Implications for the law of effect. *Journal of Experimental Psychology. Animal Behavior Processes*, *27*, 354–372.

Geary, D. C. (1994). *Children's mathematical development: Research and practical applications*. Washington, DC: American Psychological Association.

Geary, D. C., Hoard, M. K., Nugent, L., & Bailey, D. H. (2013). Adolescents' functional numeracy is predicted by their school entry number system knowledge. *PLoS ONE*, *8*(1), e54651. http://dx.doi.org/10.1371/journal.pone.0054651.

Gelman, R. (1972). The nature and development of early number concepts. In H. W. Reese (Ed.), *Advances in child development* (pp. 3, 115–3, 167). New York: Academic Press.

Gelman, R. (1993). A rational-constructivist account of early learning about numbers and objects. In D. Medin (Ed.), *Learning and motivation* (pp. 30, 61–30, 96). New York: Academic Press.

Gelman, R., & Gallistel, C. R. (1978). *The child's understanding of number.* Cambridge, MA: Harvard University Press.

Halberda, J., & Feigenson, L. (2008). Developmental change in the acuity of the "number sense": The approximate number system in 3-, 4-, 5-, and 6-year-olds and adults. *Developmental Psychology, 44*(5), 1457–1465.

Hernstein, R. J. (1961). Relative and absolute strength of response as a function of frequency of reinforcement. *Journal of the Experimental Analysis of Behavior, 4,* 267–272.

Huntley-Fenner, G., & Cannon, E. (2000). Preschoolers' magnitude comparisons are mediated by a preverbal analog mechanism. *Psychological Science, 11,* 147–152.

Huntley-Fenner, G., Carey, S., & Solimando, A. (2002). Objects are individuals but stuff doesn't count: Perceived rigidity and cohesiveness influence infants' representations of small numbers of discrete entities. *Cognition, 85,* 203–221.

Hyde, D. C., & Spelke, E. S. (2011). Neural signatures of number processing in human infants: Evidence for two core systems underlying non-verbal numerical cognition. *Developmental Science, 14*(2), 360–371.

Izard, V., Sann, C., Spelke, E. S., & Streri, A. (2009). Newborn infants perceive abstract numbers. *Proceedings of the National Academy of Sciences of the United States of America, 106,* 10382–10385.

Jordan, K. E., & Brannon, E. M. (2006). The multisensory representation of number in infancy. *Proceedings of the National Academy of Sciences of the United States of America, 103,* 3486–3489.

Jordan, K. E., Clark, K., & Mitroff, S. R. (2010). See an object, hear an object file: Object correspondence transcends sensory modality. *Visual Cognition, 18,* 492–503.

Kahneman, D., Treisman, A., & Gibbs, B. J. (1992). The reviewing of object files: Object-specific integration of information. *Cognitive Psychology, 24,* 174–219.

Kobayashi, T., Hiraki, K., & Hasegawa, T. (2005). Auditory-visual intermodal matching of small numerosities in 6-month-old infants. *Developmental Science, 8,* 409–419.

Kobayashi, T., Hiraki, K., Mugitani, R., & Hasegawa, T. (2004). Baby arithmetic: One object plus one tone. *Cognition, 91,* B23–B34.

Koechlin, E., Dehaene, S., & Mehler, J. (1997). Numerical transformations in five-month-old human infants. *Mathematical Cognition, 3,* 89–104.

Le Corre, M., & Carey, S. (2007). One, two, three, four, nothing more: An investigation of the conceptual sources of the verbal counting principles. *Cognition, 105,* 395–438.

Leslie, A. M., Gallistel, C. R., & Gelman, R. (2007). Where integers come from. In P. Caruthers, S. Laurence, & S. Stich (Eds.), *The innate mind, Vol. 3: Foundations and the future* (pp. 109–138). New York: Oxford University Press.

Leslie, A. M., Gelman, R., & Gallistel, C. R. (2008). The generative basis of natural number concepts. *Trends in Cognitive Sciences, 12*(6), 213–218.

Lipton, J. S., & Spelke, E. S. (2003). Origins of number sense: Large number discrimination in human infants. *Psychological Science, 14,* 396–401.

Macchi-Cassia, V., Picozzi, M., Girelli, L., & de Hevia, D. D. (2012). Increasing magnitude counts more: Asymmetrical processing of ordinality in 4-month-old infants. *Cognition, 124*(2), 183–193.

McCrink, K., & Spelke, E. S. (2010). Core multiplication in childhood. *Cognition, 116,* 204–216.

McCrink, K., & Wynn, K. (2004). Large-number addition and subtraction by 9-month-old infants. *Psychological Science, 15,* 776–781.

McCrink, K., & Wynn, K. (2007). Ratio abstraction by 6-month-old infants. *Psychological Science, 18,* 740–746.

Meck, W. H., & Church, R. M. (1983). A mode control model of counting and timing processes. *Journal of Experimental Psychology. Animal Behavior Processes*, *9*, 320–334.

Mitroff, S. R., Wynn, K., Scholl, B. J., Johnson, S. P., & Shuwairi, S. M. (2004). 'Bouncing vs. streaming' as a measure of infants' dynamic object individuation. Poster presented at the annual meeting of the International Society on Infant Studies, Chicago, IL.

Mix, K., Huttenlocher, J., & Levine, S. (2002). Multiple cues for quantification in infancy: Is number one of them? *Psychological Bulletin*, *128*, 278–294.

Mou, Y., & vanMarle, K. (2013a). Two core systems of numerical representation in infants. *Developmental Science*. http://dx.doi.org/10.1016/j.dr.2013.11.001.

Mou, Y., & vanMarle, K. (2013b). Abstract matching: Six-month-old infants' intermodal representations of ratio. Poster presented at the Society for Research in Child Development biennial meeting in Seattle, WA.

Penn, D. C., Holyoak, K. J., & Povinelli, D. J. (2008). Darwin's mistake: Explaining the discontinuity between human and nonhuman minds. *Behavioral and Brain Sciences*, *31*, 109–178.

Piaget, J. (1952). *The child's conception of number*. London, England: Routledge & Kegan Paul.

Picozzi, M., de Hevia, M. D., Girelli, L., & Macchi-Cassia, V. (2010). Seven-month-olds detect ordinal numerical relationships within temporal sequences. *Journal of Experimental Child Psychology*, *107*(3), 359–367.

Plaisier, M. A., Tiest, W. M. B., & Kappers, A. M. L. (2010). Range dependent processing of visual numerosity: Similarities across vision and haptics. *Experimental Brain Research*, *204*, 525–537.

Pylyshyn, Z. W., & Storm, R. W. (1988). Tracking multiple independent targets: Evidence for a parallel tracking mechanism. *Spatial Vision*, *3*, 179–197.

Rips, L. J., Asmuth, J., & Bloomfield, A. (2006). Giving the boot to the bootstrap: How not to learn the natural numbers. *Cognition*, *101*, B51–B60.

Rips, L. J., Asmuth, J., & Bloomfield, A. (2008a). Do children learn the integers by induction? *Cognition*, *106*, 940–951.

Rips, L. J., Bloomfield, A., & Asmuth, A. (2008b). From numerical concepts to concepts of number. *Behavioral and Brain Sciences*, *31*, 623–642.

Ritchie, S. J., & Bates, T. C. (2013). Enduring links from childhood mathematics and reading achievement to adult socioeconomic status. *Psychological Science*, *24*, 1301–1308.

Rousselle, L., Palmers, E., & Noel, M.-P. (2004). Magnitude comparison in preschoolers: What counts? Influence of perceptual variables. *Journal of Experimental Child Psychology*, *87*(1), 57–84.

Scholl, B. J. (2001). Objects and attention: The state of the art. *Cognition*, *80*, 1–46.

Scholl, B. J., & Leslie, A. M. (1999). Explaining the infant's object concept: Beyond the perception/cognition dichotomy. In E. Lepore & Z. Pylyshyn (Eds.), *What is cognitive science?* (pp. 26–73). Oxford: Blackwell.

Schöner, G., & Thelen, E. (2006). Using dynamic field theory to rethink infant habituation. *Developmental Review*, *113*(2), 273–299.

Seok, J. H., & vanMarle, K. (2014). *Rate learning in infancy*, Manuscript in preparation.

Siegel, L. S. (1974). Development of number concepts: Ordering and correspondence operations and the role of length cues. *Developmental Psychology*, *10*, 907–912.

Simon, T. J. (1997). Reconceptualizing the origins of number knowledge: A 'non-numerical' account. *Cognitive Development*, *12*, 349–372.

Simon, T. J., Hespos, S. J., & Rochat, P. (1995). Do infants understand simple arithmetic? A replication of Wynn (1992). *Cognitive Development*, *10*, 253–269.

Sophian, C., & Adams, N. (1987). Infants' understanding of numerical transformations. *British Journal of Developmental Psychology, 5*(3), 257–264.

Spelke, E. S. (2000). Core knowledge. *American Psychologist, 55*, 1233–1243.

Spelke, E. S. (2011). Natural number and natural geometry. In E. Brannon, & S. Dehaene (Eds.), *Attention and performance Vol.24. Space, time and number in the brain: Searching for the foundations of mathematical thought* (pp. 287–317). Oxford, UK: Oxford University Press.

Spelke, E. S., & Tsivkin, S. (2001). Initial knowledge and conceptual change: Space and number. In M. Bowerman & S. Levinson (Eds.), *Language acquisition and conceptual development* (pp. 70–100). Cambridge, UK: Cambridge University Press.

Starkey, P., Spelke, E. S., & Gelman, R. (1983). Detection of intermodal numerical correspondences by human infants. *Science, 222*, 179–181.

Starkey, P., Spelke, E., & Gelman, R. (1990). Numerical abstraction by human infants. *Cognition, 36*, 97–127.

Strauss, M. S., & Curtis, L. E. (1981). Infant perception of numerosity. *Child Development, 52*, 1146–1152.

Suanda, S. H., Tompson, W., & Brannon, E. M. (2008). Changes in the ability to detect ordinal numerical relationships between 9 and 11 months of age. *Infancy, 13*, 308–337.

vanMarle, K. (2013). Different mechanisms underlie infants' representation of small and large numbers in an ordinal choice task. *Journal of Experimental Child Psychology, 114*, 102–110.

vanMarle, K., Chu, F., Li, Y., & Geary, D. C. (2014a). Acuity of the approximate number system and preschoolers' quantitative development. *Developmental Science.* http://dx.doi.org/10.1111/desc.12143.

vanMarle, K., Chu, F., Mou, Y., & Geary, D. C. (2014b). *Attaching meaning to the number words: Contributions of the object tracking and approximate number systems,* Manuscript submitted for publication.

vanMarle, K., Mou, Y., & Seok, J. (2014c). *Infants' ordinal judgments show ratio signature for large numbers,* Manuscript in preparation.

vanMarle, K., & Scholl, B. J. (2003). Attentive tracking of objects versus substances. *Psychological Science, 14*(5), 498–504.

vanMarle, K., Seok, J., & Mou, Y. (2014d). *Overcoming the boundary effect: When can infants successfully compare small and large sets?* Manuscript in preparation.

vanMarle, K., & Wynn, K. (2006). Six-month-old infants use analog magnitudes to represent duration. *Developmental Science, 9*, F41–F49.

vanMarle, K., & Wynn, K. (2009). Infants' auditory enumeration: Evidence for analog magnitudes in the small number range. *Cognition, 111*, 302–316.

vanMarle, K., & Wynn, K. (2011). Tracking and quantifying objects and non-cohesive substances. *Developmental Science, 14*(3), 502–515.

Walsh, V. (2003). A theory of magnitude: Common cortical metrics of time, space and quantity. *Trends in Cognitive Sciences, 7*, 483–488.

Whalen, J., Gallistel, C. R., & Gelman, R. (1999). Nonverbal counting in humans: The psychophysics of number representation. *Psychological Science, 10*, 130–137.

Wood, J., & Spelke, E. (2005). Infants' enumeration of actions: Numerical discrimination and its signature limits. *Developmental Science, 8*, 173–181.

Wynn, K. (1992a). Addition and subtraction by human infants. *Nature, 358*, 749–750.

Wynn, K. (1992b). Children's acquisition of the number words and the counting system. *Cognitive Psychology, 24*, 220–251.

Wynn, K. (1996). Infants' individuation and enumeration of actions. *Psychological Science, 7*, 164–169.

Wynn, K., & Chiang, W. (1998). Limits of infants' knowledge of objects: The case of magical appearance. *Psychological Science, 9*, 448–455.

Xu, F. (2003). Numerosity discrimination in infants: Evidence for two systems of representations. *Cognition, 89*, B15–B25.

Xu, F., & Spelke, E. S. (2000). Large number discrimination in 6-month-old infants. *Cognition, 74*, B1–B11.

Xu, F., Spelke, E. S., & Goddard, S. (2005). Number sense in human infants. *Developmental Science, 8*, 88–101.

Chapter 8

Intuitive Nonsymbolic Arithmetic

Koleen McCrink
Barnard College, Columbia University, New York, NY, USA

INTRODUCTION

The ability to represent quantities and numbers of objects—one of the foundational aspects of our cognitive experience—is supported early in development by a mechanism commonly referred to as the approximate number system (ANS), part of untrained "number sense" (Dehaene, 1997). The ANS is distinct from the rote, symbolic schoolroom knowledge one typically thinks of when hearing the term "math." This system takes input from the perceptual systems to yield nonsymbolic, abstract, and approximate mental representations of amounts in the world. The number sense is not exclusive to humans, with animals as varied as mallard ducks (*Anas platyrhynchos* L.: Harper, 1982), chickens (*Galas galas*: Rugani, Vallortigara, & Regolin, 2013), pigeons (*Columba livia*: Brannon, Wusthoff, Gallistel, & Gibbon, 2001), rats (*Rattus novegicus*: Capaldi & Miller, 1988), elephants (*Loxodanta Africana*: Perdue, Talbot, Stone, & Beran, 2012), nonhuman primates (*Macaca mulatta*: Beran, 2007, 2012; Cantlon & Brannon, 2007), and spiders (*Evarcha culicivora*: Nelson & Jackson, 2012) able to represent and discriminate magnitudes in a variety of experimental paradigms and real-life situations (see also Agrillo et al., this volume; Beran et al., this volume; Geary et al., this volume; Pepperberg, this volume; Vallortigara, this volume).

Using the ANS is intuitive and occurs without training; it is active from the moment of birth in humans, with even newborn infants able to discriminate two magnitudes when the difference between them is large enough (Izard, Sann, Spelke, & Streri, 2009). Representations garnered from this system are fundamentally abstract; they can arise from arrays of objects (Cordes, Gelman, Gallistel, & Whalen., 2001; Starkey & Cooper, 1980, Starkey, Spelke, & Gelman 1990; van Loesbrook & Smitsman, 1990; Xu & Spelke, 2000), sounds (Lipton & Spelke, 2004; Meck & Church, 1983; vanMarle & Wynn, 2006, 2009), and events (Platt & Johnson, 1971; Sharon & Wynn,

Mathematical Cognition and Learning, Vol. 1. http://dx.doi.org/10.1016/B978-0-12-420133-0.00008-9

1998; Wood & Spelke, 2005), and relate to each other independently of modality (Barth et al., 2003, Feigenson, 2011).

There are many kinds of amounts to be represented in the world: an array of five squares, for example, also yields representations of the quantity of colored area, the perimeter of the squares, and the density of the set (dimensions sometimes referred to as spatial or continuous extent, in addition to numerical magnitude—though c.f. Feigenson, 2007, and Walsh, 2003, for arguments that spatial and numerical dimensions are not fundamentally distinct; see also Lourenco, this volume). Magnitude representations generated by the ANS are thought to be plainly numerical and abstract, separate from the perceptual qualities of that representation (e.g., the numerical magnitude "8" is the only meaningful way 8 peanuts and 8 airplanes are identical). A large set of objects with magnitude x of one type and a large set of objects of magnitude x of a different type yield the same magnitude-dependent neuronal activity in the parietal area of the brain that houses the ANS (Cantlon, Brannon, Carter, & Pelphrey, 2006.) The ANS is most commonly discussed in the context of large magnitude representations (>3), since representation of the quantity of smaller magnitudes is thought to engage multiple overlapping systems, some of which are object- or individual-based (the object-tracking system, after Kahneman, Treisman, & Gibbs, 1992; see Feigenson, Dehaene, & Spelke, 2004; vanMarle, this volume) and not inherently numerical.

The inexactitude of the ANS is well documented in the developmental literature. Frequently termed "noisy," the ability to discriminate two mental representations of magnitude is dependent not on the absolute difference but rather on the ratio between amounts (the so-called Weber ratio, after the early experimental psychologist Ernst Weber; Halberda & Odic, this volume). This Weber ratio limit (the point at which two magnitudes become indiscriminatable) decreases over the course of development (Halberda & Feigenson, 2008), is improved by education (Piazza, Pica, Izard, Spelke, & Dehaene, 2013; Pica, Lemer, Izard & Dehaene, 2004), and varies from individual to individual (Halberda, Ly, Wilmer, Naiman, & Germine, 2012; Halberda, Mazzocco, & Feigenson, 2008). Infant researchers have mapped out the rapidly improving ability to discriminate magnitudes across early development, using what are known as "habituation" and "violation-of-expectation" paradigms that capitalize on a fundamental propensity to attend differentially to novel items in the environment. In the habituation paradigm, infants are shown one stimuli type repeatedly and then presented with a novel stimulus at test; if they detect the difference, they look longer at the new stimulus. In the violation-of-expectation paradigm, infants are shown multiple sets of stimuli, some of which adults consider to be correct (expected) and some incorrect (unexpected). If they are attuned to these differences, they will look differently to the two stimuli types.

Izard et al., (2009) familiarized newborn infants to a magnitude of 18 sounds and found that they attended differentially at test only when the new magnitude differed by a ratio of 3:1 (i.e., a magnitude of 6 objects and sounds after having been familiarized to 18 sounds). Six-month-old infants, by contrast, can detect a magnitude difference of 2:1 (vanMarle & Wynn, 2006; Xu & Spelke, 2000). In one such study, 6-month-olds who had habituated to 8 objects dishabituated (i.e., attended longer to) to 16 objects, but not 12 (Xu & Spelke, 2000). By 9 months, infants reliably distinguish 12 from 18, indicative of a 1.5:1 Weber ratio function (Lipton & Spelke, 2003, 2004). Over the course of development from infancy into adulthood, humans typically achieve roughly a 1.15:1 discrimination ability, peaking around 30 years of age, although this is highly variable across individuals (Halberda et al., 2012; Pica et al., 2004). Formal education plays a part in this eventual decrease, as illustrated by studies that show a slight but significant discrimination deficit in indigenous populations with little or no formal schooling (Pica et al., 2004), and increased ANS acuity dependent on years of schooling, but not age (McCrink, Spelke, Dehaene & Pica, 2013; Piazza et al., 2013).

Some theorists consider the generation of number representations— alongside representations of objects, agents, and geometric relations—to be part of the suite of evolved core cognitive capacities that provide a foundation for later thinking (Kinzler & Spelke, 2007). A tenet of this overarching theory is that new and flexible knowledge systems build on the representations generated by the core systems. Given the preceding findings that many species can reliably represent and discriminate nonsymbolic magnitudes, the question then turns to the functional significance of these representations of magnitudes.

In his seminal *The Organization of Learning*, Gallistel (1990) outlined the manner in which many nonhuman animals use different types of quantity information, including numerical magnitude, to solve a variety of foraging quandaries and laboratory tasks (see also Gallistel, 2008; Gallistel & Gibbon, 2000; Gallistel et al., 2007; Leon & Gallistel, 1998). The animals must integrate all the information in order to arrive at an optimal solution, and this integration is inherently abstract and can involve some types of nonsymbolic arithmetic. For example, work by Clayton and Dickinson (see Clayton, Emery, & Dickinson, 2006, for a review; example via Gallistel, 2012) suggests that food-caching scrub jays (*Aphelocoma Californica*) must store multiple types of information about many different caches of food (where they made it, when they made it, who was watching, what was in the cache, and whether or not the cache has been emptied), and integrate and store this information differentially for each cache, in order to survive.

If one views cognition through an evolutionary framework in which humans' cognitive mechanisms underwent selective pressures in the same way their physical characteristics did (Buss, 2005; Cosmides & Tooby, 1987; Hirschfield & Gelman, 1994), it seems likely that humans—even those without much meaningful life experience, such as infants—will have a version of these arithmetic and integrative capacities as well. In this spirit, the frameworks of evolutionary psychology and core cognition suggest the existence of implicit mathematical operations such as ordering, adding, and dividing, using magnitudes generated by an approximate number system. Below, evidence for these untrained and intuitive foundational mathematical operations (ordering, adding, subtracting, multiplying, and dividing; after Gallistel & Gelman, 1992) is discussed. (For the sake of a more complete review, the emphasis will be on studies that specifically look at large-number representations, but classic or relevant studies that look at small-number or spatial extent computations will also be noted where appropriate.)

ORDERING MAGNITUDES

Although ordering is not always in the canon of arithmetic operations one finds in a traditional mathematics textbook, it is a fundamentally relational aspect of comparing two or more magnitudes to each other. Which is more, which less, and where does another amount fall along this same abstract dimension of more-ness and less-ness? An early study on the topic by Cooper (1984) looked at ordering of small numbers of objects in 10–12-, and 14–16-month-olds. The infants were shown a visual stimulus of a pair of sets of colored rectangles, one set presented after the other (such as an array of 1 rectangle followed by an array of 4 rectangles), that exhibited either a greater-than or less-than relation of two small amounts (i.e., magnitudes in the absolute value range of 1 to 4, thus potentially activating either or both the approximate number system and object-tracking system). At test, they saw four different trial types: one familiar pair directly seen during habituation, one novel pair with the habituated ordinal relation, one novel pair with a reversed relation, and one novel pair in which the magnitudes in the pair were equal. Older infants (>14 months of age) looked significantly longer to any new ordinal relation, whereas younger infants (<12 months of age) looked longer only to the equal-pair test item, suggesting that the appreciation of order develops over infancy from knowledge of equal/unequal to a notion of increasing/decreasing. Cooper (1984) used both small values (1, 2, and 3) and large values (4); thus, these results may not generalize to those studies that examine solely large approximate number representations. Indeed, vanMarle (2013) found that 10–12-month-old infants who were tasked with crawling toward greater or lesser amounts of food succeeded in tasks that utilized small number comparison (1 vs. 2) or large number comparison (4 vs. 8), but failed when the two number types were mixed (2 vs. 4).

This ordering ability has also been examined in the realm of the ANS over large numbers of objects, both with infants (Brannon, 2002; Suanda, Tompson, & Brannon, 2008) and with nonhuman animals (Brannon & Terrace, 2000; Cantlon & Brannon, 2006; Pepperberg, 2006; Rugani, Regolin, & Vallortigara, 2007). Brannon (2002) habituated 9- and 11-month-old infants to ascending or descending displays of dot arrays, some of which fell within the small-number range (up to 3 objects) and some within the large-number range (>3 objects). The displays contained differing magnitudes (e.g., 1–2–4, or 2–4–8), and each sequence contained the same type of ordering (ascending or descending). They were then presented with test items of new values, arranged in either familiar or unfamiliar order relative to habituation. Eleven-month-olds—but not 9-month-olds—looked longer at the new order at test, regardless of whether non-numerical cues (such as area and contour length) were controlled.

The ordering abilities of the 11-month-old infants in this study are in contrast to the failure documented by Cooper (1984). This is likely due to several differences between the study procedures: (1) in the Brannon study, the infants saw more magnitudes with values in the large number realm, and this activated their ANS; (2) the values were in triplicate across the exemplar (i.e., 1–2–4) instead of just a pair (i.e., 1–2); and (3) in the Brannon (2002) study, habituation was more varied, and that may have led to a stronger scaffold on which to form a concept of number order (Needham, Dueker, & Lockhead, 2005). Nine-month-old infants in the Brannon study were actually able to successfully detect the ordinality of a set of squares that either increased or decreased in area, which suggests the developmental change between 9 and 11 months of age is specific to understanding of number magnitudes per se. Nine-month-olds are sensitive to ordinality in some domains but are unable to appreciate ordinality in number; this appears to develop between 9 and 11 months of age. Even this interpretation may be generous, however, since 9-month-old infants in a similar study failed to discriminate ordinal properties of area and in fact needed multiple redundant cues across quantity (e.g., simultaneously increasing area, perimeter, and number) to perform the ordering operation (Suanda et al., 2008).

Consistent with this finding, Picozzi, de Hevia, Girelli, and Cassia (2010) found that even 7-month-old infants can order numerical sequences (i.e., collections of objects) if each of the exemplars had its own unique perceptual information (e.g., specific color) and if the relation *between* each of the habituation sequences was consistent with the relation showed *within* the sequence (e.g., if the ascending condition was presented 6–12–24, followed by 9–18–36, followed by 12–24–48.) Additionally, there is evidence that infants can extract an ordering concept from a set of stimuli and apply it to both numerical and non-numerical domains. de Hevia and Spelke (2010) habituated 8-month-old infants to arrays of objects (dots, triangles, or rectangles) that either increased or decreased in number, controlling for other non-numerical cues. At test, the infants saw sequences of horizontal lines that either increased or decreased in length. Infants looked longer at test trials that presented a new order relative to habituation; this suggests a spontaneous transfer of ordinality concepts across different quantity dimensions. The competence shown by these 8-month-olds in de Hevia and Spelke (2010)—but not by the 9-month-olds tested by Suanda et al., (2008)—suggests that amodal transfer of ordinality across quantity dimensions (number to length) comes in earlier than the ability to represent the ordinality of numerical stimuli within just one specific quantity type (number to number).

Overall, these results illustrate an intuitive and nonsymbolic understanding of order in both numerical and non-numerical domains of quantity in the first year of life. However, a stringent evolutionary perspective may predict the presence of ordering from as early as can be tested in human infancy; the literature instead points to a gradual development of this ability that is varied with respect to the number of objects to be represented and the way in which ordinality is highlighted in the stimuli.

ADDING AND SUBTRACTING NONSYMBOLIC MAGNITUDES

The ability to spontaneously add and subtract nonsymbolic magnitudes has been found in both humans (Barth, Kanwisher & Spelke, 2003; Barth, La Mont, Lipton, Dehaene, Kanwisher, & Spelke, 2006; Barth, La Mont, Lipton & Spelke, 2005; Cordes, Gallistel, Gelman, & Latham, 2007; Knops, Zitzmann, & McCrink, 2013; McCrink, Dehaene, & Dehaene-Lambertz, 2007; Wynn, 1992) and some nonhuman animals (Beran, Evans, Leighty, Harris, & Rice, 2008; Brannon et al., 2001; Cantlon & Brannon, 2007). In a seminal study, Wynn (1992) showed 5-month-old infants a nonsymbolic addition or subtraction scenario, in which small dolls were placed onto (in the 1+1 addition condition) or removed from (in the 2–1 subtraction condition) a stage behind an occluder. When the occluder was dropped, either a correct or incorrect number of dolls was revealed to the infant. In both the addition and subtraction scenarios, infants looked longer to incorrect outcomes compared to correct outcomes, a finding that has been replicated several times in a variety of paradigms (Koechlin, Dehaene, & Mehler, 1997; Simon, Hespos, & Rochat, 1995; Uller, Carey, Huntley-Fenner, & Klatt, 1999).

One challenge to the conclusions of these studies is that the infants are not necessarily calculating the number of objects expected in the outcomes, but rather perceptual variables that correlate with the number of objects, such as the total expected amount of area or perimeter after the addition or subtraction of a doll (Clearfield & Mix, 1999, 2001; Feigenson, Carey, & Spelke, 2002). In one study that tested this theory, infants who saw outcomes that controlled for total amount of area (e.g., an incorrect outcome of 1 large doll in a 1+1 small doll scenario) failed to look longer at the numerically incorrect outcome (Feigenson et al., 2002). Similarly, other studies demonstrate that infants could be tracking the placement and presence of individual dolls instead of attending to numerosity (Simon et al., 1995).

To address these critiques, McCrink and Wynn (2004) presented 9-month-old infants with computerized nonsymbolic addition and subtraction scenarios using large sets of objects, controlled for area and contour length. Because even adults cannot track individual entities of more than four objects in a scene (see Scholl, 2001, for a review), and because cues to continuous extent were unavailable or uninformative at outcome, success in this task required infants' engagement of their ANS to correctly detect addition and subtraction. Infants were assigned to either an addition scenario, in which they saw 5 objects of constantly changing size become occluded and 5 more added behind the occluder, or a subtraction scenario, in which they saw 10 objects become occluded and 5 taken from behind the occluder. Both groups saw outcomes of 5 objects and 10 objects when the occluder dropped at test. The infants in the addition scenario looked longer to the outcome of 5 objects, whereas infants in the subtraction scenario looked longer to the outcome of 10 objects—a "flip-flop" in looking time that suggested that the previous addition or subtraction of objects influenced the expected quantity after the occluder was dropped and that the infants were expecting the correct arithmetic process (i.e., they looked longer at incorrect outcomes).

Follow-up studies (Barth et al., 2005; Knops et al., 2013; McCrink & Wynn, 2009) have replicated the finding that children with no formal experience in symbolic addition and subtraction can perform these operations on groups of items. However, it is unclear whether the infants in McCrink and Wynn (2004) were genuinely computing estimated outcomes to these nonsymbolic problems. Instead, the ANS may operate using general arithmetic logic resulting in a simple heuristic of "if adding, accept more than the initial amount" and "if subtracting, accept less than the initial amount."

Recent work in my laboratory suggests this is the case. Using a similar paradigm to McCrink and Wynn (2004, 2009), I presented 9-month-old infants with nonsymbolic addition and subtraction problems of 6+4=10, 20, and 40 objects, and 30–10=20, 10, and 5 objects. I predicted that if infants are representing the approximate solution to these problems, they will look the least to the correct outcomes (as they are the least unexpected), slightly more to the outcomes closest to the correct outcome (since they are incorrect, but the Weber ratio of the incorrect:correct is relatively low at 2:1), and most to the outcomes farthest from the correct outcome (which have a relatively large Weber ratio of 4:1). Instead, infants in both the addition and subtraction scenarios failed to discriminate their response to any of the outcomes, and looked similarly to the correct (e.g., 6+4=10) and both incorrect outcomes (e.g., 6+4=20, 40). This indicates that the infants were not representing an approximate outcome of 10 (for addition) or 20 (for subtraction); rather, because all outcomes obeyed the logical principles heuristic, they were encoded and responded to similarly as "acceptable." To be clear, this does not minimize the mathematical prowess of the infants in these studies; it is still inherently numerical and reflects an inherent understanding of the nature of the operation, but it is fundamentally a different process than outcome calculation (even if the calculations were approximate rather than exact).

The reliance on this heuristic changes by the time children are entering into more formal educational contexts; by that point they do appear to generate an actual outcome to problems that require the use of ANS representations (Barth et al., 2005, 2006; Gilmore, McCarthy, & Spelke, 2007; Knops et al., 2013). For example, Knops et al., (2013) showed 7-year-old children addition or subtraction scenarios of two large-number dot arrays. The children were then presented with multiple different outcomes from which to choose the correct one; other options included arrays that were much smaller, slightly smaller, slightly larger, or much larger than the correct outcome. Knops et al., (2013) observed that children's choices of the correct answer varied with the Weber ratio. As predicted by the ANS properties, children more frequently selected outcomes that were a shorter mental "distance" from the correct outcome as compared with outcomes that were a greater mental distance from the correct response. Thus, ANS processing likely accounted for their responses. For both the addition and subtraction trials, the children appeared to have an estimated outcome in mind and could discriminate the correct outcome from an incorrect outcome that fell within the range that the logical principles heuristic allowed.

Unlike the infants, they would not be likely to view an array of 40 as an acceptable outcome to the problem of 6+4, even though "more than the initial amount" would heuristically be acceptable. Further, children can perform this addition and subtraction over abstract, cross-modal ANS representations (Barth et al., 2005). Barth et al. found that children were able to add two nonsymbolic magnitudes (arrays of dots) and compare this outcome to a nonsymbolic magnitude in a different modality (e.g., a series of tones). Again, the signature Weber ratio modulation of performance was found; children performed more poorly when the comparison array of tones was close to the correct outcome (a Weber ratio of 1.25:1) than when it was distant (a Weber ratio of 1.75:1). The closer an incorrect array of tones was in mental "distance" to the correct response, the more likely it was to be accepted as correct. When these results are taken together with the infant results, one observes an appreciation for the logic of addition and subtraction present early in infancy that develops over early childhood into an increasingly precise ability to estimate outcomes to a variety of nonsymbolic addition and subtraction problems.

MULTIPLICATION AND DIVISION OF NONSYMBOLIC QUANTITY

Ratio and Proportion Tasks

In adults, symbolic multiplication and division problems have traditionally evoked neural responses that are distinct from those found when the ANS is activated in comparison to addition/subtraction tasks (Cohen & Dehaene,

2000; Dehaene & Cohen, 1997; Lampl, Eshel, Gilad, & Sarova-Pinhas, 1994; Van Harskamp & Cipolotti, 2001), leading to some question as to whether these operations are supported by the ANS in the same way that ordering, addition, and subtraction are supported. However, when more-sensitive methods that do not emphasize previously memorized rote problems are used with adults, ANS activation is found. For example, Jacob and Neider (2009a, 2009b), using fMRI imaging and a neural adaptation procedure in which activation of a particular brain area is measured while stimuli are repeatedly shown to the subject, presented subjects with nonsymbolic and symbolic proportions of one type (2/4, 3/6, 4/8, which all reduce to 1/2) while occasionally showing oddball proportions (e.g., 3/5).

Proportional understanding is frequently used as a proxy for nonsymbolic division because reduction of a proportion is inherently reliant on a process that relates two amounts by means of a division calculation. The intraparietal sulcus (IPS, a well-known brain area active during ANS representation [Dehaene, 1997; Naccache & Dehaene, 2001; Thioux, Pesenti, Costes, De Volder, & Seron, 2008]) rebounded in activation to these oddball proportions, and the degree of rebound activation was based on the degree of difference between these proportions and those used in the adaptation phase (e.g., 3/4 showed less rebound activation than 4/5 for the 1/2 standard), consistent with a Weber ratio modulation found in most ANS tasks (Jacob & Neider, 2009a, 2009b). Additionally, this effect was found irrespective of the format of the ratios; it was found for spatial proportions such as line length as well as for symbolic numerical proportions (Jacob & Nieder, 2009a), and even when the standard stimuli were Arabic numerals but deviants were presented as words (Jacob & Neider, 2009b).

Currently, few studies have assessed abstract division and multiplication so elegantly in untrained or very young populations. In fact, numerous studies suggest that division, at least, is a difficult concept for young children in formal schooling. The studies that have been done on division have largely focused on formal, symbolic division problems, and the results consistently indicate that children have difficulty with these types of problems as well as with ratios and fraction problems in general (Carpenter, Corbitt, Kepner, Lindquist, & Reys, 1981; Dixon & Moore, 1996; Fischbein, 1990; Kieren, 1988; Mack, 1990; Moore, Dixon, & Haines, 1991; Nunes, Schliemann, & Carraher, 1993; Piaget & Inhelder, 1956; Post, 1981; Reyna & Brainerd, 1994; Singer, Kohn, & Resnick, 1997). The set–subset relation, a key component to fractional reasoning, can be particularly difficult to master; in one classic study, children up to the age of 7 struggled with knowing that an array composed of 6 roses and 2 daisies has more flowers than roses (Inhelder & Piaget, 1964).

Moreover, there is an inverse relation between number of divided units and the size of those units, and this relation can be challenging for young children to grasp (Frydman & Bryant, 1988; Sophian, Garyantes, & Chang, 1997;

Spinillo & Bryant, 1991). In one such study, 5–7-year-old children were presented with a number of "pizza bits" that must be distributed among multiple sharers, with the goal of obtaining the largest share of bits for a "pizza monster" (Sophian et al., 1997). The children had difficulty realizing that the more sharers in the equation, the smaller the portions, and often predicted that there would be a direct (not inverse) relation between the number of partitions and the number of items in each partition. Finally, preschoolers who see an amount divided into half and are provided with the number of objects in one of the halves are unable to infer that the other half is equivalent in number (Frydman & Bryant, 1988). This suggests a logical failure to appreciate the relation between parts in a division setting.

In contrast, young children reason much more competently about proportions and ratios in tasks that emphasize non-numerical quantities such as overall area or space (Duffy, Huttenlocher, & Levine, 2005; Goswami, 1989; Jeong, Levine, & Huttenlocher, 2007; Mix, Levine, & Huttenlocher 1999; Spinillo & Bryant, 1991; Sophian, 2000; though c.f. Piaget & Inhelder, 1956). In one such study, Goswami (1989) found that by 6 years of age children who were given multiple examples of a particular proportion of shaded:unshaded area in a shape were able to pick a test item that showed the same proportion. In a similar paradigm, Sophian (2000) presented 4- and 5-year-old children with a sample stimulus that exhibited a particular proportional relation (e.g., an animal with a relatively small body and large head) and asked them to choose the animal that was "just like" it from a pair of test items, one that exemplified the same proportion and one that did not. At all ages tested, children successfully performed this analogical task across a range of stimulus types and configurations.

Many of these successful proportional analogies have come about only when the stimuli exhibited non-numerical quantities of a spatial nature. Indeed, children seem to be actively impaired in proportion-analogy reasoning in which they must utilize exact number instead of continuous quantity (Boyer, Levine, & Huttenlocher, 2008; Jeong, Levine, & Huttenlocher, 2007; Spinillo & Bryant, 1991; Singer-Freeman & Goswami, 2001). For example, Boyer et al., (2008) provided first- and third-graders with a standard that exemplified a particular proportion (e.g., a beaker one-third full of juice) and instructed them to choose from a pair of test stimuli the picture that was the "right mix" of juice. Children were able to perform the task readily when the choice was over a continuous amount of liquid, but when the standard was notched into countable units of juice and water, even older children failed to intuit and apply the correct relation between the standard and test item. Dissociations such as these have led some theorists to posit that the strategies children use to reason about number, such as counting, interfere with intuitive notions of proportion (Gelman, Cohen, & Hartnett, 1989; Mix, Levine, & Huttenlocher, 1999).

In contrast, my colleagues and I have found evidence for intuitive proportional understanding when ANS representations (that are not symbolic, precise, or readily counted) are invoked in the experimental task. In a set of experiments with Karen Wynn, I tested the ability of infants to extract ratios from a set of stimuli and apply this ratio to new stimuli (McCrink & Wynn, 2007). Six-month-old infants were habituated to slides of arrays of yellow and blue objects. Half of the infants saw slides that conveyed exemplars that reduced to 1 yellow object: 4 blue objects (in the 1:4 condition), and half saw slides with exemplars that reduced to 1 yellow object: 2 blue objects (in the 1:2 condition); the absolute number of each object type and overall number of objects differed between the slides (e.g., for the 1:2 condition—4 yellow: 8 blue, then 10 yellow: 20 blue, etc.). At test, infants saw either new exemplars of the habituated ratio or an exemplar of an entirely new ratio. Perceptual factors such as area and contour length were strictly controlled for in both habituation and test movies, meaning that the calculation must be performed over the variable of numerical quantity and not non-numerical quantity.

Infants who were habituated to a 1:2 ratio looked significantly longer at test when presented with a 1:4 ratio versus a 1:2 ratio, whereas infants who were habituated to a 1:4 ratio showed the opposite pattern. A second experiment examined whether this ratio-extraction ability exhibited poorer performance when the ratio values to be discriminated were arithmetically closer, indicative of the modulation by Weber ratio seen in many ANS tasks. Infants were presented with exemplars of 1:2 or 1:3 and then tested with new versions of these ratios at test. In this experiment, the infants did not distinguish between the habituated and new ratio, indicating that the system underlying this process is similarly "noisy" to absolute number discrimination at this age and cannot discriminate ratios that differ only by a factor of 1.5 (but can do so for ratios that differ by a factor of 2.0). This ratio extraction ability also appears to be cross-modal.

Mou and vanMarle (2013) showed 6- and 10-month-old infants slides of stimuli exemplifying a 2:1 and 4:1 ratio side-by-side while playing a sequence of tones with either 2:1 or 4:1 ratio information in the background. The infants looked significantly longer to the slide that matched the auditory ratio information they were hearing. Further, this process was again modulated by the Weber ratio, as 6-month-old infants (but not 10-month-olds) were unable to perform a similar task when the ratios to be discriminated were 2:1 vs. 3:1. The ability of infants to perform this cross-modal task, with decreasing performance as the Weber ratio decreases, is strong evidence that the ability to encode proportions in infancy is supported by the ANS.

Another type of task that is suggestive of sensitivity to proportional relations in infancy is probability reasoning tasks (Denison & Xu, 2010; Teglas, Girotto, Gonzalez, & Bonatti, 2007; Teglas et al., 2011; Xu & Denison, 2009, Xu & Garcia, 2008). In one such study, Teglas et al., (2007; also see

Teglas et al., 2011) showed 12-month-old infants either a yellow ball coming out of a container that contains 1 blue ball and 3 yellow balls (a probable event) or a yellow ball coming out of a container that contains 3 blue balls and 1 yellow ball (an improbable event). The infants looked significantly longer to improbable events than probable ones, indicating sensitivity to the proportional base rate of occurrence of each single event.

Xu, Denison, and colleagues (Denison & Xu, 2010; Xu & Denison, 2009, Xu & Garcia, 2008) have established that infants are sensitive to the likelihood of an event occurring (e.g., finding a lollipop) based on how common the event is in the context of a wider group of possibilities. Xu and Garcia (2008), for instance, showed 11-month-old infants the contents of a box of colored ping-pong balls and then tested their expectations of what a sample from this box would look like; infants reliably looked longer at an unrepresentative sample. They also performed the converse reasoning and looked longer to a box's unexpected contents given a previously drawn sample (see also Denison, Reed, & Xu, 2013, for a similar finding with 6-month-olds). Denison and Xu (2010) also found evidence for probabilistic reasoning in infancy; in this study, 12- to 14-month-old infants were shown a large-number population—thought to evoke the ANS—consisting of mostly either desirable or undesirable objects (e.g., a preferred lollipop), and reliably crawled to the sample drawn from the population that had a higher ratio of desirable objects. These results are analogous to nonhuman animals' sensitivity of similar types of information in the context of foraging.

Serial Multiplication and Division in Infants and Children

One critical question that arises in the context of the study of the ANS is the practicality of this line of research. Does the ANS in any substantive way matter to how children learn math in a formal, educational setting? Some recent research suggests this is the case. Several interventions have been aimed at using ANS and "number sense" training to promote numerical understanding and skills in children with dyscalculia and other mathematical deficits, with varying degrees of success (Star & Brannon, this volume; Ramani & Seigler, 2008; Wilson, Revkin, Cohen, Cohen, & Dehaene, 2006). There is a consistent relation between an individual's ANS acuity and his or her performance in school math, even as early as the preschool years (Halberda, Mazzocco, & Feigenson, 2008; Libertus, Feigenson & Halberda, 2011; Libertus, Odic, & Halberda, 2012; Mazzocco, Feigenson, & Halberda, 2011). Further, children appear to bring their ANS representations to the table when dealing with symbolic arithmetic problems they cannot solve precisely. Gilmore, McCarthy, and Spelke (2007) presented kindergarten-aged children with exact symbolic addition and subtraction problems, and found that although they could not answer the problems exactly, the children recruited their ANS to devise approximate solutions to the problems.

Given these findings, it would be useful to have a serial, explicit paradigm for nonsymbolic multiplication that mirrors the step-by-step format of beginner multiplication and division problems in a formal setting. In this way, the relation between a multiplicative or divisive transformation in the realm of the ANS can be made more transparent to the task of whole-number precise multiplication or division when first learning these terms. In one such study, McCrink and Spelke (2010; using a paradigm adapted from Barth, Baron, Spelke, & Carey, 2009) presented 5- and 6-year-old children with a single trial in which they saw a single computerized object transform via a "magic wand" into either 2 objects (Times 2 condition) or 4 objects (Times 4 condition), or 2 objects transform into 5 objects (Times 2.5 condition). After this introduction trial, they were presented with large-number sets of objects that were occluded and transformed. An additional array came down onscreen next to the occluded array, and children needed to indicate which array (the transformed or the comparison) was more numerous. (For a schematic, see Figure 8-1.)

In all multiplication conditions, the children reliably performed above chance, indicating they had correctly inferred the hidden product of this nonsymbolic calculation. Their performance decreased as the multiplicative factor increased and was lower when the ratio of the comparison array:correct outcome decreased—both signs of the ANS at work. The children performed above chance on the very first trial at testing, suggesting that this calculation was a fairly straightforward one for them to make. Further, a companion set of experiments I conducted in collaboration with Spelke, using a similar setup,

Introduction to Multiplicative Factor:"Look, here's a magic multiplying wand. Let's watch what it can do. Wow! It made more rectangles."

Training Trials (12 total):"We have *this* many blue rectangles. Now our magic wand is going to multiply them! Which side has more? Good job/ Oops! That was the right/ wrong answer."

Testing Trials (16 total):"We have *this* many blue rectangles. Now our magic wand is going to multiply them! Which side has more?"

FIGURE 8-1 A schematic of the Times 4 condition from McCrink and Spelke (2010).

found that children of the same age can nonsymbolically halve and quarter the magnitude of a large set, with overall very similar performance to doubling and quadrupling. This led to the idea that the ANS supports an operation of *core scaling*, in which the magnitude of a set of objects can be scaled up or down flexibly using the same underlying scaling process. Because I have not yet tested whether children's competence on this nonsymbolic task is related to their competence on more formal arithmetic problems, we cannot know whether early nonsymbolic scaling supports later formal division and multiplication. However, given the findings of Gilmore et al., (2007), one testable prediction would be that that nonsymbolic scaling can be used to approximate outcomes to whole-number symbolic multiplication and division problems.

Interestingly, even though the psychophysical profiles of performance are similar, the explicit concepts children have about these operations are incomplete at best. For example, children seem to appreciate that addition and subtraction are the logical inverse of each other, even in a nonsymbolic context. Gilmore and Spelke (2008) found that 5-year-old children with no formal training in inversion reasoning were more accurate on addition and subtraction problems that contained inversion reasoning (e.g., 4+8–8) compared to those problems that did not (4+10–6), when presented in either a symbolic or nonsymbolic (ANS supported) format. Work by Barth and myself (McCrink & Barth, 2013) explored whether this was the case with nonsymbolic multiplication and division as well. We presented 7-year-old children who had no formal knowledge of inversion reasoning with a serial chain of multiplication or division events, some of which exemplified inversion principles (10∗4/4) and some of which did not (10∗2/4). Children performed similarly, irrespective of whether or not the problem type lent itself to inversion reasoning. It is notable, however, that they performed similarly *well*—largely comparable to single-step nonsymbolic multiplication or division—which is the first evidence for nonsymbolic multistep multiplication and division problem solving in the literature to date.

The ability to solve a nonsymbolic division problem was further tested in children of the Mundurucu tribe (McCrink et al., 2013), a group of indigenous people native to the Amazonian area of Brazil whose language possesses no precise number words for magnitudes greater than 5. This is an important test of the theory of the intuitive and untrained version of core scaling, because although the U.S.-based children had no formal education in multiplication and division, they are deeply enculturated to symbolic number concepts in informal contexts. In this experiment, Mundurucu-speaking children (aged 7 to 11 years) were able to successfully halve a large-number magnitude, and do so on the very first test trial, like the U.S. sample. A subset of the children spoke Portuguese and had precise number words and up to 3 years of formal schooling; this formal education increased their performance relative to children who did not have formal education (although even uneducated children in the sample performed above chance on the task).

Overall, performance of the Mundurucu children with some schooling was similar to the U.S. sample, although they were less likely to use non-numerical strategies to solve the problems (i.e., computing how large the comparison array was relative to the range of previously presented arrays, and using this information to gauge the likelihood that it was larger than the outcome); this likely is an outcome of their lack of testing experience or experimental savvy with following directions closely. A larger number of years in school—but not their age—was correlated with better performance on the task, lending support for the idea that precise number words, in tandem with practice on formal mathematical concepts, sharpen the acuity of the ANS in middle childhood (see also Piazza et al., 2013). Given that this relationship is correlational only, it is difficult to know whether it is truly causal; there may be some underlying variable that goes along with schooling—such as executive function development—that is the primary mediator between schooling and ANS acuity change.

In my laboratory, we have extended this work to look at whether this nonsymbolic core scaling ability is found as early as infancy. In a study currently being conducted, we provided 24 5–9-month-old infants with a series of computerized displays to probe their sensitivity to multiplicative or divisive relations. The infants were assigned to either a doubling condition or a halving condition. Infants assigned to the doubling condition see an introduction trial in which a happy face bounces down from offscreen and sits in the middle while a computerized wand waves over it, transforming it with a popping sound into two happy faces. They are then trained further in this doubling transformation via familiarization to five doubling scenarios: 4 doubled to 8, 8 to 16, 10 to 20, 12 to 24, and 16 to 32. They then see two test pairs, each of which is composed of one incorrect transformation and one correct transformation (6*2=12 or 24, 14*2=14 or 28). Infants in the halving condition see the complementary transformation, in which two happy faces are transformed into a single happy face, followed by the halving of sets of 64, 56, 44, 32, and 16. At test, they see 48/2=12 or 24, and 28/2=28 or 14.

The exact test values are worth noting here, as they get at two different questions. First, the 14*2=14 or 28 and 28/2=28 or 14 trials test whether infants are attentive to the principle of the logic of the transformation—that multiplication results in a bigger amount, and division a smaller amount. Second, the 6*2=12 or 24 and 48/2=24 or 12 trials require that the infants compute the outcome approximately, not just detect the general direction or logic of a transformation. The revealed outcomes for infants in both conditions (12, 14, 24, and 28) are the same, which allows us to conclude that different-looking time patterns are due to sensitivity of the operation the infant viewed beforehand. Infants look reliably longer to the incorrect outcomes than the correct outcomes, and do so in both types of test trials, implying not only an appreciation that a transformation occurred but also having had an idea of approximately how many objects this transformation should yield.

In a companion experiment, infants of the same age range were provided with the same magnitudes and transformation processes (doubling and halving), but continuous extent variables (e.g., area, contour length, density, and item size) were controlled across training and testing trials. The infants in this experiment looked similarly to all outcomes at test, indicating they (a) did not compute an approximate answer to the visual equation being presented and (b) failed to detect the presence of the scaling transformation in general. This dissociation of performance between the two experiments mirrors that found in the ordering literature (Picozzi et al., 2010; Suanda et al., 2008); infants here are performing a scaling computation only when multiple spatial cues to quantity (and not just numerical magnitude represented abstractly by the ANS) are available.

CONCLUSIONS

The evidence reviewed here provides clear support for the theory that, from very early on in development, infants are sensitive to the effects of arithmetic operations that involve nonsymbolic number representations. Infants, children, adults, and nonhuman animals can productively combine and relate two magnitudes in an inexact manner. In humans, these skeletal abilities undergo substantial development across infancy and childhood, and on into adulthood, as children become enculturated and their abilities to represent both the magnitudes and their operational outcomes sharpen.

The exact relation between education and magnitude representation change remains unclear. Simply because the two are related does not mean the causal arrow points directly; work by Halberda et al., (2012) has established through large-scale Internet testing that ANS acuity sharpens even after formal education ends, and this acuity is more predictive of informal—and not formal—math abilities (Libertus, Feigenson, & Halberda, 2013). These findings suggest a potentially important role for functional math abilities as either the driver of ANS acuity change, or one that works in tandem with formal math abilities to sharpen the ANS. The focus on function, as opposed to formal education, fits with a theory of evolutionary cognition that emphasizes practical and facile everyday use to enact change in the precision of an underlying core ability such as number representation. It also makes the prediction that there will be individuals who, despite not having formal education, will possess highly accurate ANS representations; one potential test case could be fishermen found in Brazil, whose livelihood depends on frequent use of intuitive and informal arithmetic operations to arrive at decisions regarding the preparation of that day's catch (Nunes, Schliemann, & Carraher, 1993). If such facile use of informal arithmetic prompts ANS acuity change, this population may outperform those who have advanced education on carefully constructed ANS representation tasks.

The finding that even young infants can intuit scaling relationships, and that this process is found throughout childhood (Barth et al., 2009;

McCrink & Spelke, 2010) as well as with populations who have no exact large number words or enculturated emphasis on formal arithmetic (McCrink et al., 2013), opens the door for using this robust nonsymbolic core scaling as a scaffold into symbolic multiplication and division. One telling aspect to the research reported here on infant core scaling is the experimental failure obtained when the serial a*b=c paradigm was implemented with continuous extent controlled; as observed in the ordering literature (Picozzi et al., 2010; Suanda et al., 2008), multiple cues to quantity (area, perimeter, and density in tandem with overall numerical magnitude) were necessary for this young group to extract the scaling relationship. This may be relevant when viewing how one would bring this paradigm into a more formal setting; by using sets of pictures that emphasize the overall change in all types of quantity (either bigger by a fixed transformation factor, or smaller), and placing symbols alongside these pictures, the process of multiplying and dividing whole numbers can be related to nonsymbolic quantity representations instead of instantiation via opaque rote memorization strategies.

The finding that young infants' sensitivity to arithmetic operations encompasses multiple types of quantity information, in addition to numerical magnitude information generated by the ANS per se, speaks to the evolutionary theory of numerical cognition. Although number and extent are sometimes dissociated in the naturalistic world (e.g., one apple is more desirable than 20 apple seeds), they mostly co-occur; maximizing a process like foraging requires attentiveness to multiple quantity dimensions and an ability to integrate information across these dimensions. Although it is doubtlessly worthy to understand the developmental profile of different types of quantity processing, this evolutionary emphasis on functionality of cognition as a driver of core capacities is a valuable reminder that scientific exploration of humans' "cognitive niche" (Tooby & Devore, 1987) is bound to yield a multifaceted picture of any important psychological mechanism.

REFERENCES

Barth, H., Baron, A., Spelke, L., & Carey, S. (2009). Children's multiplicative transformations of discrete and continuous quantities. *Journal of Experimental Child Psychology, 103*, 441–454.

Barth, H., Kanwisher, N., & Spelke, E. (2003). The construction of large number representations in adults. *Cognition, 86*, 201–221.

Barth, H., La Mont, K., Lipton, J., Dehaene, S., Kanwisher, N., & Spelke, E. (2006). Nonsymbolic arithmetic in adults and young children. *Cognition, 98*, 199–222.

Barth, H., La Mont, K., Lipton, J., & Spelke, E. (2005). Abstract number and arithmetic in preschool children. *Proceedings of the National Academy of Sciences, 102*(39), 14117–14121.

Beran, M. J. (2007). Rhesus monkeys (*Macaca mulatta*) enumerate large and small sequentially presented sets of items using analog numerical representations. *Journal of Experimental Psychology. Animal Behavior Processes, 33*, 42–54.

Beran, M. J. (2012). Quantity judgments of auditory and visual stimuli by chimpanzees (*Pan troglodytes*). *Journal of Experimental Psychology. Animal Behavior Processes, 38*, 23–29.

Beran, M. J., Evans, T. A., Leighty, K. A., Harris, E. H., & Rice, D. (2008). Summation and quantity judgments of sequentially presented sets by capuchin monkeys (*Cebus apella*). *American Journal of Primatology, 70*, 191–194.

Boyer, T., Levine, S. C., & Huttenlocher, J. (2008). Development of proportional reasoning: Where young children go wrong. *Developmental Psychology, 44*(5), 1478–1490.

Brannon, E. (2002). The development of ordinal numerical knowledge in infancy. *Cognition, 83*, 223–240.

Brannon, E., & Terrace, H. S. (2000). Representation of the numerosities 1–9 by rhesus macaques (*Macaca mulatta*). *Journal of Experimental Psychology. Animal Behavior Processes, 26*, 31–49.

Brannon, E., Wusthoff, C., Gallistel, C. R., & Gibbon, J. (2001). Subtraction in the pigeon: Evidence for a linear subjective number scale. *Psychological Science, 12*(3), 238–243.

Buss, Da vi d M. (2005). *The handbook of evolutionary psychology*. Hoboken: Wiley.

Cantlon, J. F., & Brannon, E. (2006). Shared system for ordering small and large numbers in monkeys and humans. *Psychological Science, 17*(5), 401–406.

Cantlon, J. F., & Brannon, E. M. (2007). Basic math in monkeys and college students. *PLoS Biology, 5*(12), e328.

Cantlon, J. F., Brannon, E., Carter, E., & Pelphrey, K. (2006). Functional imaging of numerical processing in adults and 4-year-old children. *PLoS Biology, 4*(5), e125.

Capaldi, E., & Miller, D. (1988). Counting in rats: Its functional significance and the independent cognitive processes that constitute it. *Journal of Experimental Psychology. Animal Behavior Processes, 14*(1), 3–17.

Carpenter, T., Corbitt, M. K., Kepner, H. S., Jr., Lindquist, M. M., & Reys, R. E. (1981). Decimals: Results and implications from national assessment. *Arithmetic Teacher, 28*(8), 34–37.

Clayton, N., Emery, N., & Dickinson, A. (2006). The prospective cognition of food caching and recovery by Western scrub-jays. *Comparative Cognition and Behavior Reviews, 1*, 1–11.

Clearfield, M. W., & Mix, K. S. (1999). Number versus contour length in infants' discrimination of small visual sets. *Psychological Science, 10*, 408–411.

Clearfield, M. W., & Mix, K. S. (2001). Amount versus number: Infants' use of area and contour length to discriminate small sets. *Journal of Cognition and Development, 2*, 243–260.

Cohen, L., & Dehaene, S. (2000). Calculating without reading: Unsuspected residual abilities in pure alexia. *Cognitive Neuropsychology, 17*, 563–583.

Cooper, R. G. (1984). Early number development: Discovering number space with addition and subtraction. In C. Sophian (Ed.), *The origins of cognitive skill* (pp. 157–192). Hillsdale, NJ: Erlbaum.

Cordes, S., Gallistel, C. R., Gelman, R., & Latham, P. (2007). Nonverbal arithmetic in humans: Light from noise. *Perception and Psychophysics, 69*(7), 1185–1203.

Cordes, S., Gelman, R., Gallistel, C. R., & Whalen, J. (2001). Variability signatures distinguish verbal from nonverbal counting for both large and small numbers. *Psychonomic Bulletin and Review, 8*(4), 698–707.

Cosmides, L., & Tooby, J. (1987). From evolution to behavior: Evolutionary psychology as the missing link. In J. Dupre (Ed.), *The latest on the best: Essays on evolution and optimality* (pp. 277–306). Cambridge, MA: MIT Press.

de Hevia, M. D., & Spelke, E. S. (2010). Number-space mapping in human infants. *Psychological Science, 21*, 653–660.

Dehaene, S. (1997). *The number sense: How the mind creates mathematics*. New York: Oxford University Press.

Dehaene, S., & Cohen, L. (1997). Cerebral pathways for calculation: Double dissociation between rote verbal and quantitative knowledge of arithmetic. *Cortex, 33*(2), 219–250.

Denison, S., Reed, C., & Xu, F. (2013). The emergence of probabilistic reasoning in very young infants: Evidence from 4.5- and 6-month-olds. *Developmental Psychology, 49*(2), 243–249.

Denison, S., & Xu, F. (2010). Twelve- to 14-month-old infants can predict single-event probability with large set sizes. *Developmental Science, 13*, 798–803.

Dixon, J. A., & Moore, C. F. (1996). The developmental role of intuitive principles in choosing mathematical strategies. *Developmental Psychology, 32*, 241–253.

Duffy, S., Huttenlocher, J., & Levine, S. (2005). How infants encode spatial extent. *Infancy, 8*(1), 81–90.

Feigenson, L. (2007). The equality of quantity. *Trends in Cognitive Sciences, 11*(5), 185–187.

Feigenson, L. (2011). Predicting sights from sounds: 6-month old infants' intermodal numerical abilities. *Journal of Experimental Child Psychology, 110*(3), 347–361.

Feigenson, L., Carey, S., & Spelke, E. S. (2002). Infants' discrimination of number vs. continuous extent. *Cognitive Psychology, 44*, 33–66.

Feigenson, L., Dehaene, S., & Spelke, E. S. (2004). Core systems of number. *Trends in Cognitive Sciences, 8*(7), 307–314.

Fischbein, E. (1990). Intuition and information processing in mathematical activity. *International Journal of Educational Research, 14*, 31–50.

Frydman, O., & Bryant, P. (1988). Sharing and the understanding of the number equivalence by young children. *Cognitive Development, 3*, 323–329.

Gallistel, C. R. (1990). *The organization of learning*. Cambridge, MA: MIT Press.

Gallistel, C. R. (2008). The neural mechanisms that underlie decision making. In P. W. Glimcher, C. F. Camerer, E. Fehr & R. A. Poldrack (Eds.), *Neuroeconomics: Decision making and the brain* (pp. 419–424). New York: Elsevier/Academic.

Gallistel, C. R. (2012). Machinery of cognition. In P. Hammerstein, J. Stevens & J. Lupp (Eds.), *Strüngmann Forum Report: vol. 11*. Cambridge, MA: MIT Press.

Gallistel, C. R., & Gelman, R. (1992). Preverbal and verbal counting and computation. *Cognition, 44*, 43–74.

Gallistel, C. R., & Gibbon, J. (2000). Time, rate, and conditioning. *Psychological Review, 107*, 289–344.

Gallistel, C. R., King, A. P., Gottlieb, D., Balci, F., Papachristos, E. B., Szalecki, M., & Carbone, K. S. (2007). Is matching innate? *Journal of the Experimental Analysis of Behavior, 7*(2), 161–199.

Gelman, R., Cohen, M., & Hartnett, P. (1989). To know mathematics is to go beyond the belief that "Fractions are not numbers." *Proceedings of Psychology of Mathematics Education, 11*.

Gilmore, C., McCarthy, S., & Spelke, E. (2007). Symbolic arithmetic knowledge without instruction. *Nature, 447*, 589–591.

Gilmore, C. K., & Spelke, E. (2008). Children's understanding of the relationship between addition and subtraction. *Cognition, 107*, 932–945.

Goswami, U. (1989). Relational complexity and the development of analogical reasoning. *Cognitive Development, 4*, 251–268.

Halberda, J., & Feigenson, L. (2008). Developmental change in the acuity of the "Number Sense": The approximate number system in 3-, 4-, 5-, 6-year-olds and adults. *Developmental Psychology, 44*(5), 1457–1465.

Halberda, J., Ly, R., Wilmer, J., Naiman, D., & Germine, L. (2012). Number sense across the lifespan as revealed by a massive Internet-based sample. *Proceedings of the National Academy of Sciences, 109*(28), 11116–11120.

Halberda, J., Mazzocco, M., & Feigenson, L. (2008). Individual differences in nonverbal number acuity predict maths achievement. *Nature, 455*, 665–668.

Harper, D. (1982). Competitive foraging in mallards: 'Ideal free' ducks. *Animal Behaviour, 30*(2), 575–584.

Hirschfield, L., & Gelman, S. (1994). *Mapping the mind: Domain specificity in cognition and culture.* Cambridge, UK: Cambridge University Press.

Inhelder, B., & Piaget, J. (1964). *The early growth of logic in the child: Classification and seriation.* (E. A. Lunzer & D. Papert, Trans.). New York: Harper & Row.

Izard, V., Sann, C., Spelke, E. S., & Streri, A. (2009). Newborn infants perceive abstract numbers. *Proceedings of the National Academy of the Sciences, 106*, 10382–10385.

Jacob, S., & Nieder, A. (2009a). Tuning to non-symbolic proportions in the human frontoparietal cortex. *European Journal of Neuroscience, 30*, 1432–1442.

Jacob, S., & Nieder, A. (2009b). Notation-independent representation of fractions in the human parietal cortex. *Journal of Neuroscience, 29*, 4652–4657.

Jeong, Y., Levine, S. C., & Huttenlocher, J. (2007). The development of proportional reasoning: Effect of continuous vs. discrete quantities. *Journal of Cognition and Development, 8*(2), 237–256.

Kahneman, D., Treisman, A., & Gibbs, B. J. (1992). The reviewing of object files: Object-specific integration of information. *Cognitive Psychology, 24*, 175–219.

Kieren, T. E. (1988). Personal knowledge of rational numbers: Its intuitive and formal development. In J. Hiebert & M. Behr (Eds.), *Number concepts and operations in the middle grades* (pp. 162–181). Hillsdale, NJ: Lawrence Erlbaum.

Kinzler, K. D., & Spelke, E. S. (2007). Core systems in human cognition. *Progress in Brain Research, 164*, 257–264.

Knops, A., Zitzmann, S., & McCrink, K. (2013). Examining the presence and determinants of operational momentum in childhood. *Frontiers in Psychology, 4*, 325.

Koechlin, E., Dehaene, S., & Mehler, J. (1997). Numerical transformations in five-month-old human infants. *Cognition, 3*, 89–104.

Lampl, Y., Eshel, Y., Gilad, R., & Sarova-Pinhas, I. (1994). Selective acalculia with sparing of the subtraction process in a patient with left parietotemporal hemorrhage. *Neurology, 44*(9), 1759–1761.

Leon, M., & Gallistel, C. R. (1998). Self-stimulating rats combine subjective reward magnitude and subjective reward rate multiplicatively. *Journal of Experimental Psychology. Animal Behavior Processes, 24*(3), 265–277.

Libertus, M. E., Feigenson, L., & Halberda, J. (2011). Preschool acuity of the approximate number system correlates with school math ability. *Developmental Science, 14*(6), 1292–1300.

Libertus, M., Feigenson, L., & Halberda, J. (2013). Numerical approximation abilities correlate with and predict informal but not formal mathematics abilities. *Journal of Experimental Child Psychology, 116*, 829–838.

Libertus, M., Odic, D., & Halberda, J. (2012). Intuitive sense of number correlates with scores on college-entrance examination. *Acta Psychologica, 141*, 373–379.

Lipton, J. S., & Spelke, E. S. (2003). Origins of number sense: Large number discrimination in human infants. *Psychological Science, 14*(5), 396–401.

Lipton, J. S., & Spelke, E. S. (2004). Discrimination of large and small numerosities by human infants. *Infancy, 5*(3), 271–290.

Mack, N. (1990). Learning fractions with understanding: Building on informal knowledge. *Journal for Research in Mathematics Education, 21*, 16–32.

Mazzocco, M. M. M., Feigenson, L., & Halberda, J. (2011). Preschoolers' precision of the approximate number system predicts later school mathematics performance. *PloS One*, *6*(9), e23749.

McCrink, K., & Barth, H. (2013). Non-symbolic multiplication, division, and inversion in young children. Poster presented at the Society for Research in Child Development, Seattle, WA.

McCrink, K., Dehaene, S., & Dehaene-Lambertz, G. (2007). Moving along the number line: The case for operational momentum. *Perception and Psychophysics*, *69*(8), 1324–1333.

McCrink, K., & Spelke, E. S. (2010). Core multiplication in childhood. *Cognition*, *116*(2), 204–216.

McCrink, K., Spelke, E. S., Dehaene, S., & Pica, P. (2013). Non-symbolic halving in an Amazonian indigene group. *Developmental Science*, *16*(3), 451–462.

McCrink, K., & Wynn, K. (2004). Large-number addition and subtraction by 9-month-old infants. *Psychological Science*, *15*, 776–781.

McCrink, K., & Wynn, K. (2007). Ratio abstraction by 6-month-old infants. *Psychological Science*, *18*, 740–746.

McCrink, K., & Wynn, K. (2009). Operational momentum in large-number addition and subtraction by 9-month-old infants. *Journal of Experimental Child Psychology*, *104*, 400–408.

Meck, W. H., & Church, R. M. (1983). A mode control model of counting and timing processes. *Journal of Experimental Psychology. Animal Behavior Processes*, *9*(3), 320–334.

Mix, K., Levine, S. C., & Huttenlocher, J. (1999). Early fraction calculation ability. *Developmental Psychology*, *35*(5), 164–174.

Moore, C. F., Dixon, J. A., & Haines, B. A. (1991). Components of understanding in proportional reasoning: A fuzzy set representation of developmental progressions. *Child Development*, *62*, 441–459.

Mou, Y., & vanMarle, K. (2013). Abstract matching: Six-month-old infants' intermodal representations of ratio. Poster presented at the Society for Research in Child Development biennial meeting in Seattle, WA.

Naccache, L., & Dehaene, S. (2001). The priming method: Imaging unconscious repetition priming reveals an abstract representation of number in the parietal lobes. *Cerebral Cortex*, *11*, 966–974.

Needham, A., Dueker, G. L., & Lockhead, G. (2005). Infants' formation and use of categories to segregate objects. *Cognition*, *94*, 215–240.

Nelson, X. J., & Jackson, R. R. (2012). The role of numerical competence in a specialized predatory strategy of an araneophagic spider. *Animal Cognition*, *15*, 699–710.

Nunes, T., Schliemann, A., & Carraher, D. (1993). *Street mathematics and school mathematics*. Cambridge: Cambridge University Press.

Pepperberg, I. M. (2006). Cognitive and communicative abilities of Grey parrots. *Applied Animal Behaviour Science*, *100*, 77–86.

Perdue, B. M., Talbot, C. F., Stone, A., & Beran, M. J. (2012). Putting the elephant back in the herd: Elephant relative quantity judgments match those of other species. *Animal Cognition*, *15*, 955–961.

Piaget, J., & Inhelder, B. (1956). *The child's conception of space*. New York: Norton.

Piazza, M., Pica, P., Izard, V., Spelke, E. S., & Dehaene, S. (2013). Education increases the acuity of the non-verbal approximate number system. *Psychological Science*, *24*(6), 1037–1043.

Pica, P., Lemer, C., Izard, V., & Dehaene, S. (2004). Exact and approximate arithmetic in an Amazonian indigene group. *Science*, *306*(5695), 499–503.

Picozzi, M., de Hevia, M. D., Girelli, L., & Cassia, V. M. (2010). Seven-month-olds detect ordinal numerical relationships within temporal sequences. *Journal of Experimental Child Psychology*, *107*, 359–367.

Platt, J. R., & Johnson, D. M. (1971). Localization of position within a homogeneous behavior chain: Effects of error contingencies. *Learning and Motivation, 2,* 386–414.

Post, T. R. (1981). Fractions: Results and implications from national assessment. *Arithmetic Teacher, 28,* 26–31.

Ramani, G. B., & Siegler, R. S. (2008). Promoting broad and stable improvements in low-income children's numerical knowledge through playing number board games. *Child Development, 79,* 375–394.

Reyna, V. F., & Brainerd, C. J. (1994). The origins of probability judgment: A review of data and theories. In G. Wright & P. Ayton (Eds.), *Subjective probability* (pp. 239–272). New York: Wiley.

Rugani, R., Regolin, L., & Vallortigara, G. (2007). Rudimental numerical competence in 5-day-old domestic chicks *(Gallus gallus)*: Identification of ordinal position. *Journal of Eperimental Psychology: Animal Behavior Processes, 33,* 21–31.

Rugani, R., Vallortigara, G., & Regolin, L. (2013). Numerical abstraction in young domestic chicks *(Gallus gallus)*. *PloS One, 8*(6), e65262. http://dx.doi.org/10.1371/journal. pone.0065262.

Scholl, B. J. (2001). Objects and attention: The state of the art. *Cognition, 80*(1/2), 1–46.

Sharon, T., & Wynn, K. (1998). Individuation of actions from continuous motion. *Psychological Science, 9,* 357–362.

Simon, T. J., Hespos, S. J., & Rochat, P. (1995). Do infants understand simple arithmetic? A replication of Wynn (1992). *Cognitive Development, 10,* 253–269.

Singer, J. A., Kohn, A. S., & Resnick, L. B. (1997). Knowing about proportions in different contexts. In T. Nunes & P. Bryant (Eds.), *Learning and teaching mathematics* (pp. 115–132). East Sussex, UK: Psychology Press.

Singer-Freeman, K., & Goswami, U. (2001). Does half a pizza equal half a box of chocolates? Proportional matching in an analogy task. *Cognitive Development, 16*(3), 811–829.

Sophian, C. (2000). Perceptions of proportionality in young children: Matching spatial ratios. *Cognition, 75,* 145–170.

Sophian, C., Garyantes, D., & Chang, C. (1997). When three is less than two: Early developments in children's understanding of fractional quantities. *Developmental Psychology, 33,* 731–744.

Spinillo, A., & Bryant, P. (1991). Children's proportional judgments: The importance of 'half.' *Child Development, 62*(3), 427–440.

Starkey, P., & Cooper, R. (1980). Perception of number by human infants. *Science, 210*(4473), 1033–1035.

Starkey, P., Spelke, E. S., & Gelman, R. (1990). Numerical abstraction by human infants. *Cognition, 36,* 97–127.

Suanda, S. H., Tompson, W., & Brannon, E. M. (2008). Changes in the ability to detect ordinal numerical relationships between 9 and 11 months of age. *Infancy, 13,* 308–337.

Teglas, E., Girotto, V., Gonzalez, M., & Bonatti, L. L. (2007). Intuitions of probabilities shape expectations about the future at 12 months and beyond. *Proceedings of the National Academy of Sciences, 104,* 19156–19159.

Teglas, E., Vul, E., Girotto, V., Gonzalez, M., Tenenbaum, J. B., & Bonatti, L. L. (2011). Pure reasoning in 12-month-old infants as probabilistic inference. *Science, 332,* 1054–1059.

Thioux, M., Pesenti, M., Costes, N., De Volder, A., & Seron, X. (2008). Task-independent semantic activation for numbers and animals. *Cognitive Brain Research, 24*(2), 284–290.

Tooby, J., & Devore, I. (1987). The reconstruction of hominid behavioral evolution through strategic modeling. In W. Kinzey (Ed.), *The Evolution of Human Behavior: Primate Models.* Albany, NY: SUNY Press.

Uller, C., Carey, S., Huntley-Fenner, G., & Klatt, L. (1999). What representations might underlie infant numerical knowledge? *Cognitive Development, 14,* 1–36.

Van Harskamp, N. J., & Cipolotti, L. (2001). Selective impairments for addition, subtraction and multiplication. Implications for the organisation of arithmetical facts. *Cortex, 37,* 363–388.

vanLoosbroek, E., & Smitsman, A. W. (1990). Visual perception of numerosity in infancy. *Developmental Psychology, 26,* 916–922.

vanMarle, K. (2013). Infants use different mechanisms to small and large number ordinal judgments. *Journal of Experimental Child Psychology, 114,* 102–110.

vanMarle, K., & Wynn, K. (2006). 6-month-old infants' use analog magnitudes to represent duration. *Developmental Science, 9*(5), F41–F49.

vanMarle, K., & Wynn, K. (2009). Infants' auditory enumeration: Evidence for analog magnitudes in the small number range. *Cognition, 111,* 302–316.

Walsh, V. (2003). A theory of magnitude: Common cortical metrics of time, space and quantity. *Trends in Cognitive Sciences, 7,* 483–488.

Wilson, A., Revkin, S., Cohen, D., Cohen, L., & Dehaene, S. (2006). An open trial assessment of "The Number Race," an adaptive computer game for remediation of dyscalculia. *Behavioral and Brain Functions, 2*(19).

Wood, J. N., & Spelke, E. S. (2005). Infants' enumeration of actions: Numerical discrimination and its signature limits. *Developmental Science, 8,* 173–181.

Wynn, K. (1992). Addition and subtraction by human infants. *Nature, 358,* 749–750.

Xu, F., & Denison, S. (2009). Statistical inference and sensitivity to sampling in 11-month-old infants. *Cognition, 112,* 97–104.

Xu, F., & Garcia, V. (2008). Intuitive statistics by 8-month-old infants. *Proceedings of the National Academy of Sciences, 105,* 5012–5015.

Xu, F., & Spelke, E. S. (2000). Large number discrimination in 6-month-old infants. *Cognition, 74,* B1–B11.

Chapter 9

Analog Origins of Numerical Concepts

Jessica F. Cantlon

Department of Brain & Cognitive Sciences, University of Rochester, New York, NY, USA

INTRODUCTION

In this review, I describe evidence consistent with the hypothesis of continuity in numerical processes from "primitive" quantitative abilities to higher-level mathematical competence. "Primitive" quantitative abilities are those that many animals use to estimate the magnitude of an object or event—for instance, its distance, duration, or number (among others). Humans and animals can make estimated judgments of these quantities rapidly, without verbally counting. The constraints on how human and animal minds estimate quantities such as these are similar (Gallistel & Gelman, 1992). For example, quantity estimation exhibits cognitive processing limitations that can be predicted by Weber's law, specifically that quantity discrimination is determined by the objective ratio between the magnitudes of two compared sets (Halberda & Odic, this volume). This ratio-based signature indicates that quantities, such as estimated number, are represented in an *analog format*, akin to the way in which a machine represents intensities in currents or voltages (Gallistel & Gelman, 1992).

Evidence that adult humans, children, infants, and nonhuman animals all have access to an analog system of quantity estimation has been interpreted as evidence that they have a common evolutionary and developmental origin and a common foundation in the mind and brain. Further evidence suggests a cognitive and neural relation between these analog numerical representations and symbolic numerical concepts such as the verbal counting system and Arabic numerals in humans. The evidence of continuity across species, stages of human development, brain and behavior, and number notation systems supports the hypothesis that the analog system is a primitive source of quantity information from which our formal mathematical abilities originated (Star & Brannon, this volume).

Mathematical Cognition and Learning, Vol. 1. http://dx.doi.org/10.1016/B978-0-12-420133-0.00009-0
225

Preverbal children and nonhuman animals possess the ability to estimate quantities, such as the approximate number of objects in a set, without counting them verbally. Instead of counting, children and animals can mentally represent quantities approximately, in an analog format. Studies from our group and others have shown that human adults, children, and nonhuman primates share cognitive algorithms for encoding numerical values as analogs, comparing numerical values, and performing simple arithmetic (Cantlon, Platt, & Brannon, 2009; Feigenson, Dehaene, & Spelke, 2004; Gallistel, 1989; Meck & Church, 1983; McCrink, this volume; vanMarle, this volume). Developmental studies indicate that these analog numerical representations interact with children's developing symbolic knowledge of numbers and mathematics (Gelman & Gallistel, 1978; Halberda, Mazzocco, & Feigenson, 2008; van Marle, Chu, Li, & Geary, in press; but see De Smedt, Noël, Gilmore, & Ansari, 2013, and Geary, 2013, for review). Furthermore, the brain regions recruited during approximate number representations are shared by adult humans, nonhuman primates, and young children who cannot yet count to 30, suggesting a primitive basis in the brain (Ansari, 2008; Dehaene, Piazza, Pinel, & Cohen, 2003; Nieder, 2005). Finally, it has recently been demonstrated that neural regions involved in analog numerical processing also are involved in symbolic numerical processing and are related to formal mathematics achievement (Cantlon, Libertus, et al., 2009; Emerson & Cantlon, 2012; Holloway & Ansari, 2010;). This chapter takes stock of the evidence that a primitive analog number system exists, that it forms the cognitive and neural basis of symbolic number concepts, and that it impacts some aspects of formal mathematics achievement in humans.

THE ANALOG NUMBER SYSTEM

There is substantial evidence that humans and other animals share a nonsymbolic, analog system of numerical concepts. Discussion of a cross-species analog system for representing numerical magnitude dates back many decades and was perhaps first and most thoroughly articulated in Gallistel's *The Organization of Learning* (1990; see also Gallistel, 1989). Since that time, evidence has accumulated that nonhuman animals, and specifically nonhuman primates, share three essential analog numerical processing mechanisms with modern humans: an ability to *represent* numerical values (e.g., Beran, 2001; Brannon & Terrace, 1998; Cantlon & Brannon, 2006; Cantlon & Brannon, 2007b; Evans, Beran, Harris, & Rice, 2009; Nieder, 2005), a general mechanism for mental *comparison* (e.g., Cantlon & Brannon, 2005), and *arithmetic* algorithms for performing addition and subtraction (e.g., Beran, 2001, 2004; Beran & Beran, 2004; Cantlon & Brannon, 2007a). These findings complement and extend a long history of research on the numerical abilities of nonhuman animals (see Beran, Gulledge, & Washburn, 2006; Emmerton, 2001, for review). The degree to which humans and nonhuman primates share numerical abilities is evidence that those abilities might derive from a common ancestor, in the same way that common morphology like the presence of 10 fingers and toes in two different primate species points to a common morphological heritage.

Representation

When adult humans and rhesus monkeys (*Macaca mulatta*) are given a task in which they have to rapidly compare two visual arrays and touch the array with the smaller numerical value (without counting the dots), their performance reliably yields the pattern shown in Figure 9-1: accuracy decreases as the ratio between the numerical values in the two arrays approaches 1 (Cantlon & Brannon, 2006; see Gallistel & Gelman, 1992; Dehaene, 1992, for review). The explanation of this performance pattern is that both groups are representing the numerical values in an analog format (Figure 9-2).

In an analog format, number is represented only approximately and it is systematically noisy (Dehaene, 1992; Gallistel & Gelman, 1992). More precisely, the probability of error in the subjective representation of a number increases with the objective number of items that are coded by that representation. This pattern is due to the spacing or spread of the distributions that represent a given number (Dehaene, 1992; Gallistel & Gelman, 1992; see Cantlon, Cordes, Libertus, & Brannon, 2009). For instance, analog processing of two items might only weakly activate representations for the adjacent quantities one and three, whereas processing of a larger value such as eight items could moderately to strongly activate the adjacent representations of seven and nine. Consequently, the probability of confusion (i.e., the overlap between distributions) between any two objective numbers increases as their values increase. This means that the probability of having an accurate subjective representation of a numerical

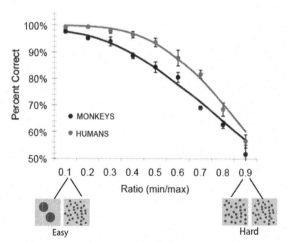

FIGURE 9-1 Accuracy on a numerical discrimination task for monkeys and humans plotted by the numerical ratio between the stimuli. *(From Cantlon & Brannon, 2006.)*

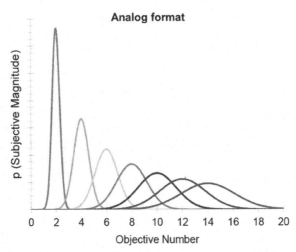

FIGURE 9-2 An analog represention of numerical value represents an objective numerical value with a probability distribution that scales with the size of the objective value. *(From Cantlon, Cordes, Libertus, & Brannon, 2009.)*

value decreases with its objective magnitude. This behavioral pattern is most evident in the comparison of two quantities and is succinctly quantified by the ratio between the values being compared. Two different pairs of numerical values that have the same ratio (e.g., 2 & 4, 4 & 8) have the same amount of overlap or the same probability of confusion. As numerical pairs get larger and closer together, their ratio increases (approaching 1.0) and so does the probability that they will be confused (leading to more errors). For example, one might be 80% accurate at choosing the larger number when the numerical choices are 45 versus 70 (45/70 = a 0.64 ratio) but might perform at chance when the choices are 45 versus 50 (45/50 = a 0.9 ratio). This effect is known as Weber's law (Halberda & Odic, this volume).

The solid lines in Figure 9-1 (from Cantlon & Brannon, 2006) represent predicted data (solid lines) from a model of number representation under Weber's law (Pica, Lemer, Izard, & Dehaene, 2004) and show that the predictions of this analog numerical model fit the actual performance data well. These and other empirical data from monkeys and humans and the fit of the analog model demonstrate that although humans can represent numerical values precisely using words and Arabic numerals, they still have an approximate, analog numerical system that functions essentially in the same way as in monkeys (see also Agrillo et al., this volume; Beran et al., this volume; Geary et al., this volume; Pepperberg, this volume; Vallortigara, this volume).

Comparison

The ratio effect, described by Weber's law, indicates that numerical values can be represented in an analog format. However, that does not tell us anything about the process by which two numerical values are compared. We have identified a signature of mental comparison in monkeys that is commonly observed when adult humans make judgments of magnitudes: the semantic congruity effect (Cantlon & Brannon, 2005; Holyoak, 1977). The effect is observed in adult humans' response times whenever they compare things along a single dimension. For instance, when people are presented with pairs of animal names and asked to identify the larger or smaller animal based on their memory of such animals, they show a semantic congruity effect in their response time: people are faster to choose the smaller of a pair of small items (e.g., ant vs. rat) but faster to choose the larger of a pair of large items (e.g., horse vs. cow). This effect suggests that the physical size of the animal pairs interacts with the verbal phrasing of the instructions (i.e., whether "which is larger?" or "which is smaller?") in subjects' judgments.

In humans, the semantic congruity effect is observed for judgments of many dimensions including judgments of numerical values, represented by Arabic numerals. We found that this effect is also observed in monkeys when they compare numerical values from arrays of dots. Monkeys performed a task in which they had to choose the larger numerical value from two visual arrays when the background color of the computer screen was blue, but when the screen background was red, they had to choose the smaller numerical value of the two arrays. As shown in Figure 9-3 (from Cantlon & Brannon, 2005), both monkeys showed a cross-over pattern of faster RTs when choosing the smaller of two small values compared to the larger of two small values and the opposite pattern for large values. The semantic congruity effect is the signature of a mental comparison process wherein context-dependent mental reference points are established (e.g., 1 for "choose smaller" and 9 for "choose larger"), and RT is determined by the distance of the test items from the reference points: this has been modeled as the time it takes for evidence to accrue in the comparison of each item to the reference point (Holyoak, 1977). In humans, the semantic congruity effect is observed for a variety of mental comparisons from both perceptual and conceptual stimuli: brightness, size, distance, temperature, ferocity, numerals, and so forth. Our data from nonhuman primates indicate that the mental comparison process that yields the semantic congruity effect is a primitive, generalized, nonverbal mental comparison process for judging quantities and other one-dimensional properties.

In fact, the ability to compare quantities, and the proposed algorithm underlying that ability, could be so primitive that it extends to nonprimate animals (Gallistel, 1989). A recent study by Scarf and colleagues (Scarf, Hayne, & Colombo, 2011) showed that pigeons can compare numerical values, and in doing so, they represent a simple numerical rule that can be applied to novel numerical values. Pigeons' accuracy on that ordinal numerical task is comparable to that of monkeys tested on an identical task (Brannon & Terrace, 1998). Evidence of similar quantitative abilities across diverse species could indicate that analog quantitative processing emerges to support foraging, for example, under a variety of conditions. The similarities

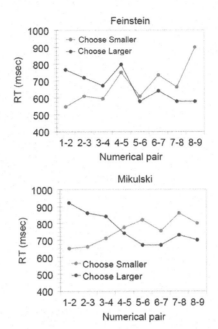

FIGURE 9-3   The semantic congruity effect in the response times of two different monkeys (Feinstein and Mikulski) on a numerical comparison task where they sometimes chose the larger numerical value from two arrays (blue) and other times chose the smaller value (red). The cross-over pattern reflects the effect of semantic congruity. *(From Cantlon & Brannon, 2005.)*

across species could occur because these competencies were present in a common ancestor or emerged independently through convergent evolution.

Arithmetic

Arithmetic is the ability to mentally combine or decompose values to create a new value or values without having directly observed that new value. Researchers have found that monkeys possess a capacity for basic, nonverbal addition that parallels human nonverbal arithmetic in a few key ways. First, monkeys and humans show a ratio effect when performing rapid nonverbal addition, similar to the ratio effect described earlier (Cantlon & Brannon, 2007a). In the addition task, monkeys and humans viewed two sequentially presented arrays of items that varied in terms of the number of elements in each array. After the sequence of arrays was presented, subjects were given two choice arrays: one that represented the arithmetic sum of the two arrays and another that differed from the sum by some amount. Humans were prevented from verbally counting by being forced to respond too rapidly to count. Monkeys and humans alike accurately selected the arithmetic sum over the distractor value and their degree of accuracy depended on the ratio between the numerical values of the choice arrays. The fact that monkeys and humans are capable of nonverbal arithmetic and show a common ratio effect suggests that the analog numerical system affords them arithmetic computation abilities (see also McCrink, this volume; vanMarle, this volume).

In other studies, it has been shown that nonhuman primates can track and represent quantities as items are added to or subtracted from sets during visual occlusion (Beran, 2001; 2004; Boysen & Berntson, 1989), and that they can remember the location and value of the transformed set over an extended period of time (Beran & Beran, 2004). That is, monkeys can form and retain a mental representation of the transformed set. Finally, monkeys recognize ordinal rules such as "ascending" and "descending" and can use those rules to express arithmetic relationships among quantities (Bongard & Nieder, 2010; Cantlon & Brannon, 2005). Together, the existing data reflect the capacity of the analog system to execute a subset of computations that are analogous to human formal arithmetic. The breadth of arithmetic operations available to the analog numerical system is not yet known. Moreover, it is not yet known how the arithmetic operations afforded by the analog numerical system are related to symbolic arithmetic.

The overarching conclusion from this line of research is that the abilities to represent and compare quantities and perform simple arithmetic computations reflect a cognitive system for numerical reasoning that is primitive and based on analog magnitude representations (Gallistel & Gelman, 1992). If analog numerical cognition is truly "primitive" and homologous across primate species, then it should be rooted in the same physical (neural) system in monkeys and humans.

NEURAL BASIS OF ANALOG NUMBER

There is evidence from multiple sources that analog numerical processing recruits a common neural substrate in monkeys, adult humans, and young children (Figure 9-4). In monkeys that are trained to match visual arrays of dots based on number, single neurons along the intraparietal sulcus (IPS) respond maximally to a preferred numerical value, and their firing rate decreases as the number that is presented gets numerically farther from that preferred value (Nieder & Miller, 2004). Moreover, neurons tuned to larger numerical values exhibit noisier responses than do neurons tuned to smaller values (Nieder & Merten, 2007). That is, neurons tuned to larger numerical values have a coarser response; they respond to a wider set of adjacent numerical values than neurons tuned to smaller values. The finding that neural responses in the IPS are modulated by both numerical distance and numerical magnitude represents a neural version of the numerical ratio effect, or Weber's law. In a recent extension of this research, it was found that IPS neurons exhibit a number-selective and ratio-dependent tuning pattern spontaneously, when the animal has no prior experience discriminating number in an experimental task (Viswanathan & Nieder, 2013; see also Roitman, Brannon, & Platt, 2007).

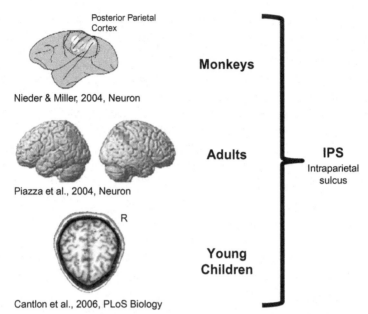

FIGURE 9-4 Monkeys, adults, and children exhibit similar activation in the intraparietal sulcus (IPS) during analog numerical processing. *(Redrawn from Nieder & Miller, 2004; Piazza, Izard, Pinel, Le Bihan, & Dehaene, 2004; Cantlon, Brannon, Carter, & Pelphrey, 2006.)*

This means that the number-selective tuning of neurons in the IPS is likely a cause of numerical representation rather than a consequence. In individual monkeys, the neural representation of numerical sensitivity in the IPS scales with the behavioral sensitivity to number and is thought to directly reflect analog numerical processing by the IPS (see Nieder, 2012, for discussion).

A similar pattern of ratio-dependent numerical tuning has been observed with fMRI in the human IPS. Piazza and colleagues (Piazza et al., 2004) found a neural adaptation effect for numerical values in the IPS that depended on the ratio between the adapted numerical value and a deviant numerical value. Adults were presented with visual arrays of dots that typically had a constant number (e.g., 8 dots). After several exposures to the constant number of dots, neural responses in the IPS were found to decrease, or adapt. Then a novel, or "deviant," number of dots was presented (e.g., 16 dots) and neural responses increased. The degree to which neural responses increased to deviant stimuli depended on the ratio between the adapted numerical value and the deviant numerical values. Our group also observed neural adaptation in the IPS for visual arrays ranging from 8 to 64 items in preschool children who could not yet verbally count to 30 (Cantlon et al., 2006). Subsequent fMRI with 4- to 6-year-olds also found evidence of ratio-dependent neural responses to numerical stimuli in the IPS (Cantlon, Libertus, et al., 2009). Ratio-dependent neural responses to numerical stimuli can be observed in posterior dorsal regions of the brain as early as 6–7 months of age using electroencephalogram (EEG) recordings (Hyde & Spelke, 2011; Libertus, Pruitt, Woldorff, & Brannon, 2009). Together, these studies with monkeys, children, and adults reflect a common source for analog numerical representation that bridges species as well as stages of human development and is thus independent of language and formal mathematics experience.

The neural representation of numerical value in the IPS could be the basis of a conceptual understanding of number. Parallel neural data from humans and nonhuman primates indicate that the neural representation of number in the IPS is abstract or, "supramodal." In humans, neuroimaging data have revealed regions of the IPS that represent numerical values regardless of whether the stimulus is presented in the auditory or visual modality (Eger, Sterzer, Russ, Giraud, & Kleinschmidt, 2003; Piazza, Mechelli, Price, & Butterworth, 2006). In one study, adults were given a task in which sequences of red and green light flashes (visual) or high and low tones (auditory) were presented on different trials. In the visual condition, subjects reported which color light was more frequent, whereas in the auditory condition, they reported which pitch was more frequent. Regions of the IPS and prefrontal cortex (PFC) showed a heightened neural response during the numerical estimation of light and pitch relative to a memory control task.

A comparable task was used with nonhuman primates during single-cell recordings from the IPS and lateral PFC to determine if a single neuron represents numerical value across sensory inputs (Nieder, 2012). Monkeys were given a task in which they released a lever to indicate when two sequences of stimuli were numerically equivalent. The sequences were composed either of light flashes (visual) or tones (auditory). Both the IPS and PFC contained neural populations that responded to numerical values presented visually and auditorily. The strongest test of modality-independence, however, is whether a single neuron represents the same numerical value in both modalities. IPS recordings did reveal neurons that responded to a specific number of items in a sequence regardless of whether the sequence was presented in the auditory or visual modality. This result represents strong evidence that neurons in the IPS *can* represent number across sensory modalities. However, the frequency and breadth of such "supramodal" neurons was found to be limited in the IPS especially in comparison to number-selective neurons in the PFC. Nieder found that the frequency of modality-independent number-selective neurons was greater in the PFC compared to the IPS and that the range of numerical values represented as modality-independent was also greater in the PFC. The relative contribution of the IPS versus PFC to modality-independent analog numerical processing is currently unknown in humans. Yet it is clear from the human and monkey data that both regions are capable of representing numerical value from various sensory inputs.

Taken together, the neural data from monkeys and humans support the conclusions derived from the behavioral data; specifically, there is continuity between humans and nonhuman animals in the mechanisms underlying non-symbolic, analog numerical representations. And, this continuity is supported at least in part by the IPS and portions of the PFC.

ANALOG ORIGINS OF NUMBER SYMBOLS

A long history of studies with preverbal human infants has shown that they too possess an ability to quantify objects with approximate, analog representations (Feigenson et al., 2004; vanMarle, this volume). There is general agreement that the analog system for numerical reasoning is primitive in human development. Indeed, human infants exhibit signatures of analog numerical processing from birth (Izard, Sann, Spelke, & Streri, 2009). The fact that newborn infants possess a common system of analog numerical representation with nonhuman primates is consistent with the previous argument for evolutionary continuity in the mechanisms underlying representations of number. A question at the evolutionary level of explanation is whether those analog numerical mechanisms facilitated the invention of number words and symbols (e.g., counting) in human cultural evolution (Geary, 2007). Evidence that symbolic number representation retains some of the signatures of analog numerical processing supports that hypothesis (Starr & Brannon, this volume).

One important source of cognitive evidence for continuity between analog and symbolic numerical processes is the semantic distance effects produced during Arabic numeral and number word processing, as is found for nonsymbolic processing of quantities (see the "Comparison" section). A classic paper by Moyer and Landauer (1967) showed that when adult humans perform simple judgments of Arabic numerals, such as choosing the larger value, the speed of their response is modulated by the numerical difference between the values. For example, subjects were faster to choose the larger numeral when the numerical difference was 3 (e.g., 2 vs. 5, 4 vs. 7, etc.) than when the difference was 2 (e.g., 2 vs. 4, 3 vs. 5, etc.). Similar effects of distance are observed for spoken and written number words (Dehaene, 1992; Dehaene & Akhavein, 1995). Numbers that are close in value are more difficult to discriminate than those that are farther apart. The fact that these distance effects emerge in subjects' response times during judgments of numerals and number words suggests engagement of the analog representational system. Unlike analog numerical representation, a strictly digital representation of numerical symbols would not show these distance effects during comparisons because performance would be uniformly perfect, like a calculator. That humans' responses are modulated by the magnitude of difference between the digits is evidence that the judgments are grounded in analog units.

A second source of cognitive evidence that symbolic numerical processing was initially derived from evolved analog numerical "machinery" is the influence of spatial dimensions on nonsymbolic and symbolic numerical judgments. The best evidence for a common constraint of spatial processing on analog and symbolic numerical representations comes from number-size congruity effects. When adults are asked to judge the numerically larger of two visual arrays of elements, performance is better when number and array size (cumulative area) are congruent between the arrays than when they are incongruent (e.g., Gebuis, Kadosh, de Haan, & Henik, 2009; Hurewitz, Gelman, & Schnitzer, 2006). Congruent trials are those in which the larger numerical value also has the greater cumulative surface area compared to the smaller value. On incongruent trials the relation between number and cumulative area is reversed, presenting the smaller numerical value with a larger cumulative area. These number-size congruity effects occur despite the fact that size (cumulative area) is a task-irrelevant dimension. Number-size congruity effects are also found in adults' judgments of symbolic numbers. When Arabic numerals are pitted against physical size (e.g., 9 5), judgments of which numeral is larger are faster when the larger numerical value is presented in the larger font size (e.g., Henik & Tzelgov, 1982). The data from number-size congruity effects indicate that adults cannot inhibit spatial magnitude estimation (i.e., estimating area or size) during numerical judgments, whether the judgments are nonsymbolic (analog) or symbolic (see also Lourenco, this volume). Yet, spatial processing is not formally necessary in order to make judgments of numbers.

The fact that spatial input cannot be avoided or ignored when humans make either symbolic or nonsymbolic numerical judgments implicates a common processing constraint on both numerical processes. Moreover, this interaction between space and number appears to be psychologically primitive in that it is also observed in nonhuman animals, human infants, and young children, at least for analog numerical judgments (e.g., Cantlon & Brannon, 2007b; Cantlon, Safford, & Brannon, 2010; de Hevia & Spelke, 2009; Lourenco & Longo, 2010; Rousselle & Noël, 2008). A possible evolutionary explanation of this phenomenon is that during foraging, for instance, food intake can be maximized by using an integrated representation of number and cumulative area.

Further cognitive evidence supporting a primitive link between analog and symbolic numerical processing comes from studies of counting development and number symbol understanding in children. Note that an evolutionary account of the relation between analog numerical processing and symbolic number does not *require* a developmental or lifespan relation between analog and symbolic numerical processing. Analog numerical mechanisms could have provided an evolved platform for the conceptualization of counting and symbolic number in early humans with symbolic numerical mechanisms growing more distinct from analog numerical structures as they became (literally) written into human culture. In this view, human symbolic mathematics in many ways may be distinct from the basic analog core (Penn, Holyoak, & Povinelli, 2008; Geary, 2007). Qualitative changes to initially primitive numerical mechanisms are hypothesized to occur within the human lifespan even for basic symbolic number representation (Carey, 2004; 2010). Thus, developmental data can provide a source of confirmation for shared representational structures between the analog and symbolic number systems, as would be expected for an evolved system; but at the same time we would expect discontinuities between the functioning of this system and children's acquisition of symbolic mathematics (Nieder, 2009).

At the developmental level of explanation, a fundamental question is how a child's early understanding of numerical symbols interfaces with preverbal analog representations of number. Of particular interest is how children initially map numerical meanings to the first few symbolic number words (Gelman & Butterworth, 2005; Gelman & Gallistel, 1978; Le Corre & Carey, 2007; Piazza, 2010; Wynn, 1990). There is currently a debate over the types of preverbal numerical representations that form the initial basis of children's verbal counting. However, regardless of how this initial mapping transpires, behavioral evidence suggests that as children learn words in the counting sequence, they map them to approximate, analog representations of number (Gilmore, McCarthy, & Spelke, 2007; Lipton & Spelke, 2005; Wynn, 1992). Lipton and Spelke (2005) found that 4-year-old children could look at a briefly presented array of 20 dots and, if they could count to 20, they could verbally report (without counting) that there were 20 dots in the array and their errors were systematically distributed around 20 (i.e., their errors exhibited a numerical ratio effect). If they could not yet count to 20, however, they responded with random number labels.

Thus, as soon as children learn a particular verbal count word in the sequence, they know the approximate quantity to which it corresponds without counting, suggesting that number words are mapped to the analog numerical code. These data have been taken to indicate that analog numerical representations are used to assign semantic meanings to numerical symbols during early human development. There is also evidence that children who have learned to count verbally, but have not yet learned to add and subtract, psychologically "piggy back" on analog arithmetic representations as they transition to an understanding of exact symbolic arithmetic (Gilmore et al., 2007). The general conclusion that then emerges is that the cognitive faculties that children initially use for nonsymbolic, analog numerical operations (and which they share with nonhuman animals) provide scaffolding for verbal counting in early childhood.

Evidence that judgments of numerical symbols, such as number words and Arabic numerals, are inherently spatial and grounded in an analog scale further supports the hypothesis that number symbols are at least partially constructed from primitive, analog mechanisms. The data from children suggest that an evolutionary relation between analog and symbolic numerical representations continues to provide input to the development of numerical concepts within the modern human lifespan.

NEURAL SUBSTRATE OF HUMAN NUMBER SYMBOLS

Adding further support for the argument that symbolic number concepts are derived from analog structures is found in studies that have revealed that the neural substrate of numerical symbol representation is the same as that of analog numerical processing. Moreover, the data that exist suggest that this relation between the neural substrates of analog and symbolic numerical processing is stable and continuous throughout human development.

As described earlier, the nonhuman primate IPS represents analog numerical values at the level of single neurons and shows numerical ratio effects in its neural signatures, and parallel findings are observed at the cortical level in the human IPS using neuroimaging methods. As with analog judgments, symbolic numerical judgments of numerals and number words also elicit neural activity in the IPS in humans. Several neuroimaging studies have uncovered number-specific IPS activation during symbolic addition, symbolic number comparison, and even when symbolic numbers are presented subliminally (e.g., Dehaene et al., 1999; Eger et al., 2003; Naccache & Dehaene, 2001; Pinel, Piazza, Le Bihan, & Dehaene, 2004; see Ansari, 2008; Dehaene et al., 2003, for review). Although the degree of neural overlap between analog and symbolic numerical representations is currently debated, there is agreement that the representation of number is based in parietal cortex for all notations (Cohen Kadosh, Cohen Kadosh, Kaas, Henik, & Goebel, 2007).

Recent data have also shown within-subject functional overlap in IPS responses for analog and symbolic numerical judgments. For instance, Piazza, Pinel, Le Bihan, and Dehaene (2007) employed fMRI with adults and found numerical distance effects in the response of the IPS across numerical notations. Arabic numerals or collections of dots were presented repeatedly, resulting in a decrease in IPS response. The IPS response rebounded with presentation of a numeral or collections of dots that differed in quantity from the original, and the extent of rebound was modulated by the numerical distance between the initially presented quantities and the new ones. For example, if subjects repeatedly saw the numeral "8," the IPS response was stronger to subsequently presented numerical values such as 4 or 12 than to the same value "8" regardless of whether those values were presented as numerals or collections of dots. The strength of the IPS response to subsequent values varied with numerical distance such that, in this example, the IPS response to 4 or 12 was stronger than the response to 7 or 9. Regions of the IPS can thus be shown to represent numerical values across analog and symbolic notations. Moreover, the amplitude of the IPS response during judgments of number symbols such as Arabic numerals varies with the numerical distance between values, as is found for analog judgments (Dehaene et al., 1999; Naccache & Dehaene, 2001; Piazza et al., 2007; Pinel et al., 2004).

It is somewhat surprising that numerical judgments of words and numerals elicit semantic distance effects in dorsal parietal cortex because semantic distance and similarity effects for other types of category judgments such as animals, objects, and places are observed in the ventral temporal cortex in adults and children (Cantlon, Pinel, Dehaene, & Pelphrey, 2011; Grill-Spector, Golarai, & Gabrieli, 2008). Thus, the semantic processing of symbolic numbers is unique from other types of semantic judgments that have been assessed in that it is associated with the dorsal stream functions of parietal cortex. The explanation of this unique neural pattern for the semantic processing of number symbols could be that the evolution of analog numerical processing in the parietal cortex provided the basis of symbolic numerical representation. Dehaene and Cohen (2007) refer to this as "cultural recycling" of cortical functions. More generally, this is a possible instance of *exaptation* (Gould & Lewontin, 1979), the evolutionary process of repurposing of a biological structure to serve a new function, or *preadaptation* (see Geary, 2007, for discussion).

Beyond showing common neural signatures of semantic processing in the IPS, symbolic and analog numerical processes also bear a common influence of spatial processing. The influence of spatial representations on numerical processes can be traced to parietal cortex. In one study (Tudusciuc & Nieder, 2007), monkeys were trained to perform a line length-matching task (i.e., a size judgment) and a nonsymbolic numerical matching task (i.e., numerosity judgment). During stimulus presentation as well as during a subsequent delay, single neurons in the IPS responded to the dimensions of numerosity and length. Although some neurons responded only to numerosity and others only to line length, a subset of cells (~20%) responded to both line length and numerical value. A similar pattern of neural overlap between number and size has been observed in human judgments of Arabic numerals. In an fMRI study, adults were given a task in which two Arabic numerals were presented on a monitor and pressed a button corresponding to the numeral with the larger font size, numerical value, or brightness (Pinel et al., 2004). Although the instructions of which dimension to judge varied throughout the session, the stimulus set was identical for all three judgment types. Neural activity associated with judgments of the number and size of the Arabic numerals significantly overlapped in the IPS. The degree of neural overlap between number and size varied with the magnitude of the behavioral congruity effect that was observed for number and size. There were also subregions of the IPS that responded only to number or only to size.

Based on these and other data, several researchers have proposed a "distributed but overlapping" representation of number and size at the neural level (Cantlon, Platt, & Brannon, 2009; Pinel et al., 2004; Tudusciuc & Nieder, 2007). Together, the data from monkey IPS neurons and human IPS cortex suggest that this principle of "distributed but overlapping" size and number representation applies to both the analog and symbolic numerical systems (see also Lourenco, this volume). This is evidence of a common constraint of spatial processing at the neural level for analog and symbolic numerical judgments.

The evidence that the semantic processing of symbolic numerical values originates from a common neural source with analog numerical processing suggests representation by a common mechanism. Indeed, studies of numerical processing in young children have shown that nonsymbolic and symbolic numerical stimuli elicit activation of the IPS throughout development. The IPS exhibits conjunctive activation to symbolic (numerals) and nonsymbolic (dot arrays) numerical judgments by at least 6 or 7 years of age (Cantlon, Platt, & Brannon, 2009; Holloway & Ansari, 2010). However, there are notable developmental changes in the IPS response to symbolic numbers over development. For example, the IPS gradually refines its response to symbolic numerical stimuli between the ages of 6 and 18 years in that develops a more robust, adult-like numerical distance effect over time (Ansari, Garcia, Lucas, Hamon, & Dhital, 2005; Bugden, Price, McLean, & Ansari, 2012; Cantlon, Platt, & Brannon, 2009; Kaufmann et al., 2006, Rivera, Reiss, Eckert, & Menon, 2005; see Ansari, 2008, for review). Moreover, the lateralization and spatial extent of IPS activation in response to symbolic numbers has been reported to mature gradually (see Ansari, 2008, for review).

The existing data thus support arguments for the development of symbolic numerical processing from a core substrate that processes nonsymbolic number. However, the existing studies are incomplete. So far, the only symbolic numerical stimuli that have been tested in neuroimaging paradigms with children are Arabic numerals, and the youngest age tested is 6 years. The real question is whether a common IPS substrate represents symbolic number words from the earliest stages of verbal counting. Verbal count words are the first numerical symbols learned by human children; Arabic numerals are learned later. Thus, although current neuroimaging data are consistent with claims of common neural mechanisms for symbolic and nonsymbolic numerical judgments throughout the lifespan, it is currently unclear whether this is the case when children are first learning number words and numerals.

Some research suggests a unique role for the PFC in the development of symbolic numerical concepts in humans. In humans, PFC activation is sometimes observed along with IPS activation during symbolic numerical tasks, especially during the early phases of learning (Ansari et al., 2005; Cantlon, Libertus, et al., 2009; Emerson & Cantlon, 2012; Piazza et al., 2007). A pattern of greater activation of prefrontal sites in children compared to adults has been observed for symbolic numerical and basic mathematical tasks (Ansari et al., 2005; Cantlon, Libertus, et al., 2009; Rivera et al., 2005). For example, in Cantlon, Libertus, et al. (2009), the inferior frontal gyrus responded more strongly than the IPS during symbolic numerical tasks in children, a pattern opposite to that observed in adults. Ansari et al. (2005) found a similar frontal-to-parietal shift in the strength of symbolic number-related activity in 9-year-old children (see also Rivera et al., 2005). We have found that unlike IPS activation, the activation of prefrontal cortex in children's symbolic numerical processing is correlated with generic performance factors such as overall response time, or "time on task" (Emerson & Cantlon, 2012; see also Schlaggar et al., 2002). Many questions remain regarding the role of the prefrontal cortex in symbolic numerical judgments. The greater recruitment of that region by children compared to adults could reflect associative learning mechanisms that link number symbols with their meanings, or it could reflect processes related to mitigating the greater difficulty children have in making symbolic numerical judgments than adults. Neural recordings from symbol-training studies with monkeys provide support for the first of those two possibilities.

Studies with nonhuman primates have suggested that they too engage PFC during numerical processing (see Nieder, 2009, for review), and that prefrontal regions play a unique role in associating analog numerical values with arbitrary symbols at the level of a single neuron (Diester & Nieder, 2007). Diester and Nieder (2008) recorded the activity of posterior parietal and prefrontal neurons in rhesus monkeys that were trained to associate Arabic numerals with nonsymbolic numerosities ranging from 1 to 4. Monkeys were trained to release a lever when an Arabic numeral was presented following a visual array containing its target number of elements. Neurons in both parietal and prefrontal cortex elicited responses specific to the numerical values of the nonsymbolic visual arrays and the trained Arabic numerals. Remarkably, a much larger percentage of prefrontal neurons were simultaneously tuned to a common specific value for the visual arrays and the associated Arabic numeral compared to parietal neurons, which did so only rarely (<2% of neurons). Thus, in monkeys, parietal and prefrontal neurons represent numerical values, but compared to parietal neurons, prefrontal neurons more frequently represent the abstract association between a nonsymbolic numerical value and its symbol. This suggests that the prefrontal cortex is involved in representing associations between analog numerical representations and their symbolic referents. This type of neural mechanism could explain the greater involvement of the prefrontal cortex in children's symbolic numerical judgments compared to adults. It could be the case that the prefrontal cortex plays a greater role in symbolic numerical judgments while the meanings of number symbols are becoming solidified in the minds of young children.

Humans have formal mathematics systems that express numerical rules and relations that are beyond the capacity of an analog system. Thus, at some point in development, human symbolic mathematics diverges from the numerical system that is shared by nonhuman primates. There are a number of factors that distinguish human numerical reasoning from that of other animals beginning from the earliest stages of human number learning. For example, human numerical representations, unlike nonhuman numerical representations, require that numerical symbols are committed to memory. Recent evidence suggests that in addition to a fronto-parietal network, numerical reasoning in human children recruits memory processes mediated by the hippocampus (Cho et al., 2012; Supekar et al., 2013). The role of the hippocampus in the development of human mathematics could be to store numerical symbols and rote memorized mathematics facts. The neural differences between human and nonhuman numerical reasoning remain to be explored, but the development of symbolic memory processes, mediated by the hippocampus, are likely a key factor that distinguishes the neural processes of human mathematics from those of other animals.

THE ANALOG SYSTEM AND FORMAL MATHEMATICS ACHIEVEMENT

A further central issue is the extent to which analog numerical abilities are related to formal mathematical learning and achievement. Researchers have begun to examine whether individual differences in formal math achievement are modulated by individual differences in functioning of the "primitive" analog numerical system (Starr & Brannon, this volume).

There are two levels at which individual differences could be used to study a potential relation between analog and symbolic numerical skills. One is a test of whether variation in the quality of analog numerical representations statistically predicts the acquisition of new numerical concepts over development (e.g., Lipton & Spelke, 2005). This approach addresses the extent to which variation in the acuity of the analog numerical system predicts the acquisition of mathematics concepts over development. For example, children with superior analog arithmetic acuity might develop symbolic addition at a younger age than their peers with lower acuity in the analog system. Age-related individual differences analyses can reveal whether children with better analog numerical structures develop new symbolic mathematics concepts more quickly than others. The relation between the development of new numerical concepts and analog representations could also be independent of children's overall mathematics achievement, especially if the analog numerical system is required for the development of some formal mathematics concepts but not others.

A second question is to what extent analog numerical aptitude and formal mathematics aptitude correlate across individuals. This approach addresses whether the general acuity of the analog and symbolic numerical systems are correlated, after controlling for age. In these analyses, controlling for age removes variability in the rate at which different types of formal mathematical concepts are initially acquired. Instead, these studies test how well children's numerical concepts function given the mathematics concepts they should have already acquired at that age. These two approaches address very different types of questions. Most individual differences studies have investigated the second question—the relations between acuity of the analog system and performance on the symbolic mathematics tasks.

Several studies have found that individual differences in mathematics achievement are predicted by differences in analog numerical sensitivity

(Bugden & Ansari, 2011; Halberda et al., 2008; Holloway & Ansari, 2009; Libertus, Feigenson, & Halberda, 2011; Libertus et al., 2012; Libertus, Feigenson, & Halberda, 2013). Studies with children indicate that acuity of the analog numerical system correlates with school-based mathematics achievement scores, controlling for other abilities that predict learning (e.g., working memory, verbal intelligence) and that mathematics achievement is more closely correlated with analog numerical abilities than it is with other formal abilities, such as vocabulary. For example, adolescents' analog numerical ability (measured by the Numerical Weber Fraction in ninth grade) correlated with their mathematics achievement measured in early childhood (Halberda et al., 2008; TEMA-2 MAS score from third grade). This and similar findings indicate that the "primitive" ability to accurately estimate numerical values from sets of objects is related to mathematics achievement.

In contrast to these findings, some studies have failed to find a relation between analog numerical ability and formal mathematics achievement in children and adults (e.g., Holloway & Ansari, 2009; see De Smedt et al., 2013, for review). There are three possible sources for the different findings. First, differences in the paradigms and statistics used to test the relation between analog number ability and formal mathematics achievement are likely to be important. For example, studies that use the difference in accuracy or response time between close and far numerical values as a metric of the "numerical distance effect" and a proxy for acuity of the analog system fail to take overall performance levels into account and thus neglect an important component of the variance in subjects' analog numerical abilities. The statistic that has shown the most reliable correlation with symbolic mathematics thus far is the Weber fraction (De Smedt et al., 2013). The Weber fraction represents the proportion difference needed to discriminate numerical values and is the best representation of a subject's analog numerical discrimination ability (Halberda & Odic, this volume). However, even when the Weber fraction is used as the metric of analog numerical sensitivity, the results are mixed (De Smedt et al., 2013).

The second issue in assessing the relation between the analog numerical system and symbolic mathematics is that the preferred assessment of mathematics achievement is a global test of school-based mathematics (Halberda et al., 2008). As highlighted in a recent study by vanMarle and colleagues (in press), the analog numerical system might only be related to a subset of the symbolic mathematics concepts that children learn. Thus, a global test of mathematics achievement could underestimate the relation between analog numerical representations and specific symbolic mathematics concepts.

A third issue concerns the role of confounding variables in testing individual differences correlations. Some studies have argued for a mediating role of executive function and working memory in the development of nonsymbolic numerical sensitivity and formal mathematical reasoning (Bull & Scerif, 2001; Mazzocco, Bhatia, & Lesniak-Karpiak, 2006; Mazzocco & Kover, 2007) or that a relation between nonsymbolic and symbolic numerical knowledge disappears when variables such as these are controlled (Mussolin, Nys, & Leybaert, 2012). Correlations among multiple interrelated performance measures can be difficult to disentangle (e.g., Friedman & Wall, 2005). Thus, it is currently unclear what set of conditions has led to the conflicting conclusions of studies investigating individual differences correlations between analog number ability and formal mathematics achievement.

Whatever the result, it is important to point out that all individual differences correlations are testing associations in the *state* of the cognitive systems as opposed to the relation in their *processes*. Correlations across individual subject scores, even those that control for third variables, can reveal whether two systems have a similar performance status (e.g., high or low performance, good fitness or poor fitness), but they cannot reveal whether they have interdependent processes or operations. Interdependent processes or operations are best revealed by experiments in which interactions between cognitive representations are examined using task and stimulus manipulations (see Starr & Brannon, this volume). The finding that two cognitive systems have correlated scores across individuals is consistent with arguments that their processes are correlated, but it is not evidence they are related due to the same processes.

Another source of information about the relations between analog numerical abilities and formal mathematics achievement is neuroimaging data. Neuroimaging studies can provide an independent assessment of the relation by testing whether individual variations in those abilities are linked via a common neural substrate. As with behavioral correlations between analog numerical ability scores and mathematics achievement scores, the results from studies on a neural relation have been mixed (see De Smedt et al., 2013; Kaufmann, Wood, Rubinsten, & Henik, 2011, for review). Some recent studies report correlations between mathematics achievement and activity in the neural substrates underlying approximate numerical processing (e.g., Bugden et al., 2012; Cantlon & Li, 2013; Emerson & Cantlon, 2012). Although the number of studies that have investigated these links is too small to draw firm conclusions, the studies that have been conducted use the amplitude of the neural response as the metric of neural performance. Neural amplitude is an important metric of brain development, but the information contained in the amplitude is limited (Cantlon & Li, 2013). Two additional neural measures have recently been used to examine the relationship between neural development and mathematics abilities in children: (1) a child's neural time course correlation with adults' and (2) functional connectivity.

Our lab recently implemented an analysis that measures neural activity while children view naturalistic education stimuli, such as counting and reading lessons on *Sesame Street* (Cantlon & Li, 2013). We collected fMRI scans at 2-second intervals as children viewed the educational video and used the whole neural time course from the session to measure the correlation between each child's neural time course and those of adults who viewed the same video. We used the strength of the child-to-adult correlation in each brain region as a metric of neural maturity for each child. Consistent with prior evidence that the IPS is the basis for at least some of children's initial learning of mathematics, we showed that children's neural activity in the IPS as they view educational television predicts their mathematics achievement, controlling for their verbal ability. Other brain regions did not exhibit a relation with mathematics achievement independently of verbal ability. Figure 9-5 shows brain regions where children's neural activity during educational video viewing statistically predicted their mathematics performance. We further showed that the regions of the IPS that bore a relationship between neural activity and mathematics achievement also responded selectively during children's approximate numerical matching judgments of digits-to-dot arrays. The result shows a relationship in the IPS between neural processes that are important for mathematics development and those that are recruited to estimate numerical values from arrays of dots during childhood.

The second neural measure that has revealed links between the analog numerical system and formal mathematics achievement is functional connectivity. As described in the previous section, developmental neuroimaging data have revealed the mutual involvement of the IPS and prefrontal cortex in numerical judgments, and this observation has led to the hypothesis that interactions between frontal and parietal regions are important for the development of uniquely human symbolic numerical coding. In a recent neuroimaging study, we found that neural synchrony, or "functional connectivity," between frontal and parietal regions that are recruited when children estimate numerical equivalence between digits and arrays of dots predicts their age-related mathematics achievement. That is, individual variability in the functional connectivity of the frontal and parietal regions that process approximate numerical values correlates with individual variability in mathematics test scores independently of verbal test scores (Emerson & Cantlon, 2012). A similar finding was reported by Cho and colleagues (Cho et al., 2012). The

Overlap of Activations

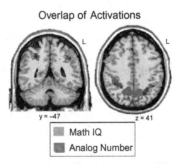

FIGURE 9-5 Overlap of IPS activation in 4- to 8-year-old children between regions in which neural activity is related to math IQ (i.e., mathematics achievement) and regions in which neural activity is related to approximate numerical processing. *(From Cantlon & Li, 2013.)*

implication is that neural interactions between the brain regions that underlie analog numerical processing are related to the development of formal mathematical concepts in children.

Data from behavioral and neural individual differences analyses suggest a relation between analog numerical processing and formal mathematics learning and achievement. Although the results of these studies are somewhat mixed, the possibility that analog numerical processes are related to formal mathematics knowledge seems likely given evidence described in previous sections. Yet, this relation may not be uniform over the course of development because cognitive and neural mechanisms have the potential to undergo qualitative changes. Theories and data that address the interactions between formal symbolic mathematical cognition, analog number representations, and domain-general mechanisms such as executive function and intelligence are needed to more fully understand how the analog number system and domain general abilities differentially contribute to or interact in the learning of formal mathematics. Individual differences correlations are an important source of information for testing relations among different types of numerical abilities, but experimental manipulations are also needed to test for overlap in cognitive processing (see Park & Brannon, 2013). Data that address a neural relation between analog number and mathematics are currently sparse and tend to focus on response amplitude as a primary measure of neural performance. Whole time course measures such as "neural maturity," or a child's neural time course correlation with adults', and anatomical and functional connectivity can provide independent insights into the neural basis of mathematics knowledge.

CONCLUSIONS

The goal of this chapter has been to examine the origins and organization of numerical abilities ranging from analog quantification to more formal arithmetic and mathematics achievement. The general hypothesis is that the uniquely human ability to perform complex and sophisticated symbolic mathematics can be traced back to a simpler computational system that is shared among many animals: the analog numerical system. Humans and nonhuman animals possess a common system for making numerical judgments via analog representations. Throughout development, analog numerical representations interact with the human ability to represent numerical values symbolically, suggesting a relation between "primitive" and modern numerical systems in humans. Data from neural analyses of numerical processing support this conclusion and provide independent confirmation that these are, in fact, related if not overlapping systems.

The analog numerical system precedes the symbolic system and potentially scaffolds the development of the more complex symbolic system. Of course, unlike other animals, humans ultimately attain a level of symbolic mathematics that far exceeds the mathematics concepts that are sparked by the analog numerical system. Questions remain regarding the precise taxonomy of the development and organization of numerical information. The analog numerical system could serve to bootstrap early childhood symbolic mathematics (e.g., counting), whereas later mathematics concepts (e.g., fraction multiplication) could be acquired completely independently of the analog system (Nieder, 2009). However, the general nature of the relation between "primitive" and modern numbers appears to derive from evolutionary constraints on the structure of numerical concepts in the mind and brain as well as the conceptual and neural foundation that evolution has provided for the development of numerical thinking in humans.

ACKNOWLEDGMENTS

The author is grateful to Katherine Blakely for editorial assistance. The author acknowledges funding from NICHD (R01HD064636), the Alfred P. Sloan Foundation, and the James S. McDonnell Foundation.

REFERENCES

Ansari, D. (2008). Effects of development and enculturation on number representation in the brain. *Nature Reviews. Neuroscience, 9*, 278–291.

Ansari, D., Garcia, N., Lucas, E., Hamon, K., & Dhital, B. (2005). Neural correlates of symbolic number processing in children and adults. *Neuroreport, 16*, 1769–1773.

Beran, M. J. (2001). Summation and numerousness judgments of sequentially presented sets of items by chimpanzees (*Pan troglodytes*). *Journal of Comparative Psychology, 155*, 181–191.

Beran, M. J. (2004). Chimpanzees (*Pan troglodytes*) respond to nonvisible sets after one-by-one addition and removal of items. *Journal of Comparative Psychology, 118*, 25–36.

Beran, M. J., & Beran, M. M. (2004). Chimpanzees remember the results of one-by-one addition of food items to sets over extended time periods. *Psychological Science, 15*, 94–99.

Beran, M. J., Gulledge, J. P., & Washburn, D. A. (2006). Animals count: What's next? Contributions from the Language Research Center to primate numerical cognition research. In D. A. Washburn (Ed.), *Primate perspectives on behavior and cognition* (pp. 161–173). Washington, DC: APA Press.

Bongard, S., & Nieder, A. (2010). Basic mathematical rules are encoded by primate prefrontal cortex neurons. *Proceedings of the National Academy of Sciences of the United States of America, 107*, 2277–2282.

Boysen, S. T., & Berntson, G. G. (1989). Numerical competence in a chimpanzee (*Pan troglodytes*). *Journal of Comparative Psychology, 103*(1), 23.

Brannon, E. M., & Terrace, H. S. (1998). Ordering of the numerosities 1 to 9 by monkeys. *Science, 282*, 746–749.

Bugden, S., & Ansari, D. (2011). Individual differences in children's mathematical competence are related to the intentional but not automatic processing of Arabic numerals. *Cognition, 118*, 32–44.

Bugden, S., Price, G. R., McLean, D. A., & Ansari, D. (2012). The role of the left intraparietal sulcus in the relationship between symbolic number processing and children's arithmetic competence. *Developmental Cognitive Neuroscience, 2*, 448–457.

Bull, R., & Scerif, G. (2001). Executive functioning as a predictor of children's mathematics ability: Inhibition, switching, and working memory. *Developmental Neuropsychology, 19*, 273–293.

Cantlon, J. F., & Brannon, E. M. (2005). Semantic congruity affects numerical judgments similarly in monkeys and humans. *Proceedings of the National Academy of Sciences of the United States of America, 102*, 16507–16511.

Cantlon, J. F., & Brannon, E. M. (2006). Shared system for ordering small and large numbers in monkeys and humans. *Psychological Science, 17*, 401–406.

Cantlon, J. F., & Brannon, E. M. (2007a). Basic math in monkeys and college students. *PLoS Biology, 5*(12), e328.

Cantlon, J. F., & Brannon, E. M. (2007b). How much does number matter to a monkey (*Macaca mulatta*)? *Journal of Experimental Psychology. Animal Behavior Processes, 33*, 32.

Cantlon, J. F., Brannon, E. M., Carter, E. J., & Pelphrey, K. A. (2006). Functional imaging of numerical processing in adults and 4-year-old children. *PLoS Biology, 4*(5), e125.

Cantlon, J. F., Cordes, S., Libertus, M. E., & Brannon, E. M. (2009). Comment on "Log or Linear? Distinct intuitions of the number scale in Western and Amazonian Indigene Cultures." *Science, 323*, 38b.

Cantlon, J. F., & Li, R. (2013). Neural activity during natural viewing of Sesame Street statistically predicts test scores in early childhood. *PLoS Biology, 11*, e1001462.

Cantlon, J. F., Libertus, M. E., Pinel, P., Dehaene, S., Brannon, E. M., & Pelphrey, K. A. (2009). The neural development of an abstract concept of number. *Journal of Cognitive Neuroscience, 21*, 2217–2229.

Cantlon, J. F., Pinel, P., Dehaene, S., & Pelphrey, K. A. (2011). Cortical representations of symbols, objects, and faces are pruned back during early childhood. *Cerebral Cortex, 21*, 191–199.

Cantlon, J. F., Platt, M. L., & Brannon, E. M. (2009). Beyond the number domain. *Trends in Cognitive Sciences, 13*, 83–91.

Cantlon, J. F., Safford, K. E., & Brannon, E. M. (2010). Spontaneous analog number representations in 3-year-old children. *Developmental Science, 13*, 289–297.

Carey, S. (2004). Bootstrapping and the origin of concepts. *Daedalus, 133*, 59–68.

Carey, S. (2010). *The origins of concepts*. New York: Oxford University Press.

Cho, S., Metcalfe, A. W., Young, C. B., Ryali, S., Geary, D. C., & Menon, V. (2012). Hippocampal–prefrontal engagement and dynamic causal interactions in the maturation of children's fact retrieval. *Journal of Cognitive Neuroscience, 24*, 1849–1866.

Cohen Kadosh, R., Cohen Kadosh, K., Kaas, A., Henik, A., & Goebel, R. (2007). Notation-dependent and -independent representations of numbers in the parietal lobes. *Neuron, 53*, 307–314.

de Hevia, M. D., & Spelke, E. S. (2009). Spontaneous mapping of number and space in adults and young children. *Cognition, 110*, 198–207.

De Smedt, B., Noël, M., Gilmore, C., & Ansari, D. (2013). The relationship between symbolic and non-symbolic numerical magnitude processing skills and the typical and atypical development of mathematics: A review of evidence from brain and behavior. *Trends in Neuroscience & Education, 2*, 48–55.

Dehaene, S. (1992). Varieties of numerical abilities. *Cognition, 44*, 1–42.

Dehaene, S., & Akhavein, R. (1995). Attention, automaticity, and levels of representation in number processing. *Journal of Experimental Psychology. Learning, Memory, and Cognition, 21*, 314.

Dehaene, S., & Cohen, L. (2007). Cultural recycling of cortical maps. *Neuron, 56*, 384–398.

Dehaene, S., Spelke, E., Pinel, P., Stanescu, R., & Tsivkin, S. (1999). Sources of mathematical thinking: Behavioral and brain-imaging evidence. *Science, 284*(5416), 970–974.

Dehaene, S., Piazza, M., Pinel, P., & Cohen, L. (2003). Three parietal circuits for number processing. *Cognitive Neuropsychology, 20*, 487–506.

Diester, I., & Nieder, A. (2007). Semantic associations between signs and numerical categories in the prefrontal cortex. *PLoS Biology, 5*(11), e294.

Diester, I., & Nieder, A. (2008). Complementary contributions of prefrontal neuron classes in abstract numerical categorization. *The Journal of Neuroscience, 28,* 7737–7747.

Eger, E., Sterzer, P., Russ, M. O., Giraud, A. L., & Kleinschmidt, A. (2003). A supramodal number representation in human intraparietal cortex. *Neuron, 37,* 719–726.

Emerson, R. W., & Cantlon, J. F. (2012). Early math achievement and functional connectivity in the fronto-parietal network. *Developmental Cognitive Neuroscience, 2,* S139–S151.

Emmerton, J. (2001). Birds' judgments of number and quantity. In R. G. Cook (Ed.), *Avian visual cognition.* [On-line]. Available: www.pigeon.psy.tufts.edu/avc/emmerton/.

Evans, T. A., Beran, M. J., Harris, E. H., & Rice, D. (2009). Quantity judgments of sequentially presented food items by capuchin monkeys (*Cebus apella*). *Animal Cognition, 12,* 97–105.

Feigenson, L., Dehaene, S., & Spelke, E. (2004). Core systems of number. *Trends in Cognitive Sciences, 8,* 307–314.

Friedman, L., & Wall, M. (2005). Graphical views of suppression and multicollinearity in multiple linear regression. *The American Statistician, 59,* 127–136.

Gallistel, C. R. (1989). Animal cognition: The representation of space, time and number. *Annual Review of Psychology, 40,* 155–189.

Gallistel, C. R. (1990). *The organization of learning.* Cambridge, MA: MIT Press.

Gallistel, C. R., & Gelman, R. (1992). Preverbal and verbal counting and computation. *Cognition, 44,* 43–74.

Geary, D. C. (2007). Educating the evolved mind: Conceptual foundations for an evolutionary educational psychology. In J. S. Carlson & J. R. Levin (Eds.), *Educating the evolved mind* (pp. 1–99). Greenwich, CT: Information Age, (Vol. 2, Psychological perspectives on contemporary educational issues).

Geary, D. C. (2013). Early foundations for mathematics learning and their relations to learning disabilities. *Current Directions in Psychological Science, 22,* 23–27.

Gebuis, T., Kadosh, R. C., de Haan, E., & Henik, A. (2009). Automatic quantity processing in 5-year olds and adults. *Cognitive Processing, 10,* 133–142.

Gelman, R., & Butterworth, B. (2005). Number and language: How are they related? *Trends in Cognitive Sciences, 9,* 6–10.

Gelman, R., & Gallistel, C. R. (1978). *The child's understanding of number.* Cambridge, MA: Harvard University Press.

Gilmore, C. K., McCarthy, S. E., & Spelke, E. S. (2007). Symbolic arithmetic knowledge without instruction. *Nature, 447,* 589–591.

Gould, S. J., & Lewontin, R. C. (1979). The spandrels of San Marco and the Panglossian paradigm: A critique of the adaptationist programme. *Proceedings of the Royal Society B, 205,* 581–598.

Grill-Spector, K., Golarai, G., & Gabrieli, J. (2008). Developmental neuroimaging of the human ventral visual cortex. *Trends in Cognitive Sciences, 12,* 152–162.

Halberda, J., Mazzocco, M. M., & Feigenson, L. (2008). Individual differences in non-verbal number acuity correlate with maths achievement. *Nature, 455,* 665–668.

Henik, A., & Tzelgov, J. (1982). Is three greater than five: The relation between physical and semantic size in comparison tasks. *Memory & Cognition, 10,* 389–395.

Holloway, I. D., & Ansari, D. (2009). Mapping numerical magnitudes onto symbols: The numerical distance effect and individual differences in children's mathematics achievement. *Journal of Experimental Child Psychology, 103,* 17–29.

Holloway, I. D., & Ansari, D. (2010). Developmental specialization in the right intraparietal sulcus for the abstract representation of numerical magnitude. *Journal of Cognitive Neuroscience, 22,* 2627–2637.

Holyoak, K. J. (1977). The form of analog size information in memory. *Cognitive Psychology, 9,* 31–51.

Hurewitz, F., Gelman, R., & Schnitzer, B. (2006). Sometimes area counts more than number. *Proceedings of the National Academy of Sciences of the United States of America, 103,* 19599–19604.

Hyde, D. C., & Spelke, E. S. (2011). Neural signatures of number processing in human infants: Evidence for two core systems underlying numerical cognition. *Developmental Science, 14,* 360–371.

Izard, V., Sann, C., Spelke, E. S., & Streri, A. (2009). Newborn infants perceive abstract numbers. *Proceedings of the National Academy of Sciences of the United States of America, 106,* 10382–10385.

Kaufmann, L., Koppelstaetter, F., Siedentopf, C., Haala, I., Haberlandt, E., Zimmerhackl, L. B., et al. (2006). Neural correlates of the number–size interference task in children. *Neuroreport, 17,* 587.

Kaufmann, L., Wood, G., Rubinsten, O., & Henik, A. (2011). Meta-analyses of developmental fMRI studies investigating typical and atypical trajectories of number processing and calculation. *Developmental Neuropsychology, 36,* 763–787.

Le Corre, M., & Carey, S. (2007). One, two, three, four, nothing more: An investigation of the conceptual sources of the verbal counting principles. *Cognition, 105,* 395–438.

Libertus, M., Feigenson, L., & Halberda, J. (2011). Preschool acuity of the approximate number system correlates with school math ability. *Developmental Science, 14,* 1292–1300.

Libertus, M. E., Feigenson, L., & Halberda, J. (2013). Is approximate number precision a stable predictor of math ability? *Learning and Individual Differences, 25,* 126–133.

Libertus, M. E., Odic, D., & Halberda, J. (2012). Intuitive sense of number correlates with math scores on college-entrance examination. *Acta Psychologica, 141*(3), 373–379.

Libertus, M. E., Pruitt, L. B., Woldorff, M. G., & Brannon, E. M. (2009). Induced alpha-band oscillations reflect ratio-dependent number discrimination in the infant brain. *Journal of Cognitive Neuroscience, 21,* 2398–2406.

Lipton, J. S., & Spelke, E. S. (2005). Preschool children's mapping of number words to nonsymbolic numerosities. *Child Development, 76,* 978–988.

Lourenco, S. F., & Longo, M. R. (2010). General magnitude representation in human infants. *Psychological Science, 21*(6), 873–881.

Mazzocco, M. M., Bhatia, N. S., & Lesniak-Karpiak, K. (2006). Visuospatial skills and their association with math performance in girls with fragile X or Turner syndrome. *Child Neuropsychology, 12,* 87–110.

Mazzocco, M. M., & Kover, S. T. (2007). A longitudinal assessment of executive function skills and their association with math performance. *Child Neuropsychology, 13,* 18–45.

Meck, W. H., & Church, R. M. (1983). A mode control model of counting and timing processes. *Journal of Experimental Psychology. Animal Behavior Processes, 9,* 320.

Moyer, R. S., & Landauer, T. K. (1967). Time required for judgments of numerical inequality. *Nature, 215,* 1519–1520.

Mussolin, C., Nys, J., & Leybaert, J. (2012). Relationships between approximate number system acuity and early symbolic number abilities. *Trends in Neuroscience and Education, 1,* 21–31.

Naccache, L., & Dehaene, S. (2001). The priming method: Imaging unconscious repetition priming reveals an abstract representation of number in the parietal lobes. *Cerebral Cortex, 11*, 966–974.

Nieder, A. (2012). Supramodal numerosity selectivity of neurons in primate prefrontal and posterior parietal cortices. *Proceedings of the National Academy of Sciences of the United States of America, 109*, 11860–11865.

Nieder, A. (2005). Counting on neurons: The neurobiology of numerical competence. *Nature Reviews. Neuroscience, 6*, 177–190.

Nieder, A. (2009). Prefrontal cortex and the evolution of symbolic reference. *Current Opinion in Neurobiology, 19*, 99–108.

Nieder, A., & Merten, K. (2007). A labeled-line code for small and large numerosities in the monkey prefrontal cortex. *The Journal of Neuroscience, 27*, 5986–5993.

Nieder, A., & Miller, E. K. (2004). A parieto-frontal network for visual numerical information in the monkey. *Proceedings of the National Academy of Sciences of the United States of America, 101*, 7457–7462.

Park, J., & Brannon, E. M. (2013). Training the approximate number system improves math proficiency. *Psychological Science, 24*, 2013–2019.

Penn, D. C., Holyoak, K. J., & Povinelli, D. J. (2008). Darwin's mistake: Explaining the discontinuity between human and nonhuman minds. *Behavioral and Brain Sciences, 31*, 109–130.

Piazza, M. (2010). Neurocognitive start-up tools for symbolic number representations. *Trends in Cognitive Sciences, 14*, 542–551.

Piazza, M., Izard, V., Pinel, P., Le Bihan, D., & Dehaene, S. (2004). Tuning curves for approximate numerosity in the human intraparietal sulcus. *Neuron, 44*, 547–555.

Piazza, M., Mechelli, A., Price, C. J., & Butterworth, B. (2006). Exact and approximate judgements of visual and auditory numerosity: An fMRI study. *Brain Research, 1106*, 177–188.

Piazza, M., Pinel, P., Le Bihan, D., & Dehaene, S. (2007). A magnitude code common to numerosities and number symbols in human intraparietal cortex. *Neuron, 53*, 293–305.

Pica, P., Lemer, C., Izard, V., & Dehaene, S. (2004). Exact and approximate arithmetic in an Amazonian indigene group. *Science, 306*, 499–503.

Pinel, P., Piazza, M., Le Bihan, D., & Dehaene, S. (2004). Distributed and overlapping cerebral representations of number, size, and luminance during comparative judgments. *Neuron, 41*, 983–993.

Rivera, S. M., Reiss, A. L., Eckert, M. A., & Menon, V. (2005). Developmental changes in mental arithmetic: Evidence for increased functional specialization in the left inferior parietal cortex. *Cerebral Cortex, 15*, 1779–1790.

Roitman, J. D., Brannon, E. M., & Platt, M. L. (2007). Monotonic coding of numerosity in macaque lateral intraparietal area. *PLoS Biology, 5*(8), e208.

Rousselle, L., & Noël, M. P. (2008). The development of automatic numerosity processing in preschoolers: Evidence for numerosity-perceptual interference. *Developmental Psychology, 44*, 544–560.

Scarf, D., Hayne, H., & Colombo, M. (2011). Pigeons on par with primates in numerical competence. *Science, 334*, 1664.

Schlaggar, B. L., Brown, T. T., Lugar, H. M., Visscher, K. M., Miezin, F. M., & Petersen, S. E. (2002). Functional neuroanatomical differences between adults and school-age children in the processing of single words. *Science, 296*, 1476–1479.

Supekar, K., Swigart, A. G., Tenison, C., Jolles, D. D., Rosenberg-Lee, M., Fuchs, L., et al. (2013). Neural predictors of individual differences in response to math tutoring in primary-grade school children. *Proceedings of the National Academy of Sciences of the United States of America, 110*, 8230–8235.

Tudusciuc, O., & Nieder, A. (2007). Neuronal population coding of continuous and discrete quantity in the primate posterior parietal cortex. *Proceedings of the National Academy of Sciences of the United States of America, 104*, 14513–14518.

vanMarle, K., Chu, F. W., Li, Y., & Geary, D. C. (in press). Acuity of the approximate number system and preschoolers' quantitative development. *Developmental Science.*

Viswanathan, P., & Nieder, A. (2013). Neuronal correlates of a visual "sense of number" in primate parietal and prefrontal cortices. *Proceedings of the National Academy of Sciences, 110*(27), 11187–11192.

Wynn, K. (1990). Children's understanding of counting. *Cognition, 36*(2), 155–193.

Wynn, K. (1992). Addition and subtraction by human infants. *Nature, 358*, 749–750.

Chapter 10

The Small–Large Divide: A Case of Incompatible Numerical Representations in Infancy

Tasha Posid and Sara Cordes
Boston College, Chestnut Hill, MA, USA

INTRODUCTION

The importance of understanding the developmental origins of our numerical abilities has recently been highlighted by studies suggesting that primitive abilities for tracking numerical information serve as the preverbal foundations for formal mathematics. Individual differences in the sensitivity of these preverbal abilities may contribute to children's initial learning of formal mathematical symbols and their meaning (e.g., Arabic numerals) and contribute to variation in mathematical outcomes in adults (e.g., Bonny & Lourenco, 2013; Geary, 2011; Geary, Hoard, Nugent, & Bailey, 2013; Halberda & Feigenson, 2008; Jordan, Kaplan, Ramineni, & Locuniak, 2009; LeFevre, Fast, Skwarchuk, Smith-Chant, & Bisanz, 2010; Libertus, Odic, & Halberda, 2012; Star & Brannon, this volume), although the relative importance of preverbal and verbal numerical processing is debated (De Smedt, Noel, Gilmore, & Ansari, 2013). Importantly, this relationship appears to be causal, at least in some domains of formal mathematics, such that arithmetic processing is improved following training on approximate numerical estimation (e.g., Hyde, Khanum, & Spelke, 2014; Park & Brannon, 2013), and is evident prior to formal mathematical experience, with preverbal numerical abilities in infancy and early childhood predicting math achievement several years later (Libertus, Feigenson, & Halberda, 2011, 2013; Starr, Libertus, & Brannon, 2013b; vanMarle, Chu, Li, & Geary, 2014). Given that math achievement upon entering school is strongly predictive of math achievement throughout later schooling (e.g., Duncan et al., 2007; Geary et al., 2013), it is critical that we understand the origins of these numerical abilities in order to target educational outcomes long before children reach the classroom.

Mathematical Cognition and Learning, Vol. 1. http://dx.doi.org/10.1016/B978-0-12-420133-0.00010-7

A substantial corpus of work accumulated over the past 40 or more years has unveiled striking similarities in the ways that numerical information is processed across phylogeny and ontogeny (see Agrillo et al., this volume; Anderson & Cordes, 2013; Beran et al., this volume; Cantlon, Platt, & Brannon, 2009; Gallistel, 1990; Geary et al., this volume; Pepperberg, this volume; Vallortigara, this volume). For example, data from infants, children, adults, and nonhuman animals consistently point to the existence of two distinct systems for tracking set sizes: one specifically dedicated to precisely tracking small sets of items (≤ 3; referred to as the object file system), and a second system responsible for representing all set sizes in an approximate manner (referred to as either the analog magnitude or approximate number system [ANS]). Despite notable parallels observed across development, robust differences in the way infants deal with the interface of these two numerical systems may point to qualitative distinctions in the way that number is processed throughout development. In this chapter, we review the evidence for these two systems across development and across species while attempting to answer open questions regarding how humans progress from a state of representational incompatibility, in which infants are generally incapable of comparing small and large sets due to qualitatively distinct representational systems, to a numerically fluid state, in which small and large set sizes are given equal treatment, just a few years later.

DISTINCT SYSTEMS OF REPRESENTATION: EVIDENCE OF CONTINUITY ACROSS DEVELOPMENT AND PHYLOGENY

Converging evidence suggests that humans and nonhuman animals have access to two distinct systems for representing number (see Feigenson, Dehaene, & Spelke, 2004, and Cordes & Brannon, 2008b, for reviews). The first is a noisy, analog magnitude system used for representing number in an approximate manner (Barth et al., 2006; Brannon & Terrace, 1998; Cantlon & Brannon, 2006; Cordes & Gelman, 2005; Cordes, Gelman, Gallistel, & Whalen, 2001; Dehaene, 1997; Gallistel & Gelman, 2000; Meck & Church, 1983; Whalen, Gallistel, & Gelman, 1999; Xu & Spelke, 2000). Importantly, the signature characteristic of the ANS is its adherence to Weber's law, such that the ease with which two set sizes are discriminated is dependent on their ratio, not absolute difference (e.g., Halberda & Feigenson, 2008; Halberda & Odic, this volume). That is, the speed and accuracy of discriminating sets of 8 and 6 items is the same as that for discriminating sets of 16 and 12 items (3:4 ratio in both cases; Barth et al., 2006). Evidence for a ratio-dependent ANS system has been found in humans throughout the lifespan, but also in nonhuman animals including primates (e.g., chimpanzees; *Pan troglodyte*; Beran & Beran, 2004; Beran, Evans, & Harris, 2008; lemurs: *Lemur catta, Eulemur mongoz, Eulemur macaco flavifrons*; Jones et al., 2013; macaques: *Macaca mulatta*; Jones et al., 2013;

orangutans; *Pongo pygmaeus*; Call, 2000; Hanus & Call, 2007), elephants (*Loxodonta Africana*; Perdue, Talbot, Stone, & Beran, 2012), birds (e.g., African Grey parrots; *Psittacus erithacus;* Al Ain, Giret, Grand, Kreutzer, & Bovet, 2009; Bogale, Kamata, Mioko, & Sugita, 2011; Zorina & Smirnova, 1996), dogs (*Canis lupus familiaris;* coyote: *Canis latrans*; Baker, Shivik, & Jordan, 2011; Ward & Smuts, 2007), bears (*Ursus americanus*; Vonk & Beran, 2012), sea lions (*Otaria flavescens*; Abramson, Hernandez-Lloreda, Call, & Colmenares, 2011), salamanders (*Caudata plethodon*; Krusche, Uller, & Dicke, 2010), and even fish (*Gambusia affinis*; Agrillo, Piffer, & Bisazza, 2010; Agrillo, Piffer, Bisazza, & Butterworth, 2012; *Xiphophorus helleri;* Buckingham, Wong, & Rosenthal, 2007; *Poecilia reticulata;* Piffer, Agrillo, & Hyde, 2011; for a review, see Anderson & Cordes, 2013; Cantrell & Smith, 2013).

Developmental data from both animals and humans suggest that the precision of ANS representations increases with age, such that the numerical ratio of discriminability approaches one over the course of development, with the greatest changes early in life (e.g., Bisazza, Piffer, Serena, & Agrillo, 2010; Halberda & Feigenson, 2008; Libertus & Brannon, 2010; Lipton & Spelke, 2003). Specifically, whereas newborn humans require as much as a 3-fold change in number to notice a change (e.g., 4 vs. 12), 6-months-olds are able to detect a 2-fold (but not a 1.5-fold) change (e.g., 4 vs. 8, 8 vs. 16, 16 vs. 32), and 9- to 10-month olds notice even smaller numerical changes (1.5-fold change; e.g., 8 vs. 12; Brannon, Abbot, & Lutz, 2004; Cordes & Brannon, 2008a; Izard, Sann, Spelke, & Streri, 2009; Lipton & Spelke, 2003, 2004; Wood & Spelke, 2005; Xu, 2003; Xu & Arriaga, 2007; Xu & Spelke, 2000; Xu, Spelke, & Goddard, 2005), with this precision continuing to increase into adulthood, such that adults discriminate a 1.14-fold (7:8) change in magnitude (Halberda & Feigenson, 2008). A similar progression in ANS acuity has been found in nonhuman animals, such that 1-day-old guppies are unable to discriminate a 2-fold difference in number (4 vs. 8), but by their 40th day of life, they do so reliably (Bisazza et al., 2010).

The second system implicated in numerical tasks, the object file system, is an exact, one-to-one representational system that can be used to track only a small number of visual items (1–3 or 4; Carey & Xu, 2001; Dehaene, 1997; Feigenson, Carey, & Hauser, 2002; Feigenson et al., 2004; Hyde & Wood, 2011; Leslie, Xu, Tremoulet, & Scholl, 1998; Simon, 1997). In contrast to the ANS, this system of parallel individuation has an absolute set size limit (e.g., Alvarez & Cavanaugh, 2004; Alvarez & Franconeri, 2007). Evidence suggests that human infants can hold exactly three or fewer items in working memory when making numerical discriminations (Carey & Xu, 2001; Feigenson, 2008; Feigenson et al., 2004; Hyde & Wood, 2011; Jordan & Brannon, 2006; Uller, Carey, Huntley-Fenner, & Klatt, 1999; Xu, 2003). Similarly, human adults simultaneously track up to 4 or 5 items before working memory becomes overly taxed (e.g., Awh, Vogel, & Oh, 2006;

Feigenson & Yamaguchi, 2009; Halberda, Simons, & Wetherhold, 2014; Klahr, 1973; Luck & Vogel, 1997; Luria & Vogel, 2011; Piazza, Giacomini, Bihan, & Dehaene, 2003; Scholl & Pylyshyn, 1999; Trick & Pylyshyn, 1994; Vogel, Woodman, & Luck, 2001; Zosh & Feigenson, 2012; Zosh, Halberda, & Feigenson, 2011). Moreover, data from nonhuman animals also reveal precise tracking of small sets of items (Agrillo et al., 2012; Bisazza et al., 2010; Hauser, Carey, & Hauser, 2000; Hunt, Low, & Burns, 2008; Piffer et al., 2011; Uller, Jaeger, Guidry, & Martin, 2003; Uller & Lewis, 2009).

Notably, the object file system is posited to be a function of the visual attention system, and as such, it is employed only when tracking visual objects (i.e., not sounds; vanMarle & Wynn, 2009, but see Mou & vanMarle, 2013). The object file system was originally conceptualized within the adult visual attention literature, where striking demonstrations revealed differences in the way adults tracked a small compared to a large number of objects. In a classic example, adult observers were briefly shown a sample array consisting of 1–12 colored squares and then saw a test array and were asked to identify whether the sample array differed from the test array in the color of one of the squares (Luck & Vogel, 1997). When arrays of 1–3 items were presented, subjects demonstrated perfect performance. However, performance declined systematically as a function of set size once the number of items in the array increased from 4 to 12. Furthermore, this decline in accuracy was not due to verbal working memory nor to any limitations in overall processing, as subjects showed no difference in performance when given more time to view the sample array. Therefore, performance accuracy varied as a function of set size specifically due to a difference in demands on working memory for dealing with small sets (<4) compared to large sets (>3; Luck & Vogel, 1997).

Importantly, although the object file system likely evolved for non-numerical processing—because it involves the tracking of individual items and, unlike the ANS, does not implicitly represent the items in a numerical fashion (i.e., as a collection with a cardinal value)—the system has been implicated in numerical tasks across the lifespan. Object file representations have been implicated in visual enumeration in adulthood, where small sets (4 or fewer items) are generally enumerated effortlessly, accurately, and quickly ("subitized"), whereas the enumeration of larger groups invokes an effortful, slower, and error-prone process of verbal counting (Balakrishnan & Ashby, 1982; Piazza, Mechelli, Butterworth, & Price, 2002; Trick, Enns, & Brodeur, 1996; Trick & Pylyshyn, 1993, 1994). Similarly, human infants also appear to show a numerical advantage in the small number range, such that infant data reveal finer-grained numerical discriminations than predicted by the ANS when sets fall exclusively within the small number range (3 or fewer). For example, although 6-month-olds robustly fail to discriminate large sets differing by only a 2:3 ratio (e.g., 4 vs 6; 8 vs. 12, 16 vs. 24; Lipton &

Spelke, 2003, 2004; Wood & Spelke, 2005; Xu, 2003; Xu & Arriaga, 2007; Xu & Spelke, 2000; Xu, Spelke, & Goddard, 2005), when sets are small (<4), they succeed in doing so under certain circumstances (2 vs. 3; e.g., Antell & Keating, 1983; Bijeljac-Babic, Bertoncini, & Mehler, 1993; Cordes & Brannon, 2009b; Jordan, Suanda, & Brannon, 2008; Kobayashi, Hiraki, & Hasegawa, 2005; see Cordes & Brannon, 2008b for review). Furthermore, similar small set numerical discrimination advantages have been demonstrated in nonhuman animals (guppies, *Poecilia reticulata*: Bisazza et al., 2010; mosquitofish, *Gambusia holbrooki:* Agrillo, Dadda, Serena, & Bisazza, 2008; chicks, *Gallus gallus*: Rugani, Regolin, & Vallortigara, 2008; dogs, *Canus lupus familiaris*: Bonanni, Natoli, Cafazzo, & Valsecchi, 2011; primates, *Macaca mulatta:* Hauser et al., 2000).

Consistent with the behavioral data, neuroscientific evidence has also revealed clear differences in how we process small and large sets. Notable differences in both the location and timing of brain activation have been demonstrated as a function set size (Ansari, Lyons, van Eimeren, & Xu, 2007; Hyde & Spelke, 2009, 2011, 2012; see Buhusi & Cordes, 2011). For example, imaging studies with human adults have revealed that processing of small nonsymbolic (i.e., arrays of 1–3 dots) sets (but not symbolic number, i.e., Arabic numerals) results in activation of an area of the brain associated with visual attention (right temporo-parietal junction, or rTPJ). In contrast, less activity in this area (presumably resulting in greater suppression of object file representations) was associated with faster numerical judgments involving large sets (Ansari et al., 2007; see also Hyde & Spelke, 2012). Moreover, event-related potential studies (ERP, tracking the timing of electrical activity in the brain) have revealed passive viewing of small sets in both human adults and infants to evoke earlier activity, with the magnitude of the activity dependent on the *absolute* size of the set (regardless of the size of previously viewed sets). In contrast, ERPs associated with viewing large sets were slightly later and dependent on the *relative* magnitude of the set (compared to other sets; Hyde & Spelke, 2009; 2011). In sum, the neuroscientific data corroborate behavioral findings to suggest small sets are processed in a very different manner than large sets across the lifespan.

EVIDENCE FOR TWO SYSTEMS IN INFANCY

As described previously, both behavioral and neuroscientific investigations indicate humans throughout the lifespan and nonhuman animals have access to two distinct systems used for representing numerical quantity: an exact object file system used to precisely track items within small sets (<4) and an approximate number system used to represent all natural numbers (van-Marle, this volume). However, arguably the strongest evidence to date of these two distinct systems emerges from work with human infants, where numerical discrimination data robustly violate Weber's law when one of the sets involved is small, in two different regards. First, as alluded to earlier, infant discriminations of small sets are reported to be more precise than dictated by Weber's law under certain circumstances. Second, discriminations of small sets from large ones yield robust failures, despite seemingly facile discrimination ratios.

Small versus Small Discriminations

Consistent with the exact nature of object file representations, data reveal infant discriminations of exclusively small sets (1–3 items) can be more precise than predicted by Weber's law. For example, 6-month-olds robustly fail to discriminate a 2:3 ratio (e.g., 6 vs. 9) of change in number for large sets, yet they notice the difference between sets of 2 and 3 items (also a 2:3 ratio, but involving exclusively small sets; Cordes & Brannon, 2009b; Jordan et al., 2008; Xu & Spelke, 2000). Importantly, however, infants' success at discriminating among small sets is found only under certain circumstances in which tasks include multimodal input or when redundant visual information is provided (Cantrell & Smith, 2013). For example, when infants are placed in a situation in which they must match the number of sounds they hear to the number of items they see, they detect a mismatch between 2 sounds and 3 objects and vice versa (Jordan et al., 2008; Kobayashi et al., 2005; Starkey, Spelke, & Gelman, 1983, 1990). Moreover, when habituated to purely visual arrays in which continuous quantities (surface area, contour) are held constant in habituation, young infants notice the difference between 2 and 3 items (Antell & Keating, 1983; Cordes & Brannon, 2009b).[1]

Importantly, however, when continuous quantities vary from trial-to-trial in habituation (thus preventing infants from using surface area as a cue for discrimination), 6-month-olds fail to detect even a 1:2 ratio change in number (1 vs. 2 items; Xu et al., 2005). This finding is particularly surprising given that similar-aged infants have no difficulty in detecting a 1:2 ratio change in number across stimuli varying in continuous quantities when sets are exclusively large (e.g., 8 vs. 16; Lipton & Spelke, 2003; Wood & Spelke, 2005; Xu, 2003; Xu & Spelke, 2000; Xu et al., 2005). In sum, unlike the ubiquitous ability to compare large sets using the ANS, infant abilities to detect numerical changes for exclusively small sets within the object file system appear to be highly contingent upon task variables.

The Small–Large Divide

Even more striking evidence in favor of the two-systems account in infancy is the finding of robust discrimination failures when comparing small (<4) and

---

1. The need for redundant visual information has provided the basis for the Signal Clarity Hypothesis (Cantrell & Smith, 2013) discussed later in this chapter.

large (≥ 4) sets, despite a favorable ratio of discriminability (e.g., Cordes & Brannon, 2009a; Feigenson & Carey, 2003, 2005; Feigenson et al., 2002; Lipton & Spelke, 2004; vanMarle, 2013; Wood & Spelke, 2005; Xu, 2003; see also Mou & vanMarle, 2013). For example, although looking time measures reveal that infants as young as 6 months reliably discriminate 1:2 ratio changes in number for large sets (e.g., 4 vs. 8, 16 vs. 32; e.g., Xu & Spelke, 2000), when presented with a similar 1:2 ratio change in number crossing the small–large divide (e.g., 2 vs. 4 or 3 vs. 6), they repeatedly fail on these discriminations (Cordes & Brannon, 2009a; Lipton & Spelke, 2004; Wood & Spelke, 2005; Xu, 2003). Similarly, paradigms involving more active infant responses (such as searching for toys placed within a box or crawling to a container with a greater number of food items) reveal successful discrimination between 1 vs. 2, 2 vs. 3, and 4 vs. 8 items, but not between 2 vs. 4 and 3 vs. 6 items (Feigenson & Carey, 2003, 2005; Feigenson & Halberda, 2004; Feigenson et al., 2002; vanMarle, 2013). In fact, using these active response measures, even when the ratio between the to-be-enumerated numbers is relatively large (e.g., 1 vs. 4), infants still fail to spontaneously detect which of two sets contains more items (Feigenson & Carey, 2003, 2005; Feigenson et al., 2002; vanMarle, 2013; but see Cordes & Brannon, 2009a).

TRACKING SMALL SETS WITH THE ANS: EXCEPTIONS TO THE RULE

Although evidence for small–large discrimination failures supports the notion of an incompatibility between ANS and object file systems in infancy (also see Mou & vanMarle, 2013), a handful of studies report successful discrimination between small and large sets (Cantrell, Boyer, Cordes, & Smith, 2014; Cordes & Brannon, 2009a; Hyde & Spelke, 2011; Starr, Libertus, & Brannon, 2013a; vanMarle & Wynn, 2009; Wynn, Bloom & Chiang, 2002). What circumstances give rise to infant abilities to cross the small–large numerical divide? And, more importantly, what do these findings tell us about these two systems in infancy?

Although the behavioral pattern of infant discriminations suggests that infants exclusively employ object files to represent small sets (up to 3 items) while solely invoking the ANS for larger sets, empirical evidence indicates that they in fact have access to ANS representations for small and large sets alike, at least under certain circumstances. For example, when presented with *nonvisual* numerical sets (i.e., sounds), infant numerical discriminations are ratio-dependent, consistent with an ANS signature, even for discriminations involving small sets. That is, 7-month-olds successfully detect the difference between 2 and 4 *tones* (1:2 ratio change), but not between 2 and 3 (2:3 ratio; vanMarle & Wynn, 2009). Importantly, this finding is consistent with claims that object files are a component of the visual attention system and thus are not invoked for tracking auditory stimuli, leaving ANS representations as the sole cognitive system available for tracking small sets. Therefore, in contrast to other small–large discrimination failures with visual sets, ratio-dependent numerical discriminations involving auditory sets strongly suggest that infants, much like older children, adults, and nonhuman animals (Brannon & Terrace, 1998; Cantlon & Brannon, 2006; Cordes et al., 2001; Meck & Church, 1983; Moyer & Landauer, 1967; Ward & Smuts, 2007), have access to ANS representations all the way down the number line.

Provided evidence that infants have access to ANS representations for small sets, it is somewhat surprising that object file representations are consistently invoked when infants confront small visual sets. If these representations result in discrimination failures, one might question why they have evolved to be the preferred system of representation for small sets early in development. An ecological account may posit that in the first year of life, the tracking of small sets is much more important to the young infant than the tracking of large sets. In natural contexts, infants may encounter large groups of objects (e.g., toys in the room), by far, the most important items to track are those that provide food and comfort to the infant—that is, the set of people in the room. Moreover, the members of this set are not necessarily interchangeable, such that, for example, Dad may provide a different source of nourishment and comfort than Mom. Therefore, being able to attend to the total number of individuals in the set (e.g., people in the room), as well as to store information such as the location and salient characteristics of each individual belonging to that set (i.e., know where Mom is sitting), may be considered much more relevant to preverbal infants than an abstract ability to compare small and large sets. Unlike the ANS, which provides only a summary representation of a set (i.e., the total number of items present), object files can be used to store not only the location of an object, but also salient characteristics of the object, even in infancy (e.g., remembering a triangle is at Location 1 and a circle is at Location 2; Feigenson, 2005; Kaldy & Leslie, 2003). Thus, it seems that the tracking demands of the infant's environment may naturally give rise to an overarching preference toward object file representations.

In addition to specific ecological demands, the significant difference in representational precision afforded by the object file system compared with the ANS system may also account for a bias toward object file representations early in development. Whereas the object file system allows for the *precise, exact* tracking of individual items, the ANS is an *approximate, noisy* system. Evidence of a numerical discrimination advantage for small sets in infancy (such that infants succeed on finer-grained discriminations in the small number range compared to the large number range) suggests that object files offer a relatively higher level of precision than the ANS in infancy (e.g., Antell & Keating, 1983; Cordes & Brannon, 2009b; Kobayashi et al., 2005; Jordan et al., 2008). Evidence of increasing precision in the ANS across development (e.g., Halberda & Feigenson, 2008; Lipton & Spelke, 2003) also indicates that, particularly in infancy, ANS representations are somewhat unreliable. Thus, object files may simply provide infants with a more reliable means for tracking objects and, as such, have evolved to be the preferred system for representation for small sets.

Consistent with this precision hypothesis are cases in which infants succeed in small–large discriminations following experimental manipulation of the relative precision of the two systems. The handful of cases in which infants have been shown to succeed on small–large discriminations of visual sets generally fall under two categories of circumstances: either (1) stimuli are presented in such a way as to allow ANS representations to provide reliable, clear information (Cantrell & Smith, 2013; Cantrell et al., 2014; Cordes & Brannon, 2009a); or (2) object file representations are made less reliable by taxing working memory (Hyde & Spelke, 2011; Starr et al., 2013a; Wynn, Bloom, & Chiang, 2002).

Signal Clarity of the ANS

When numerical information tracked via the ANS can provide a clear, reliable signal, ANS representations may predominate over the object file system. That is, young infants may fail to compare small and large sets on the basis of number when sets differ by a 2-fold ratio (i.e., 3 vs. 6 or 2 vs. 4; Cordes & Brannon, 2009a; Xu, 2003), yet the reason may be that this ratio change in number is near the limits of their ability to discriminate. Thus, while the ANS may be able to detect a change in number, the signal is fairly weak and unreliable, resulting in reliance on precise object files in this case. In contrast, 7-month-olds have been shown to successfully compare numbers across the small–large boundary when the sets vary by a 4-fold change in magnitude (i.e., 1 vs. 4 or 2 vs. 8; Cordes & Brannon, 2009a). That is, when provided a greater ratio of numerical change, the ANS unequivocally detects the change and produces a clear, strong signal that overrides the information provided by object file representations.

A similar pattern has been found for at least one nonhuman species (guppy, *Poecilia reticulata*). Piffer et al. (2011) demonstrated that guppies can succeed at a comparison between large numbers (5 vs. 10), small numbers (3 vs. 4), but not small–large comparisons close in magnitude (3 vs. 5). However, increasing the distance between the small and large numbers resulted in successful discrimination across the small–large boundary (3 vs. 6, 3 vs. 7, 3 vs. 9). Results such as these have been explained in terms of the signal strength of the ANS. In particular, it is posited that ANS magnitudes trump object file ones when the magnitude of the ratio between the small and large set exceeds some threshold criterion, such that ANS representations provide strong, clear, and precise information regarding changes in set sizes (Cordes & Brannon, 2009a).

Similarly, when numerical stimuli are designed in such a way as to present numerical information with reduced noise by providing redundant perceptual information across displays, numerical discrimination is facilitated across the small–large divide (Cantrell et al., 2014; Cantrell & Smith, 2013). Cantrell et al. (2014) habituated 9-month-old infants to arrays of 2 or 4 items in which, unlike standard practice in the field of numerical cognition (i.e., Xu & Spelke, 2000), continuous quantitative variables such as surface area, density, and contour, were held constant throughout habituation (i.e., in Xu & Spelke's design, infants were habituated to displays containing the same number of dots [e.g., 8], yet surface area of the dots varied dramatically [5-fold] across habituation; in contrast, in Cantrell et al.'s design, both surface area and number remained constant across habituation). When numerical information co-varied with continuous perceptual information in habituation, infants successfully discriminated between arrays of 2 and 4 items, and in a second experiment, between arrays of 3 and 4 items. Importantly, the change in cumulative surface area from habituation to test was significantly smaller than has been shown to be detectable by infants of this age (1:2 or 2:3 ratio; Cordes & Brannon, 2008a, 2011), indicating that increases in infant looking time must have been partly accounted for by an attention to numerical changes in their task.

In other words, consistent with the Signal Clarity Hypothesis (Cantrell & Smith, 2013), redundancy in number and other continuous dimensions across habituation reduced noise in the infants' ANS representations, allowing for the formation of a strong, clearer signal. In contrast to studies in which continuous variables varied throughout habituation (e.g., Cordes & Brannon, 2009 a, b; Xu, 2003; Xu et al., 2005), Cantrell and colleagues (2014; Cantrell & Smith, 2013) posit that infants in their study were sensitive to statistical information during habituation, resulting in a lower noise-to-signal ratio. This clearer signal allowed them to pick out the relevant numeric properties of the stimuli more quickly and discriminate much smaller ratios than previously demonstrated, especially those that cross the small–large boundary (e.g., 2 vs. 4, 3 vs. 6; Cordes & Brannon, 2009a, b; Xu, 2003; Xu et al., 2005). In a similar vein to the finding of successful discrimination of greater ratios crossing the small–large boundary (Cordes & Brannon, 2009a), the precision of infants' representations appear to vary as a function of the clarity of the information provided by the numerical displays.

Taxed Working Memory

Alternatively, when individual items within a small set become difficult to track via object files, ANS representations may become the default system for infants' numerical representations. Data indicate that the memory demands of a given task may influence recruitment of the ANS for tracking small sets (also see Mou & vanMarle, 2013). Hyde and Spelke (2009, 2011) found that infants' and adults' neural signatures of processing small (<4) versus large (≥4) numerical arrays were best characterized by the object file and approximate number systems, respectively. But, importantly, neural signatures associated with the ANS were obtained when processing small sets in the context of high memory load requirements (Hyde & Wood, 2011).

According to Hyde (2011), when items are presented under conditions that allow individuation, they are represented as distinct mental items, not as numerical magnitudes; however, when items are presented outside one's attentional limits (e.g., too many, too close together, display time too brief, high memory or attentional load, etc.), they are represented as mental magnitudes (e.g., Burr, Anobile, & Turi, 2011; Burr, Turi, & Anobile, 2010; Hyde & Wood, 2011; Piazza et al., 2002, 2003). To this end, this system is affected by the limits of attention and working memory, suggesting the object file system is automatically recruited for numerical discriminations when representing exclusively small sets for visual stimuli low in cognitive load, but does not remain a "default" when small numbers are presented outside the limits of the brain's ability to encode them precisely and accurately (Hyde, 2011; Hyde & Wood, 2011; see also Mou & vanMarle, 2013). Similar to the proposal that object files are automatically recruited for the representation of small sets unless there is a clear ANS signal for small quantities (Cantrell & Smith, 2013; Cordes & Brannon, 2009a), data from Hyde and colleagues likewise suggest the ANS may be used to represent small quantities when attentional and working memory manipulations degrade the precision of the object file system.

Consistent with this view, Starr et al. (2013a) found that 6-month-old infants successfully discriminate sets of 2 from 4 using a change-detection paradigm. This paradigm, originally designed to assess working memory in infancy (Ross-Sheehy, Oakes, & Luck, 2003), involves presenting infants two side-by-side movies simultaneously. One movie (the nonchanging display) shows images containing the same number of items (e.g., 4 items), but changing in surface area, density, contour, and configuration every 500 ms. The second movie (the changing display) is similar, except displays alternate between two set sizes (e.g., sets of 2 and 4) every 500 ms yet continuous extent continues to vary across displays, making it an unreliable cue for discrimination. The assumption of the change detection paradigm is that infants prefer to look to things that change, and thus will preferentially attend to the numerically changing display *if they detect the numerical changes involved* (Libertus & Brannon, 2010). Because change detection paradigms require infants to recognize a change in number across rapidly changing and dynamic visual images, this particular paradigm is thought to tax working memory to a greater extent than standard habituation paradigms, making it difficult to attend to individual items within each set and limiting the recruitment of object files (Starr et al., 2013a). In fact, across all numerical comparisons tested using this paradigm to date (1 vs. 3, 1 vs. 2, 2 vs. 3, 2 vs. 4, plus larger sets e.g., 10 vs. 20), infants' performance exhibits ratio-dependence (Starr et al., 2013a; Libertus & Brannon, 2010), the signature of the ANS. These findings, consistent with Hyde's (2011) hypothesis, suggest that different working memory demands may selectively recruit the ANS (during high cognitive load) over the object file system (Hyde, 2011; Hyde & Wood, 2011; Mou & vanMarle, 2013; Starr et al., 2013a; Wynn, Bloom, & Chiang, 2002).

In sum, although human infants demonstrate striking failures to discriminate between small (<4) and large (>3) sets, the few circumstances under which they have succeeded in doing so are consistent with the idea that infants employ ANS representations for small sets when (1) ANS signals are strong and (2) object file signals are weak, such as when working memory is taxed. Thus, it appears that ANS and object files can both simultaneously be used to represent small sets, but it is the relative precision (i.e., signal strength) of the two systems that determines which system will drive responses. At least in infancy, it appears that object files generally provide a stronger, clearer signal over the ANS, resulting in a greater reliance on this system for representing small sets. We next turn to the question of when and, importantly, *how*, during the course of development, we overcome this dependency on the object file system for representing small sets, such that older children reliably discriminate small from large sets without error.

OVERCOMING THE SMALL–LARGE DIVIDE

Despite the fact that infants typically fail to discriminate between small and large sets early in infancy, ample evidence demonstrates that children and adults do not. The earliest reported evidence of reliable discrimination of 2 from 4 visual items (without taxing working memory) is 3 years of age (e.g., Cantlon, Safford, & Brannon, 2010). So, how is it that, over the course of early human development, young children overcome the discrimination difficulties posed by the two-system interaction? Almost no research has examined the development of this understanding in toddler-hood (i.e., between 15 and 35 months; but see Barner, Thalwitz, Wood, Yang, & Carey, 2007; see also Mou & vanMarle, 2013, for a review). This section considers the ontogenetic continuity of this representational interaction by addressing how children eventually come to reliably discriminate small sets from large ones. We propose two noncompeting hypotheses regarding the mechanism(s) responsible for this change over the course of development: (1) children's acquisition of numerical language and (2) increasing precision in the ANS.

Children's Acquisition of Numerical Language

One observable cognitive change between infancy and early childhood is the acquisition of language, and, specifically relevant to the current discussion, the acquisition of numerical language. Several distinct lines of research suggest that children—at approximately 2 years of age—have begun to learn the number word list and have begun to understand the cardinal meanings associated with the first few words in the list (e.g., Condry & Spelke, 2008; Gallistel & Gelman, 1992; Wynn, 1990, 1992). Additionally, at this same time, children begin to appropriately use the plural form of nouns around 22–24 months of age (Barner et al., 2007). Thus, it is quite possible that this newly acquired ability to talk about small and large set sizes using a common system—number words—may facilitate thinking about these numbers as belonging to a common integrated system. In fact, the idea that the way we talk about number impacts numerical abilities is not a new one; it has long been reported that differences in the way number is referred to within a language (spoken or signed) may promote (or hinder) the acquisition of other numerical abilities (e.g., Geary, Bow-Thomas, Liu, & Siegler, 1996; Leybaert & Van Cutsem, 2002; Miller & Stigler, 1987). On this linguistic account, children may develop an ability to represent small and large sets via yet another representational system—verbal language.

One line of research suggests children's developing understanding of singular and plural sets expressed through numerical language may foster success on small–large discriminations, particularly on 1 vs. 4 comparisons. Barner et al. (2007) found that children succeed in discriminating exclusively small sets (1 vs. 3) between the ages of 14- and 18-months, but could not yet compare sets such as 1 vs. 4 at this age (e.g., Feigenson & Carey, 2003, 2005). However, around 22–24 months of age, children begin to succeed in these discriminations, with the timing of this coinciding with the onset of plural word production in spoken language. The authors posit that the ability to verbally express the distinction between a single item ("a") compared to many items ("some") provides cues to children as to how many items are being enumerated (Barner et al., 2007; Wood, Kouider, & Carey, 2009). Thus, the researchers suggest that the child's developing singular–plural morphology aids them in making a previously difficult discrimination by making the distinction between the sets more salient to them (Barner et al., 2007; Li, Ogura, Barner, Yang, & Carey, 2009). Although a positive correlation exists between the acquisition of plural nouns and children's successes in certain small–large discriminations, it is not the entire story. That is, 22–24-month-olds, who have begun to produce plural nouns, continue to fail on 2 vs. 4 numerical discriminations (Barner et al., 2007), suggesting that the singular–plural distinction alone cannot account for the acquisition of the ability to compare small and large sets.

Instead of highlighting the use of plural vs. singular nouns, children's abilities to successfully compare small and large sets during toddlerhood and early childhood may be the result of the acquisition and use of number words—and, eventually, an understanding of the cardinalities associated with those words. Children begin to produce number words at approximately 2 years of age (Wynn, 1992), yet just as young children can recite the alphabet before they can read or write, young children can count out loud before they understand what those number words really mean (e.g., Condry & Spelke,

2008; Gallistel & Gelman, 1992; Wynn, 1990). In time, though, children come to understand the meaning behind those number words that they can recite out loud, and, more importantly, they learn that reciting those words in sequence (i.e., counting) constitutes a reliable strategy for determining the cardinality of a set ("the Cardinal Principle"; Gelman & Gallistel, 1978). In this regard, verbal counting may provide children with an alternative and exact system for representing small and large numbers alike along a single continuum, allowing them to bridge the gap between salient preverbal representation systems. Thus, numerical language may provide a third integrated and reliable system for representing number that children begin to rely on when tracking differences among set sizes.

Importantly, it should be pointed out that although the acquisition of the count list does generally coincide with the onset of small–large discrimination successes (roughly around 3 years of age for both; e.g., Cantlon et al, 2010; Le Corre & Carey, 2007), more data are necessary to determine the viability of this account. If this is the case, then small–large discrimination successes should correlate with counting abilities, and moreover, providing linguistic support in the form of counting during small–large discrimination tasks should promote performance on these tasks. These are questions for future research.

Increasing Precision in the ANS across Development

Although children's acquisition of numerical language may afford them an alternative form of representation by which to learn about numbers, it may also be the case that improvements in precision in the ANS may allow them to overcome discrimination failures across small and large sets. In fact, one of the most recognized changes in the way we process numerical information between infancy and childhood—and even into adulthood—is the increasing precision of the ANS. As noted, newborns can discriminate a 3-fold change in number (e.g., Izard et al., 2009); 6-month-olds, a 2-fold change (e.g., Xu & Spelke, 2000); 9-month-olds, a 1.5-fold change (Wood & Spelke, 2005); 3-year-olds, a 1.3-fold change; and so on, such that adults generally discriminate a 1.14-change in number (Halberda & Feigenson, 2008). Importantly, consistent with the modality-independence of the ANS, a similar developmental change in precision has been observed for numerical information presented in the auditory domain (Barth, La Mont, Lipton, & Spelke, 2005; Barth et al., 2006; Barth, Spelke, & Beckmann, 2007; Lipton & Spelke, 2003, 2004). This increasing precision over the course of development, coupled with the idea that signal strength may contribute to the pattern of successes and failures observed for small–large discriminations in infancy, supports the idea that increased precision may result in stronger signal strength of the ANS, leading to a greater reliance on the ANS for comparing small and large sets in the preschool years.

Although ample evidence suggests that the ANS increases in precision into adulthood—both across sensory modalities and possibly across species (see

Anderson & Cordes, 2013, for review)—the mechanisms driving these changes are not fully understood. Research reveals that ANS training can lead to improvements in precision, suggesting more generally that children's repeated use of the ANS contributes to this improvement. ANS acuity has been demonstrated to improve with practice in adulthood and in children with mathematical difficulties (DeWind & Brannon, 2012; Wilson et al., 2006). Moreover, ANS training may even selectively improve symbolic exact addition and subtraction (Hyde, Khanum, & Spelke, 2014; Park & Brannon, 2013), indicating that enhanced ANS acuity may acutely boost performance for some types of formal mathematics problems. Training has also been shown to improve ANS acuity in guppies (Piffer, Miletto, Petrazzini, & Agrillo, 2013). Moreover, controlled experiments with guppies have also shown that ecologically salient environmental factors may similarly influence precision of the ANS across development, such that guppies raised in large social groups are able to discriminate a 1:2 ratio change in large sets earlier than those raised in pairs (Bisazza et al., 2010), suggesting that more than just acute training may impact precision of the ANS. In sum, although much research already suggests that training and other experiences can lead to enhancements in ANS acuity, more work is needed to determine the potential lasting support this may have on the developing ANS.

To conclude, evidence suggests that the ANS is less salient early in development and that early tracking of small sets in the visual domain may defer to the object file system, with the ANS emerging as the dominant system of number representation with age, maturation, and even experience. Furthermore, recent research suggests that ANS acuity is fairly malleable, such that training paradigms improve not only ANS acuity itself, but also performance on tests of symbolic arithmetic. Although data to date do not distinguish whether ANS acuity training may be the source of the developmental change observed in small–large discriminations, future research should explore this possibility.

OPEN QUESTIONS AND FUTURE DIRECTIONS

In sum, a robust pattern of failures to discriminate small from large sets has been documented, supporting claims that infants have access to two distinct systems for representing number. Evidence suggests that, like children, adults, and nonhuman animals, infants eventually overcome this incompatibility of representations when ANS representations of small sets are given priority over object file ones. Little research to date, however, has explored when and how children overcome this incompatibility, but we have proposed two potential mechanisms to account for this developmental change: namely, (1) numerical language (offering a third, alternative form of representing number along an integrated continuum) and (2) increased precision of the ANS (resulting in a stronger ANS signal and decreased reliance on object files). To conclude, we pose some open questions and future directions to further explore the circumstances contributing to small–large discrimination successes in infancy and early toddlerhood.

What Parameters Help Infants Succeed at Small–Large Comparisons?

Evidence from infants' successful discrimination of exclusively small sets may shed light on future avenues of investigation. Specifically, evidence for successful discrimination of exclusively small sets (1–3 items) from infant looking-time studies (for a review, see Cantrell & Smith, 2013) comes primarily from paradigms in which numerical information is presented cross-modally (e.g., Feron, Gontaz, & Streri, 2006; Jordan & Brannon, 2006; Kobayashi et al., 2005; Starkey, Spelke, & Gelman, 1983, 1990) or when infants use item-specific cues, such as perceptual variability, to individuate the items to be enumerated (Baker & Jordan, this volume; Feigenson, 2005). For example, researchers found that 6-month-olds expect the number of sounds of an object dropping when hitting the stage to match the number of objects seen behind an occluder (Kobayashi et al., 2005), while 5-month-olds successfully discriminated a difficult ratio (2:3) in a cross-modal transfer task between tactile and visual information (Feron et al., 2006). Similarly, 7-month-olds tracked number across the small number range when items were heterogeneous (i.e., varying in color, pattern, texture), but not homogeneous (i.e., all the same shape and color), suggesting the item variability may also facilitate infants' discrimination of small numbers (Feigenson, 2005; Tremoulet, Leslie, & Hall, 2000; Wilcox, 1990). Stimulus heterogeneity similarly has been shown to enhance 3- to 10-year-old children's numerical discrimination performance (Posid, Huguenel, & Cordes, 2014; but see Cantlon, Fink, Safford, & Brannon, 2007). Together, data suggest that cross-modal input and perceptual variability in the form of set heterogeneity may, in fact, facilitate an early attention to number and thus selectively recruit the ANS for abstract numerical representation. Future research should examine the impact of these two variables on infants' abilities to compare small and large sets. Moreover, if cross-modal input and perceptual variability promote attention to numerical differences, then large number discriminations may also be enhanced under these conditions, such that infants may successfully discriminate finer ratios than found with visual sets alone.

Is the Small–Large Incompatibility Receptive to Feedback or Training?

Finally, another avenue for future research examining the small–large divide in infancy is to examine how training of the ANS may impact young toddler abilities to discriminate small from large. If it really is the case that increased ANS acuity is the driving force behind children's increasing reliance on ANS representations (resulting in successful small–large comparisons), then earlier improvements in the ANS should promote earlier acquisition of small–large discriminations. If it is possible to shape ANS acuity as early as infancy, then this may provide a clear way to identify whether ANS acuity is the mechanism of change in the observed pattern of small–large discriminations across development.

CONCLUSIONS

We have considered the question of phylogenetic and ontogenetic continuity of the two-system hypothesis for representing quantities and specifically addressed the issue of how and under what circumstances children come to exclusively rely on the ANS for numerical comparisons. Consistent with data revealing an early reliance on the object file system in human infant studies, we posit that the ANS may be suppressed for the visual tracking of a small number of items relative to the object file system, but eventually emerges as the dominant system for tracking and representing number with increasing age, maturation, and experience. We discussed evidence suggesting that when the ANS is invoked to represent small sets, infants generally succeed in discriminating small and large sets, including much finer ratios than previously demonstrated (Cantrell et al., 2013). Finally, we proposed some open questions and future directions for distinguishing between these accounts, in order to clarify when and how infants may distinguish between small and large sets, and what mechanisms may facilitate this distinction.

This review demonstrates parallels in the way that numerical information is processed across phylogeny and ontogeny. The importance of this line of research has been highlighted recently be several studies indicating that the ability to precisely discriminate sets of items predicts math achievement across the lifespan, even when assessed as early as infancy (e.g., Halberda & Feigenson, 2008; Libertus et al., 2011, 2012, 2013; Lyons & Beilock, 2009, 2011; Starr et al., 2013b; but see De Smedt et al., 2013). Importantly, this relationship appears to be causal, such that approximate number training improves some aspects of formal mathematical processing (e.g., Park & Brannon, 2013). Therefore, numerical abilities as they appear in infancy may facilitate some aspects of children's formal mathematics learning and competence. If this is the case, understanding how infants successfully attend to number and why they may not be able to do so under certain circumstances may shed light on the origins of one potential preverbal precursor to formal mathematical abilities.

REFERENCES

Abramson, J. Z., Hernandez-Lloreda, V., Call, J., & Colmenares, F. (2011). Relative quantity judgments in South American sea lions (*Otaria flavescens*). *Animal Cognition, 14*, 695–706.

Agrillo, C., Dadda, M., Serena, G., & Bisazza, A. (2008). Do fish count? Spontaneous discrimination of quantity in female mosquitofish. *Animal Cognition, 11*, 495–503.

Agrillo, C., Piffer, L., & Bisazza, A. (2010). Large number discrimination by mosquitofish. *PLoS ONE*, *5*(12), e15232.

Agrillo, C., Piffer, L., Bisazza, A., & Butterworth, B. (2012). Evidence for two numerical systems that are similar in humans and guppies. *PLoS ONE*, *7*, 1–8.

Al Ain, S., Giret, N., Grand, M., Kreutzer, M., & Bovet, D. (2009). The discrimination of discrete and continuous amounts in African Grey parrots (*Psittacus erithacus*). *Animal Cognition*, *12*, 145–154.

Alvarez, G. A., & Cavanagh, P. (2004). The capacity of visual short-term memory is set both by visual information load and by number of objects. *Psychological Science*, *15*, 106–111.

Alvarez, G. A., & Franconeri, S. L. (2007). How many objects can you track? Evidence for a resource-limited tracking mechanism. *Journal of Vision*, *7*, 1–10.

Anderson, U., & Cordes, S. (2013). 1 < 2 and 2 < 3: Nonlinguistic appreciations of numerical order. *Frontiers in Comparative Psychology*, *4*(5). http://dx.doi.org/10.3389/fpsyg.2013.00005.

Ansari, D., Lyons, I. M., van Eimeren, L., & Xu, F. (2007). Linking visual attention and number processing in the brain: The role of the temporo-parietal junction in small and large symbolic and nonsymbolic number comparison. *Journal of Cognitive Neuroscience*, *19*, 1845–1853.

Antell, S. E., & Keating, D. P. (1983). Perception of numerical invariance in neonates. *Child Development*, *54*(3), 695–701.

Awh, E., Vogel, E. K., & Oh, S. H. (2006). Interactions between attention and working memory. *Neuroscience*, *139*, 201–208.

Baker, J. M., Shivik, J., & Jordan, K. E. (2011). Tracking of food quantity by coyotes (*Canis latrans*). *Behavioral Processes*, *88*, 72–75.

Balakrishnan, J. D., & Ashby, F. G. (1992). Subitizing: Magical numbers or mere superstition? *Psychological Research*, *54*(2), 80–90.

Barner, D., Thalwitz, D., Wood, J., Yang, S., & Carey, S. (2007). On the relation between the acquisition of singular-plural morpho-syntax and the conceptual distinction between one and more than one. *Developmental Science*, *10*(3), 365–373.

Barth, H., La Mont, K., Lipton, J., Dehaene, S., Kanwisher, N., & Spelke, E. (2006). Non-symbolic arithmetic in adults and young children. *Cognition*, *98*, 199–222.

Barth, H., La Mont, K., Lipton, J., & Spelke, E. S. (2005). Abstract number and arithmetic in pre-school children. *Proceedings of the National Academy of Sciences*, *102*, 14116–14121.

Barth, H., Spelke, E., & Beckmann, L. (2007). Nonsymbolic, approximate arithmetic in children: Abstract addition prior to instruction. *Developmental Psychology*, *44*(5), 1466–1477.

Beran, M. J., & Beran, M. M. (2004). Chimpanzees remember the results of one-by-one addition of food items to sets over extended time periods. *Psychological Science*, *15*, 94–99.

Beran, M. J., Evans, T. A., & Harris, E. H. (2008). Perception of food amounts by chimpanzees based on number, size, contour length and visibility of items. *Animal Behaviour*, *75*, 1793–1802.

Bijeljacbabic, R., Bertoncini, J., & Mehler, J. (1993). How do 4-day-old infants categorize multi-syllabic utterances. *Developmental Psychology*, *29*(4), 711–721.

Bisazza, A., Piffer, L., Serena, G., & Agrillo, C. (2010). Ontogeny of numerical abilities in fish. *PLoS ONE*, *5*, 1–9.

Bogale, B. A., Kamata, N., Mioko, K., & Sugita, S. (2011). Quantity discrimination in jungle crows, *Corvus macrorhynchos*. *Animal Behavior*, *82*, 635–641.

Bonanni, R., Natoli, E., Cafazzo, S., & Valsecchi, P. (2011). Free-ranging dogs assess the quantity of opponents in intergroup conflicts. *Animal Cognition*, *14*, 103–115.

Bonny, J. W., & Lourenco, S. F. (2013). The approximate number system and its relation to early math achievement: Evidence from the preschool years. *Journal of Experimental Child Psychology, 114*, 375–388.

Brannon, E. M., Abbot, S., & Lutz, D. J. (2004). Number bias for the discrimination of large visual sets in infancy. *Cognition, 93*, B59–B68.

Brannon, E. M., & Terrace, H. S. (1998). Ordering of the numerosities 1–9 by monkeys. *Science, 282*, 746–749.

Buckingham, J. N., Wong, B. B. M., & Rosenthal, G. G. (2007). Shoaling decisions in female swordtails: How do fish gauge group size? *Behaviour, 144*, 1333–1346.

Buhusi, C. V., & Cordes, S. (2011). Time and number: The privileged status of small values in the brain. *Frontiers in Integrative Neuroscience, 5*(67). http://dx.doi.org/10.3389/fnint.2011.00067.

Burr, D. C., Anobile, G., & Turi, M. (2011). Adaptation affects both high and low (subitized) numbers under conditions of high attentional load. *Seeing and Perceiving, 24*, 141–150.

Burr, D. C., Turi, M., & Anobile, G. (2010). Subitizing but not estimation of numerosity requires attentional resources. *Journal of Vision, 10*(6), 1–10, 20.

Call, J. (2000). Estimating and operating on discrete quantities in orangutans (*Pongo pygmaeus*). *Journal of Comparative Psychology, 114*, 136–147.

Cantlon, J. F., & Brannon, E. M. (2006). Shared system for ordering small and large numbers in monkeys and humans. *Psychological Science, 17*, 401–406.

Cantlon, J., Fink, R., Safford, K., & Brannon, E. M. (2007). Heterogeneity impairs numerical matching but not numerical ordering in preschool children. *Developmental Science, 10*(4), 431–440.

Cantlon, J. F., Platt, M., & Brannon, E. M. (2009). Beyond the number domain. *Trends in Cognitive Sciences, 13*(2), 83–91.

Cantlon, J. F., Safford, K. E., & Brannon, E. M. (2010). Spontaneous analog number representations in 3-year-old children. *Developmental Science, 13*(2), 289–297.

Cantrell, L., Boyer, T. W., Cordes, S., & Smith, L. B. (2014). *Signal clarity: An account of the variability in infant quantity discrimination tasks*. Manuscript submitted for publication.

Cantrell, L., & Smith, L. B. (2013). Open questions and a proposal: A critical review of the evidence on infant numerical abilities. *Cognition, 128*, 331–352.

Carey, S., & Xu, F. (2001). Infants' knowledge of objects: Beyond object files and object tracking. *Cognition, 80*, 179–213. Special Issue: Objects and Attention.

Condry, K. F., & Spelke, E. S. (2008). The development of language and abstract concepts: The case of natural number. *Journal of Experimental Psychology. General, 137*, 22–38.

Cordes, S., & Brannon, E. M. (2008a). The difficulties of representing continuous extent in infancy: Using number is just easier. *Child Development, 79*(2), 476–489.

Cordes, S., & Brannon, E. M. (2008b). Quantitative competencies in infancy. *Developmental Science, 11*, 803–808.

Cordes, S., & Brannon, E. M. (2009a). Crossing the divide: Infants discriminate small from large numerosities. *Developmental Psychology, 45*(6), 1583–1594.

Cordes, S., & Brannon, E. M. (2009b). The relative salience of discrete and continuous quantity in young infants. *Developmental Science, 12*, 453–463.

Cordes, S., & Brannon, E. M. (2011). Attending to one of many: When infants are surprisingly poor at discriminating an item's size. *Frontiers in Psychology, 2*, 65.

Cordes, S., & Gelman, R. (2005). The young numerical mind: When does it count? In J. Campell (Ed.), *Handbook of mathematical cognition* (pp. 127–142). London: Psychology Press.

Cordes, S., Gelman, R., Gallistel, C. R., & Whalen, J. (2001). Variability signatures distinguish verbal from nonverbal counting for both large and small numbers. *Psychonomic Bulletin & Review, 8,* 698–707.

De Smedt, B., Noel, M., Gilmore, C., & Ansari, D. (2013). How do symbolic and non-symbolic numerical magnitude processing relate to individual differences in children's mathematical skills? A review of evidence from brain and behavior. *Trends in Neuroscience and Education.* http://dx.doi.org/10.1016/j.tine.2013.06.001.

Dehaene, S. (1997). *The number sense: How the mind creates mathematics.* New York, NY: Oxford University Press.

DeWind, N. K., & Brannon, E. M. (2012). Malleability of the approximate number system: Effects of feedback and training. *Frontiers in Human Neuroscience, 6*(68), 1–10.

Duncan, G. J., Dowsett, C. J., Claessens, A., Magnuson, K., Huston, A. C., Klebanov, P., et al. (2007). School readiness and later achievement. *Developmental Psychology, 43*(6), 1428–1446.

Feigenson, L. (2005). A double-dissociation in infants' representations of object arrays. *Cognition, 95*(3), B-37–B-48.

Feigenson, L., & Carey, S. (2003). Tracking individuals via object-files: Evidence from infants' manual search. *Developmental Science, 6,* 568–584.

Feigenson, L., & Carey, S. (2005). On the limits of infants' quantification of small object arrays. *Cognition, 97,* 295–313.

Feigenson, L., Carey, S., & Hauser, M. D. (2002). The representations underlying infants' choice of more: Object files versus analog magnitudes. *Psychological Science, 13,* 150–156.

Feigenson, L., Dehaene, S., & Spelke, E. (2004). Core systems of number. *Trends in Cognitive Sciences, 8*(7), 307–314.

Feigenson, L., & Halberda, J. (2004). Infants chunk object arrays into sets of individuals. *Cognition, 91,* 173–190.

Feigenson, L., & Yamaguchi, M. (2009). Limits on infants' ability to dynamically update object representations. *Infancy, 14*(2), 244–262.

Feron, J., Gentaz, E., & Streri, A. (2006). Evidence of amodal representation of small numbers across visuo-tactile modalities in 5-month-old infants. *Cognitive Development, 21*(2), 81–92.

Gallistel, C. R. (1990). *The organization of learning.* Cambridge, MA: Bradford Books/MIT Press.

Gallistel, C. R., & Gelman, R. (1992). Preverbal and verbal counting and computation. *Cognition, 44,* 43–74.

Gallistel, C. R., & Gelman, R. (2000). Non-verbal numerical cognition: From reals to integers. *Trends in Cognitive Sciences, 4,* 59–65.

Geary, D. C. (2011). Cognitive predictors of individual differences in achievement growth in mathematics: A five-year longitudinal study. *Developmental Psychology, 47,* 1539–1552.

Geary, D. C., Bow-Thomas, C. C., Liu, F., & Siegler, R. S. (1996). Development of arithmetic competencies in Chinese and American children: Influence of age, language, and schooling. *Child Development, 67,* 2022–2044.

Geary, D. C., Hoard, M. K., Nugent, L., & Bailey, D. H. (2013). Adolescents' functional numeracy is predicted by their school entry number system knowledge. *PLoS ONE, 8*(1), e54651.

Gelman, R., & Gallistel, C. R. (1978). *The child's understanding of number.* Cambridge, MA: Harvard University Press.

Halberda, J., & Feigenson, L. (2008). Developmental change in the acuity of the "Number sense": The approximate number system in 3-, 4-, 5-, and 6-year-olds and adults. *Developmental Psychology, 44,* 1457–1465.

Halberda, J., Simons, D. J., & Wetherhold, J. (2014). *Superfamiliarity affects perceptual grouping but not the capacity of visual working memory.* Manuscript submitted for publication.

Hanus, D., & Call, J. (2007). Discrete quantity judgments in the great ape (*Pan paniscus, Pan troglodytes, Gorilla gorilla, Pongo pygmaeus*): The effect of presenting whole sets versus item-by-item. *Journal of Comparative Psychology, 121,* 241–249.

Hauser, M. D., Carey, S., & Hauser, L. B. (2000). Spontaneous number representations in semi-free-ranging rhesus monkeys. *Proceedings of the Royal Society B: Biological Sciences, 267,* 829–833.

Hunt, S., Low, J., & Burns, K. C. (2008). Adaptive numerical competency in a food-hoarding songbird. *Proceedings of the Royal Society B: Biological Sciences, 275,* 2373–2379.

Hyde, D. C. (2011). Two systems of non-symbolic numerical cognition. *Frontiers in Human Neuroscience, 5,* 1–8.

Hyde, D. C., Khanum, S., & Spelke, E. S. (2014). Brief non-symbolic, approximate number practice enhances subsequent exact symbolic arithmetic in children. *Cognition, 131,* 92–107.

Hyde, D. C., & Spelke, E. S. (2009). All numbers are not equal: An electrophysiological investigation of small and large number representations. *Journal of Cognitive Neuroscience, 21,* 1039–1053.

Hyde, D. C., & Spelke, E. S. (2011). Neural signatures of number processing in human infants: Evidence for two core systems underlying numerical cognition. *Developmental Science, 14*(2), 360–371.

Hyde, D. C., & Spelke, E. S. (2012). Spatiotemporal dynamics of processing nonsymbolic number: An event-related potential source localization study. *Human Brain Mapping, 33,* 2189–2203.

Hyde, D. C., & Wood, J. N. (2011). Spatial attention determines the nature of nonverbal number representation. *Journal of Cognitive Neuroscience, 23,* 2336–2351.

Izard, V., Sann, C., Spelke, E. S., & Streri, A. (2009). Newborn infants perceive abstract numbers. *Proceedings of the National Academy of Science, 106,* 10382–10385.

Jones, S. M., Pearson, J., DeWind, N. K., Paulsen, D., Tenekedjieva, A., & Brannon, E. M. (2013). Lemurs and macaques show similar numerical sensitivity. *Animal Cognition, 17*(3), 503–515. http://dx.doi.org/10.1007/s10071-013-0682-3.

Jordan, K. E., & Brannon, E. M. (2006). The multisensory representation of number in infancy. *Proceedings of the National Academy of Sciences of the United States of America, 103,* 3486–3489.

Jordan, N. C., Kaplan, D., Ramineni, C., & Locuniak, M. N. (2009). Early math matters: Kindergarten number competence and later mathematics outcomes. *Developmental Psychology, 45,* 850–867.

Jordan, K. E., Suanda, S. H., & Brannon, E. M. (2008). Intersensory redundancy accelerates preverbal numerical competence. *Cognition, 108,* 210–221.

Kaldy, Z., & Leslie, A. M. (2003). Identification of objects in 9-month-old infants: Integrating 'what' and 'where' information. *Developmental Science, 6*(3), 360–373.

Klahr, D. (1973). Quantification processes. In W. G. Chase (Ed.), *Visual information processing* (pp. 3–34). San Diego, CA: Academic Press.

Kobayashi, T., Hiraki, K., & Hasegawa, T. (2005). Auditory-visual intermodal matching of small numerosities in 6-month-old infants. *Developmental Science, 8,* 409–419.

Krusche, P., Uller, C., & Dicke, U. (2010). Quantity discrimination in salamanders. *Journal of Experimental Biology, 213,* 1822–1828.

Le Corre, M., & Carey, S. (2007). One, two, three, four, nothing more: An investigation of the conceptual sources of the verbal counting principles. *Cognition, 105,* 395–438.

LeFevre, J. A., Fast, L., Skwarchuk, S. L., Smith-Chant, B. L., Bisanz, J., et al. (2010). Pathways to mathematics: Longitudinal predictors of performance. *Child Development, 81*, 1753–1767.

Leslie, A. M., Xu, F., Tremoulet, P. D., & Scholl, B. J. (1998). Indexing and the object concept: Developing 'what' and 'where' systems. *Trends in Cognitive Sciences, 2*, 10–18.

Leybaert, J., & Van Custem, M. (2002). Counting in sign language. *Journal of Experimental Child Psychology, 81*(4), 482–501.

Li, P., Ogura, T., Barner, D., Yang, S., & Carey, S. (2009). Does the conceptual distinction between singular and plural sets depend on language? *Developmental Psychology, 45*(6), 1644–1653.

Libertus, M. E., & Brannon, E. M. (2010). Stable individual differences in number discrimination in infancy. *Developmental Science, 13*(6), 900–906.

Libertus, M. E., Feigenson, L., & Halberda, J. (2011). Preschool acuity of the approximate number system correlates with school math ability. *Developmental Science, 14*, 1292–1300.

Libertus, M. E., Feigenson, L., & Halberda, J. (2013). Is approximate number precision a stable predictor of math ability? *Learning and Individual Differences, 25*, 126–133.

Libertus, M. E., Odic, D., & Halberda, J. (2012). Intuitive sense of number correlates with math scores on college-entrance examination. *Acta Psychologica, 141*(3), 373–379.

Lipton, J. S., & Spelke, E. S. (2003). Origins of number sense: Large-number discrimination in human infants. *Psychological Science, 14*, 396–401.

Lipton, J. S., & Spelke, E. S. (2004). Discrimination of large and small numerosities by human infants. *Infancy, 5*, 271–290.

Luck, S. J., & Vogel, E. K. (1997). The capacity of visual working memory for features and conjunctions. *Nature, 390*, 279–281.

Luria, R., & Vogel, E. K. (2011). Visual search demands dictate reliance upon working memory storage. *The Journal of Neuroscience, 31*, 6199–6207.

Lyons, I. M., & Beilock, S. L. (2009). Beyond quantity: Individual differences in working-memory and the ordinal understanding of numerical symbols. *Cognition, 113*(2), 189–204.

Lyons, I. M., & Beilock, S. L. (2011). Numerical ordering ability mediates the relation between number-sense and arithmetic competence. *Cognition, 121*, 256–261.

Meck, W. H., & Church, R. M. (1983). A mode control model of counting and timing processes. *Journal of Experimental Psychology. Animal Behavior Processes, 9*, 320–334.

Miller, K. F., & Stigler, J. W. (1987). Counting in Chinese: Cultural variation in a basic cognitive skill. *Cognitive Development, 2*, 279–305.

Mou, Y., & vanMarle,, K. (2013). Two core systems of numerical representation in infants. *Developmental Review.* http:/dx.doi.org/10.1016/j.dr.2013.11.001.

Moyer, R., & Landauer, T. (1967). Time required for judgments of numerical inequity. *Nature, 215*, 1519.

Park, J., & Brannon, E. M. (2013). Training the approximate number system improves math proficiency. *Psychological Science, 24*(10), 2013–2019. http://dx.doi.org/10.1177/0956797613482944.

Perdue, B., Talbot, C., Stone, A., & Beran, M. (2012). Putting the elephant bask in the herd: Elephant relative quantity judgments match those of other species. *Animal Cognition, 15*(5), 955–961.

Piazza, M., Giacomini, E., Le Bihan, D., & Dehaene, S. (2003). Single-trial classification of parallel pre-attentive and serial attentive processes using functional magnetic resonance imaging. *Proceedings of the Royal Society B: Biological Sciences, 270*(1521), 1237–1245.

Piazza, M., Mechelli, A., Butterworth, B., & Price, C. J. (2002). Are subitizing and counting implemented as separate or functionally overlapping processes? *NeuroImage, 15*(2), 435–446.

Piffer, L., Agrillo, C., & Hyde, D. C. (2011). Small and large number discrimination in guppies. *Animal Cognition, 15*, 215–221.

Piffer, L., Miletto, Petrazzini, M. E., & Agrillo, C. (2013). Large number discrimination in new-born fish. *PLoS ONE, 8*(4), e62466.

Posid, T., Huguenel, B., & Cordes, S. (2014). *Stimulus heterogeneity facilitates difficult number judgments in children.* Manuscript in preparation.

Ross-Sheehy, S., Oakes, L., & Luck, S. (2003). The development of visual short-term memory capacity in infants. *Child Development, 74*(6), 1807–1822.

Rugani, R., Regolin, L., & Vallortigara, G. (2008). Discrimination of small numerosities in young chicks. *Journal of Experimental Psychology. Animal Behavior Processes, 34*(3), 388–399.

Scholl, B. J., & Pylyshyn, Z. W. (1999). Tracking multiple items through occlusion: Clues to visual objecthood. *Cognitive Psychology, 38*, 259–290.

Simon, T. J. (1997). Reconceptualizing the origins of number knowledge: A non-numerical account. *Cognitive Development, 12*, 349–372.

Starkey, P., Spelke, E. S., & Gelman, R. (1983). Detection of intermodal numerical correspondences by human infants. *Science, 222*(4620), 179–181.

Starkey, P., Spelke, E. S., & Gelman, R. (1990). Numerical abstraction by human infants. *Cognition, 36*, 97–128.

Starr, A. B., Libertus, M. E., & Brannon, E. M. (2013a). Infants show ratio-dependent number discrimination regardless of set size. *Infancy, 18*(6), http://dx.doi.org/10.1111/infa.12008.

Starr, A. B., Libertus, M. E., & Brannon, E. M. (2013b). Number sense in infancy predicts mathematical abilities in childhood. *Proceedings of the National Academy of Sciences* (Early Edition), *110*(45), 18116–18120.

Tremoulet, P. D., Leslie, A. M., & Hall, D. G. (2000). Infant individuation and identification of objects. *Cognitive Development, 15*, 499–522.

Trick, L. M., Enns, J. T., & Brodeur, D. A. (1996). Life-span changes in visual enumeration: The number discrimination task. *Developmental Psychology, 32*(5), 925–932.

Trick, L. M., & Pylyshyn, Z. W. (1993). What enumeration studies can show us about spatial attention: Evidence for limited capacity preattentive processing. *Journal of Experimental Psychology. Human Perception and Performance, 19*(2), 331–351.

Trick, L. M., & Pylyshyn, Z. W. (1994). Why are small and large numbers enumerated differently: A limited capacity preattentive stage in vision. *Psychological Review, 101*(1), 80–102.

Uller, C., Carey, S., Huntley-Fenner, G., & Klatt, L. (1999). What representations might underlie infant numerical knowledge? *Cognitive Development, 1*(3), 249–280.

Uller, C., Jaeger, R., Guidry, G., & Martin, C. (2003). Salamanders (*Plethodon cinereus*) go for more: Rudiments of number in an amphibian. *Animal Cognition, 6*, 105–112.

Uller, C., & Lewis, J. (2009). Horses (*Equus caballus*) select the greater of two quantities in small numerical contrasts. *Animal Cognition, 12*, 733–738.

vanMarle, K. (2013). Infants use different mechanisms to make small and large number ordinal judgments. *Journal of Experimental Child Psychology, 114*, 102–110.

vanMarle, K., Chu, F., Li, Y., & Geary, D. C. (2014). Acuity of the approximate number system and preschoolers' quantitative development. *Developmental Science.* http://dx.doi.org/10.1111/desc.12143.

vanMarle, K., & Wynn, K. (2009). Infants' auditory enumeration: Evidence for analog magnitudes in the small number range. *Cognition, 111*, 302–316.

Vogel, E. K., Woodman, G. F., & Luck, S. J. (2001). Storage of features, conjunctions, and objects in visual working memory. *Journal of Experimental Psychology. Human Perception and Performance, 27*, 92–114.

Vonk, J., & Beran, M. J. (2012). Bears 'count' too: Quantity estimation and comparison in black bears, *Ursus americanus*. *Animal Behavior, 84*, 231–238.

Ward, C., & Smuts, B. B. (2007). Quantity-based judgments in the domestic dog (*Canis lupus familiaris*). *Animal Cognition, 10*, 71–80.

Whalen, J., Gallistel, C. R., & Gelman, R. (1999). Non-verbal counting in humans: The psychophysics of number representation. *Psychological Science, 10*, 130–137.

Wilcox, T. (1999). Object individuation: Infants' use of shape, size, pattern, and color. *Cognition, 72*, 125–166.

Wilson, A. J., Dehaene, S., Pinel, P., Revkin, S. K., Cohen, L., & Cohen, D. (2006). Principles underlying the design of "The Number Race," an adaptive computer game for remediation of dyscalculia. *Behavioral and Brain Functions, 2*, 19. http://dx.doi.org/10.1186/1744-9081-2-1.

Wood, J. N., Kouider, S., & Carey, S. (2009). Acquisition of singular–plural morphology. *Developmental Psychology, 45*(1), 202–206.

Wood, J. N., & Spelke, E. S. (2005). Infants' enumeration of actions: Numerical discrimination and its signature limits. *Developmental Science, 8*, 173–181.

Wynn, K. (1990). Children's understanding of counting. *Cognition, 36*, 155–193.

Wynn, K. (1992). Children's acquisition of number words and the counting system. *Cognitive Psychology, 24*, 220–251.

Wynn, K., Bloom, P., & Chiang, W. C. (2002). Enumeration of collective entities by 5-month-old infants. *Cognition, 83*(3), B55–B62.

Xu, F. (2003). Numerosity discrimination in infants: Evidence for two systems of representation. *Cognition, 89*, B12–B25.

Xu, F., & Arriaga, R. I. (2007). Number discrimination in 10-month-old infants. *British Journal of Developmental Psychology, 25*, 103–108.

Xu, F., & Spelke, E. S. (2000). Large number discrimination in 6-month-old infants. *Cognition, 74*(1), 1–11.

Xu, F., Spelke, E. S., & Goddard, S. (2005). Number sense in human infants. *Developmental Science, 8*, 88–101.

Zorina, Z. A., & Smirnova, A. A. (1996). Quantitative evaluations in gray crows: Generalization of the relative attribute "larger set." *Neuroscience and Behavioral Physiology, 26*, 357–364.

Zosh, J. M., & Feigenson, L. (2012). Memory load affects object individuation in 18-month-old infants. *Journal of Experimental Child Psychology, 113*, 322–336.

Zosh, J. M., Halberda, J., & Feigenson, L. (2011). Memory for multiple visual ensembles in infancy. *Journal of Experimental Psychology. General, 140*, 141–158.

Chapter 11

The Influence of Multisensory Cues on Representation of Quantity in Children

Joseph M. Baker[1] and Kerry E. Jordan[2]
[1]Stanford University School of Medicine, Department of Psychiatry and Behavioral Sciences, Stanford, CA, USA
[2]Utah State University, Department of Psychology, Logan, UT, USA

INTRODUCTION

We live in a world in which we are constantly bombarded with sensory information of all types. For an infant, this sort of multisensory environment could potentially seem chaotic and confusing. However, inherent biases in features of environmental stimuli that capture infants' attention (and those that do not) bring some order to this potential chaos and provide the foundation for various aspects of their later learning (Gibson, 1969; Lewkowicz & Ghazanfar, 2009). A large body of foundational studies has resulted in key insights into how the infant mind processes information across multiple sensory systems, including insights into the evolutionary and developmental underpinnings of numerical cognition (Bahrick & Lickliter, 2000). Moreover, these studies have helped identify ways in which infants' and children's perception of number may be enhanced (Jordan & Baker, 2011; Jordan, Suanda, & Brannon, 2008), and have hinted at ways to benefit real-world applications such as instructional math tools (Moyer-Packenham et al., 2013).

In this chapter, we review this research, beginning with a discussion on the nature of nonverbal numerical representations, followed by a brief review of how intersensory integration emerges throughout development. Next, we discuss the effects of multisensory stimulation on learning and cognition generally, followed by a review of the beneficial effects of multisensory information on numerical competence and how these effects may arise. Finally, we conclude with a brief discussion of currently unanswered questions regarding multisensory processing and number, and possible future directions for this science.

Mathematical Cognition and Learning, Vol. 1. http://dx.doi.org/10.1016/B978-0-12-420133-0.00011-9

REPRESENTATION OF NUMBER WITHOUT LANGUAGE

How do infants and nonhuman animals represent number without language? As adults who commonly associate the property of number with number words, Arabic numerals, and mathematics, we may find it difficult to imagine how language and number can be dissociated. However, we, in fact, engage our nonverbal numerical representations often. For instance, imagine walking into a crowded lecture hall that is divided down the middle. You need to find a seat, but all of the seats in the back are taken. You quickly view both sides of the hall and head toward the one with fewer students, where you are more likely to find a seat. Chances are, unless both sides were near equally crowded, you would correctly choose the side of the hall with fewer students. But how were you able to do this if you didn't count each student individually? In this situation, you are representing the number of students in the hall as a continuous quantity, as opposed to a set of discrete entities. When judging which side of the hall has more or fewer students, you are making an approximation, much like when you judge which of two lights is brighter or which of two cups has more liquid. Importantly, nonverbal numerical representations such as these are subject to the same principles that govern representations of other continuous quantities, including predictions made by Weber's law (see Halberda & Odic, this volume).

Weber's law states that the just noticeable difference between two quantities is proportional to their original intensities. Mathematically, Weber's law is given as $\frac{\Delta I}{I} = k$, where ΔI represents the just noticeable difference between two continuous quantities, I represents the initial stimulus intensity, and k represents the constant ratio of the Weber fraction (Weber, 1996). For example, imagine that each side of the lecture hall contained 50 students. If it took an additional 10 students to sit in the left half of the hall before you could identify a difference in number between the two sides, your Weber fraction would be .2 (e.g., $10/50 = .2$). Knowing this fraction, you could estimate that you would require an additional 40 students to sit in the left half of the hall if each half originally contained 200 students (e.g., $200 \times .2 = 40$). Thus, while k stays constant, your just noticeable difference changes in proportion to the original number of students seated in the hall.

If, after your numerical judgment, you mistakenly choose the side of the lecture hall with more students (i.e., fewer open seats), it would be said that you committed an estimation error. Such errors are common because your nonverbal representation of the number of students on either side of the lecture hall is a noisy approximation of the actual number of students (Figure 11-1). Similar to the just noticeable differences described previously, estimation error is mediated by the original intensity of the continuous stimulus. For example, in a class with 50 students per side, you may find yourself off on estimations by only 5 or so students on average, meaning it may take 15 additional students on the left side of the hall for you to identify a

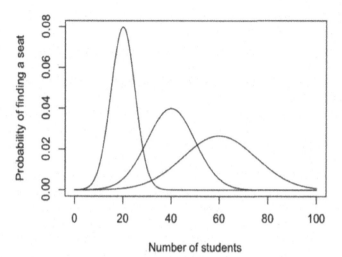

Number of students

FIGURE 11-1   Example of scalar variance. Standard deviation increases in proportion to the mean numerical value. Error increases with the total number of students.

difference instead of the predicted 10 based on Weber's law. In a larger class with closer to 200 students per side, your error may be closer to 20 students so that some days you may require a difference of 60 additional students on one side to identify a difference. Thus, your error in estimating the total number of students in a room increases in proportion with the total quantity of students.

The width of each probability function in Figure 11-1 represents the error surrounding the psychophysical representations of the scalar values along the x-axis. The increase in error that coincides with an increase in stimulus magnitude is known as scalar variance. For example, imagine a lecture hall in which one side contains 20 students, and the other contains 40 students. Notice how the probability functions for these values in Figure 11-1 overlap considerably. This suggests that, over repeated estimations, the probability of committing an estimation error is relatively high. Now, imagine the same scenario with 20 and 60 students, respectively: Here, the amount of overlap between either function is small, suggesting that estimation errors would be infrequent. Thus, the numerical distance between competing magnitudes mediates discrimination. However, the same outcomes would not be expected if the number of students per side of the lecture hall was 220 compared to 240 or 260. The reason is that scalar variance for such large values would result in significant overlap of each probability function despite the same absolute difference in numerical magnitude. Thus, nonverbal numerical discrimination is dependent on the ratio between competing magnitudes, as opposed to their absolute numerical distance. For example, the ratio between 20 and 60 students is 1:3 (.33), whereas the ratio between 220 and 260 students is 11:13 (.84). As the ratio between competing quantities approaches 1:1 (1.0), discrimination becomes more difficult. Furthermore, as numerical quantities increase in magnitude, the numerical distance between the two quantities must increase in order to maintain the ratio needed to identify a difference (Cordes, Gelman, Gallistel, & Whalen, 2001).

Evidence for the claims that human and nonhuman animals possess a common analog nonverbal system of number representation that is governed by Weber's law has been found across many different species and experimental paradigms (Agrillo et al., this volume; Beran et al., this volume; Geary et al., this volume; Pepperberg, this volume; Vallortigara, this volume; see also Baker, Rodzon, & Jordan, 2013; Beran, 2004; Beran & Beran, 2004; Beran & Rumbaugh, 2001; Brannon 2002; Brannon, Abbot, & Lutz, 2004; Brannon & Roitman, 2003; Brannon & Terrace, 1998; Cantlon, Safford, & Brannon, 2010; Feigenson, Dehaene, & Spelke, 2004; Gallistel & Gelman, 1992; Hauser, Tsao, Garcia, & Spelke, 2003; Jordan & Brannon, 2006a; Jordan & Brannon, 2006c; Jordan & Baker, 2011; Lipton & Spelke, 2003, 2004; Neider, Freedman, & Miller, 2002; Xu & Arriaga, 2007; Xu & Spelke, 2000; Xu, Spelke, & Goddard, 2005).

A particularly robust finding across human adults, human infants, and nonhuman animals is that the accuracy and time needed to respond to numerical magnitudes are modulated by their ratios (Beran, 2001, 2004; Beran, Beran, Harris, & Washburn, 2005; Brannon & Terrace, 1998, 2000; Dehaene, Dupoux, & Mehler, 1990; Judge, Evans, & Vyas, 2005; Moyer & Landauer, 1967; Nieder & Miller, 2004; Rumbaugh, Savage-Rumbaugh, & Hegel, 1987; Sekuler & Mierkiewicz, 1977; Smith, Piel, & Candland, 2003; Temple & Posner, 1998; Washburn & Rumbaugh, 1991). For example, 6-month-old infants are capable of discriminating a 1:2 (e.g., 6 vs. 12) ratio of dot arrays but fail to discriminate a 2:3 (e.g., 8 vs. 12) ratio (Lipton & Spelke, 2003; Wood & Spelke 2005; Xu & Spelke, 2000). Similarly, 3- to 5-year-olds more easily compare the numerosity of arrays occurring in a 1:2 ratio than a 2:3 ratio, although they can discriminate the latter at above chance levels (Huntley-Fenner, & Cannon, 2000). Such developmental improvements in nonverbal number discriminations are present throughout adolescence and into adulthood (Halberda & Feigenson, 2008; Halberda, Ly, Wilmer, Naiman, & Germine, 2012). Moreover, when asked to estimate the number of items in an array by pointing to the corresponding Arabic numeral on a number line, 5- to 7-year-olds demonstrated scalar variability in their responses, such that variability in responses increased proportionally with magnitude (Huntley-Fenner, 2001).[1] For adults, response times to identify the larger of two Arabic digits decreased as the numerical distance increased (i.e., decreased in ratio) (Moyer & Landauer, 1967).

---

1. Divergent evidence suggests that behavioral patterns resembling scalar variability may arise from proportion judgments made from numerical landmarks in between a number line's numerical range (e.g., 25%, 50%, etc.) (Barth & Paladino, 2011).

Using a numerical bisection task wherein participants were required to identify whether visual arrays contained a number of dots that were closer to either 2 or 8, 6-year-old children and rhesus macaques (*Macaca mulatta*) performed similarly, and as the authors argue, the performance of both groups was consistent with Weber's law (Jordan & Brannon, 2006a). Convergent results were identified using a numerical ordering task, wherein human adults and monkeys were required to choose the array that contained the smaller of two numerosities (Cantlon & Brannon, 2006). In this task, percent of trials correct, response time, and Weber fractions for humans and monkeys were strikingly similar, suggesting that a common mechanism was modulating performance in both groups. Finally, when asked to make a specified number of responses on each of a series of trials, the mean number of responses for both human adults and rats (*Rattus norvegicus*) increased systematically with the magnitude of the required number, and the variability in the accuracy of the number of responses increased linearly with the mean number of responses (Platt & Johnson, 1971; Whalen, Gallistel, & Gelman, 1999). Taken together, these and many other empirical findings have established that humans share an abstract, analog nonverbal system of number representation with many other species (Brannon, 2006; Brannon & Roitman, 2003; Feigenson, Dehaene, & Spelke, 2004).

DEVELOPMENT OF INTERSENSORY INTEGRATION

As infants are constantly being bombarded by stimulation of all kinds, their environment is fraught with a potentially confusing sensory array. Stimuli of different sizes and colors stream across their visual field, while stimuli of different pitches and volumes resonate within their auditory space. To further complicate matters, some of these stimuli occur simultaneously across sensory modalities, whereas others may partially overlap, or occur separately but still close together in time. It may seem likely that this cacophony would overwhelm infants' perceptual capabilities. To the contrary, infants are able to make sense of their world and even use the busy environment to their advantage. This remarkable ability arises in part because of infants' early propensities to integrate stimuli that occur synchronously across multiple sensory modalities. Take, for example, an infant watching a ball bounce: here, the infant sees the ball rise and fall, and hears a bounce every time the ball hits the floor. Evidence suggests that a process of intersensory integration leads the infant to perceive the number of visual bounces and the number of auditory floor hits occurring as singular, unified events (Lewkowicz, 2000).

Intersensory integration is an adaptive evolutionary tool that is present in many species (Bahrick, Lickliter, & Flom, 2004; Lewkowicz & Kraebel, 2004; Taylor & Ryan, 2013), enhances infants' cognitive abilities, and is facilitated by a number of environmental and biological factors (Lewkowicz & Ghazanfar, 2009). Arguably the most apparent factor influencing intersensory integration is the ubiquity of amodal (i.e., non-modality-specific) properties in an infant's environment. For example, when an infant plays with her toys, she sees their color and hears their sounds. Each of these attributes is conveyed only by a specific sensory modality (i.e., visual or auditory), and is thus considered unimodal. At the same time, the infant may also ascertain the size of each toy by sight and/or touch. Because the property of size is not restricted to any single sensory modality, it is considered amodal. Given their ubiquity in our multisensory environment, amodal attributes play a very important role in infants' perceptual development by facilitating intersensory integration through repeated exposure (Gibson, 1969). Infants learn to process multiple amodal attributes occurring together as a singular entity, preventing them from becoming bogged down with an overwhelming amount of discrete sensory stimulation they might otherwise perceive as unrelated. As infants learn from such repetition, they become more adept at processing amodal events throughout development.

Assisting greatly in the development of intersensory integration are the inherent developmental limitations of infancy. For instance, at birth an infant has relatively poor eyesight, which is useful for reducing competition between auditory and visual pathways as the former develops. As a result, audiovisual intersensory stimulation can occur only for objects that lie very near the infant. Throughout development, the scope of the infant's intersensory receptive field broadens and she or he is able to experience a larger array of such stimulation. This developmental progression may promote orderly integration of sensory modalities and help avoid overwhelming the infant's nervous system by reducing potential competition between developing systems (Turkewitz & Kenny, 1982). For example, for a newborn, it may be much simpler to associate the visual movement of his mother's close face with the auditory rhythm of her lullaby without being able to see other nonassociated moving stimuli in the background. Just as these simple associations are made, the visual field begins to expand and more complicated associations can be made. Moreover, theorists suggest the neural mechanisms underlying intersensory integration are closely tied to species-typical developmental experiences. As the idiom suggests, "neurons that fire together wire together"; the neural networks associated with multisensory processing and intersensory integration that are activated consistently throughout development will develop strong synaptic ties, resulting in selective elaboration of such synapses throughout development (Lewkowicz & Ghazanfar, 2009).

Evidence in support of this form of progressive intersensory integration has been demonstrated in a broad array of paradigms. For example, early in development, infants are capable of matching faces and voices based on simple intersensory cues (i.e., McGurk effect) (Kuhl & Meltzoff, 1982; Patterson & Werker, 2003; Rosenblum, Schmuckler, & Johnson, 1997; Walton & Bower, 1993). Throughout development, however, infants become more adept at perceiving higher-order cues inherent in audiovisual facial expressions such as affect and gender (Patterson & Werker, 2002; Kahana-Kalman & Walker-Andrews, 2001), while also relying less on low-level attributes such as synchrony to perceive higher-level cues (Walker-Andrews, 1986). This developmental progression toward reliance on higher-order cues is dependent on appropriate sensory input (Maurer, Mondloch, & Lewis, 2007) and is characterized by a high degree of neural plasticity that can accommodate many types of perceptual information (Horng & Sur, 2006; King, 2009; Wallace & Stein, 2007).

Alternatively, multisensory perceptual narrowing may facilitate intersensory integration. For example, after experiencing white light followed by white noise of different intensities, 3-week-old infants demonstrate a U-shaped relationship between loudness and magnitude of cardiac response (Lewkowicz & Turkewitz, 1980). In contrast, adults did not show any systematic relationship between cardiac response and loudness, suggesting that the propensity for infants to make cross-modal matches of intensity is lost throughout development. Similarly, when 4-, 6-, 8-, and 10-month-olds are habituated to side-by-side movies of rhesus monkey faces producing a "coo" call on one side and a "grunt" on the other, only the 4- and 6-month-olds are able to match the visual and auditory calls in test. Consistent with a multisensory perceptual narrowing account, the older infants had lost the ability to make such associations. However, when the synchrony between the audiovisual stimuli was removed so that the calls did not synch with the facial gestures, none of the infants were able to make the association—suggesting that successful matching in young infants was mediated by the synchrony of intersensory stimulation (Lewkowicz, Sowinski, & Place, 2008).

Similar paradigms have also been used to demonstrate infants' and monkeys' abilities to integrate concurrent visual and auditory stimuli in order to match the *number* of voices they hear to the number of faces they see. For example, Jordan and Brannon (2006a) presented 7-month-olds with simultaneous video displays wherein one video showed two human faces and the other a group of three faces that were all saying "look" repeatedly (Figure 11-2).

By systematically varying the number of voices speaking the word "two" or "three" and recording the looking time of infants to either side of the display, Jordan and Brannon were able to determine that infants preferred (i.e., viewed longer) the group of faces that matched the number of voices. A convergent pattern of results was also found with rhesus monkeys using a similar paradigm (Jordan, Brannon, Logothetis, & Ghazanfar, 2005). Here, monkeys were capable of matching the number of conspecific faces they saw with the number of voices synchronously producing a "coo" call.

Taken together, these data provide evidence that humans and nonhuman animals experience abundant sources of intersensory information throughout their environments. They develop the ability to integrate this multisensory information early in life, and even connect specific domains such as numerical representations across different sensory modalities.

Participant

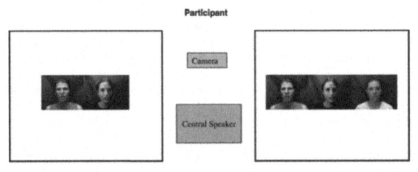

FIGURE 11-2 Trial structure from Jordan and Brannon (2006b). Infants sat in front of two screens that synchronously played a movie of two or three faces speaking the word "look" at a rate of .5s for 60s.

BENEFICIAL EFFECTS OF MULTISENSORY STIMULATION ON COGNITION AND LEARNING

The ability to integrate multisensory stimulation from our environment has a profound effect on our perception of amodal attributes and, by extension, on cognition and learning (see Lewkowicz, 2000; Lewkowicz & Lickliter, 1994; and Bahrick & Lickliter, 2000, for review). Data show that when paired synchronously in time, even changes in many seemingly unrelated unimodal properties significantly affect discrimination abilities in human infants. For instance, while there is no obvious relation between random shapes and sounds, 2-day-old infants learn to associate such pairings when they are presented synchronously in time (Slater, Quinn, Brown, & Hayes, 1999). These findings suggest that infants' learning of arbitrary audiovisual associations is constrained by the presence of redundant stimulation. Other similarly arbitrary pairings include brightness and auditory stimulation (Lewkowicz & Turkewitz, 1981), visual and auditory temporal frequencies (Gardner, Lewkowicz, Rose, & Karmel, 1986), object motion and sounds (Spelke, Born, & Chu, 1983), vocal expressions of emotion and the face (Walker-Andrews & Lennon, 1991), color/shape of objects and pitch (Bahrick, 1994), emotional expression and sound quality (Phillips, Wagner, Fells, & Lynch, 1990), and abstract visual stimuli and sounds (Wagner, Winner, Cicchetti, & Gardner, 1981).

Within infants, these studies are often conducted using basic habituation/dishabituation paradigms in which participants are habituated to an amodal property presented in a bimodal or unimodal fashion. For instance, Bahrick and Lickliter (2000) used this paradigm to investigate the effect of intersensory redundancy on infants' discrimination of rhythm. First, 5-month-old infants were habituated to a videotaped audiovisual event that depicted a bright red hammer moving up and down from a wooden surface, leading to one of four distinctive rhythms. Following habituation, each infant was assigned to an experimental or control condition wherein they were shown two visual test trials depicting a change in rhythm (experimental) or no change in rhythm (control). Infants in the experimental condition significantly dishabituated to the change in rhythm, while those in the control condition did not. However, when habituated to unimodal (e.g., visual or auditory) rhythm sequences, the infants did not dishabituate to such changes in rhythm— suggesting that the redundant information about rhythm presented across the auditory and visual sensory modalities led to more accurate encoding of rhythm during habituation, and ultimately to enhanced discrimination in test.

Furthermore, to affirm their hypothesis that synchrony between the redundant intersensory stimuli was necessary for such an effect, Bahrick and Lickliter (2000) replicated their initial study using asynchronous presentation of rhythm. This was accomplished by offsetting the auditory and visual components of the hammer sequence so that the visual motion of the hammer striking the table did not directly coincide with the tone made when the hammer hit the wooden surface. Within this preparation, the infants did not dishabituate to changes in rhythm, suggesting that synchrony between redundant intersensory stimuli is a key component behind the beneficial effects of intersensory redundancy.

Results from this study, along with the host of others demonstrating similar effects, led Bahrick and Lickliter (2000) to propose the *Intersensory Redundancy Hypothesis*, whereby intersensory redundancy recruits infant attention, causing amodal properties of events to be perceptual "foreground." Thus, intermodal redundancy operates primarily at the level of attention, such that identical information presented in a spatially coordinated and synchronous manner is highly salient to infants and directs attentional selectivity at the expense of nonredundant information. Regarding the need for synchrony, Bahrick and Lickliter (2000) argue that tight temporal coupling further causes the redundant amodal property to stand out from other nonredundant properties. Next, the attentional focus on redundant amodal stimuli leads to perceptual processing, learning, and eventually memory for amodally specified properties over other properties. Thus, when multimodal information is available, these properties will be differentiated before unimodal information. Finally, the prioritization of amodal information ensures unitary perception of single multimodal events and even constrains further processing. Information available to the different senses will thus be coordinated properly, ensuring that infants will attend to the single, unitary event producing the multimodal information.

The breadth of cognitive domains influenced by intersensory redundancy is extensive. For example, as discussed previously, 5-month-old infants were able to discriminate a novel rhythm when they were habituated to a multisensory but not unisensory rhythm (Bahrick & Lickliter, 2000). Similarly, at 3 months of age, infants can differentiate between divergent tempos after bimodal but not unimodal habituation (Bahrick, Flom, & Lickliter, 2002).

The ability to discriminate affect in response to bimodal (audiovisual) stimulation emerged by 4 months of age, while auditory discrimination emerges at 5 months and visual discrimination at 6 months (Flom & Bahrick, 2007). At 7 months, synchronized vocalizations and object motion facilitated learning of arbitrary speech-object relations (Gogate & Bahrick, 1998). Presenting light in conjunction with a sound enhances young adults' ability to detect low-intensity sounds (Lovelace et al., 2003). Finally, while the sensory processes of elderly adults often deteriorate, multisensory presentation of stimuli actually enhances their discrimination (e.g., Corso, 1971; Lichtenstein, 1992). For example, 65- to 90-year-olds were as fast responding to a two-alternative forced-choice discrimination task when provided with multisensory stimuli as 18- to 38-year-olds were when provided with unisensory stimuli (Laurienti, Burdette, Maldjian, & Wallace, 2006).

Evidence suggests that the ability to process and benefit from multisensory stimulation emerged in phylogenetically early organisms and has been sustained in higher-order organisms (e.g., Bahrick & Lickliter, 2000; Gogate & Bahrick, 1998; Lewkowicz & Kraebel, 2004; Lickliter, Bahrick, & Huneycutt, 2002; Lovelace, Stein, & Wallace, 2003; Mellon, Kraemer, & Spear, 1991; Meredith & Stein, 1983). For example, multisensory neurons in the superior colliculus of domestic cats increase firing rates in response to multimodal cues that occur together in time and space, to levels above the responses evoked by unisensory cues alone. Furthermore, multisensory compared to unisensory cues are more effective in enhancing cats' ability to detect, orient toward, and approach the cue (e.g., Meredith & Stein, 1983; Stein, Huneycutt, & Meredith, 1988). Bobwhite quail (*Colinus virginianus*) that received prenatal exposure to a patterned light paired synchronously with maternal calls were better at learning these calls than birds that received only auditory information prenatally (Lickliter et al., 2002). Similarly, preweaned rats that experienced a paired light and tone exhibited greater behavioral suppression to a light than those rats that did not experience such pairing (Mellon, Kraemer, & Spear, 1991).

In sum, multimodal information is highly salient in the environment, often causing humans and nonhuman animals to preferentially process intersensory redundancy over unimodal stimuli in their environments. This has been shown to occur in many domains, ranging from rhythm discrimination to affect recognition.

DOES MULTISENSORY INFORMATION IMPROVE NUMERICAL ABILITIES?

A similar pattern as those described previously has been demonstrated for numerical discriminations in both infants and preschool children. For example, studies have shown that 6-month-olds who were habituated to 8 or 16 dots later dishabituated to the novel of these numerosities when tested with both in alternation (Xu & Spelke, 2000). However, when habituated in the same way to 8 or 12 dots, 6-month-old infants did not dishabituate to the alternating novel numerosity during test, suggesting that at 6 months infants require a 1:2 ratio to make numerical discriminations between visual arrays (see also Xu, 2003; Xu et al., 2005). A similar pattern of ratio dependence was also observed for 6-month-old infants within an auditory numerical discrimination task (Lipton & Spelke, 2003). Here, a preferential head-orienting paradigm was used to show that infants of this age were capable of discriminating 8 vs. 16 tones but not 8 vs. 12 tones. Such results suggest that 6-month-olds also need a 1:2 ratio to successfully discriminate auditory number. Finally, Wood and Spelke (2005) demonstrated successful discrimination of a 1:2 but not 2:3 ratio of puppet jumps in 6-month-old infants. However, by 9 months of age, infants can discriminate 8 from 12 dots, 4 from 6 puppet jumps, and 8 from 12 tones (Lipton & Spelke, 2003; Wood & Spelke, 2005; Xu & Spelke, 2000). Taken together, these data suggest that 6-month-olds require a 1:2 ratio for successful numerical discrimination, regardless of the sensory modality in which stimuli are experienced. Furthermore, infants' threshold of discrimination increases from a 1:2 ratio to a 2:3 ratio between 6 and 9 months of age.

To test if this is an absolute limitation of 6-month-old infants' numerical discrimination abilities, Jordan and colleagues (2008) used a habituation/discrimination paradigm to demonstrate that intersensory redundancy enhances infants' performance to levels previously judged to emerge only around 9 months of age. Here, half of the infants were habituated to movies in which a ball dropped and bounced 8 times, and the other half were habituated to movies in which a ball dropped and bounced 12 times. In test, infants were shown alternating movies in which a ball dropped and bounced 8 or 12 times. The first of three experiments provided intersensory redundancy during numerical habituation by pairing the visual bouncing motion of the ball with a tone each time the ball struck the floor in the movie. The second experiment provided unisensory visual numerical habituation, wherein the tone was removed so that only the visual cue to the number of bounces was available. The third experiment provided non-numerical multisensory habituation, in which the visual bouncing ball stimulus was presented along with non-numerical classical music. When infants received redundant multisensory information about number throughout habituation, they significantly dishabituated to the novel bounce number in test, indicating that they were capable of discriminating a 2:3 ratio change in number. The same was not true when only visual numerical information was provided during habituation, consistent with many previous unimodal studies (above). Finally, infants did not dishabituate to the novel bounce number when they received concurrent audiovisual information that did not provide redundant information about number. Importantly, infants also fail to dishabituate to asynchronous audiovisual information within a paradigm similar to that described earlier (Bahrick & Lickliter, 2000; Flom & Bahrick, 2007).

Similar beneficial effects of intersensory redundancy on numerical discriminations have recently been shown in preschool children (Jordan & Baker, 2011). Using a within-subject delayed match-to-sample paradigm, Jordan and Baker demonstrated that 3- to 5-year-olds were significantly more accurate matching number on audiovisual trials, compared to trials in which only auditory or visual information is provided. Here, each participant completed 30 computerized trials, wherein 10 trials provided audiovisual elements to numerically match, 10 provided purely visual elements to numerically match, and 10 provided purely auditory elements to match (Figure 11-3).

Each trial specifically presented between 5 and 9 sample elements serially, too quickly for children to verbally count. The elements presented were squares of differing colors and sizes, tones, or a combination of squares and tones that occurred simultaneously. Immediately following the sample element presentation, two visual answer choices were provided at random locations on the computer screen. Each answer choice was composed of an array of differently colored squares. One answer choice (correct, matching option) always contained the same number of squares as sample elements presented in that trial, whereas the other answer choice (incorrect, distractor option) contained a differing number of squares. The ratio between the match and distractor options was varied systematically. Thus, trials in which the ratio between options was small (e.g., 5 vs. 6) were objectively more difficult

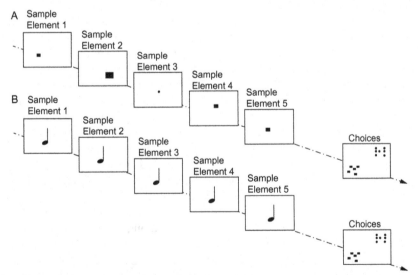

FIGURE 11-3 Trial structure from Jordan and Baker's (2011) study of intersensory redundancy and numerical discrimination. Example of visual (A) and auditory (B) trials with five sample elements each. Audiovisual trials were a combination of these two types of samples stimuli that occurred simultaneously. Dotted lines indicate time.

than trials in which this ratio was large (e.g., 5 vs. 9), which, as discussed earlier, is a key component of Weber's law.

We predicted that providing redundant multisensory numerical information would lead to overall greater performance than providing unisensory numerical information. This prediction of overall greater accuracy on audiovisual trials was upheld (Figure 11-4).

Children's performance on the audiovisual trials was significantly greater than that predicted by chance, whereas performance on both unisensory trials was at chance levels at each ratio larger than .65. Moreover, a significant main effect of trial type indicated that accuracy on audiovisual trials was greater than that on both unisensory trial types. Furthermore, ratio-dependent accuracy was identified for both unisensory trial types, as predicted by Weber's law; as the ratio between correct and distractor answer choices approached 1:1, performance on both unisensory trial types decreased to chance. However, unexpectedly, this was not the case for audiovisual trials, wherein performance remained above chance levels and steady across all ratios. This outcome was surprising, as it suggests that intersensory redundancy may in some cases override the well-established limits of ratio dependency constrained by Weber's law.

How Does Multisensory Information Improve Numerical Competence?

As argued by Gibson (1969), low-level perception of basic elements in the environment underlies much of the learning process throughout infancy. Amodal properties such as number are particularly salient in an infant's environment, and ultimately help focus infants' attention on meaningful, unitary multimodal events, while buffering them against learning of incongruent or arbitrary intersensory relations (Bahrick & Lickliter, 2000). Take, for example, the infants in Jordan and colleagues' (2008) Experiment 1: Here, multiple perceptual properties outside of number were competing for the infants' attention such as the color of the ball, the height of each bounce, etc. Because information about number was being provided synchronously across multiple sensory modalities, whereas information about these other properties was provided through only one modality, number likely became the most salient property within the scene (although this remains to be tested, and for exceptions, see Lewkowicz & Schwartz, 2002). It is possible that such selective recruiting of attention to the amodal property of number leads to more effective encoding of the number of bounces that occurred, which would in turn lead to greater overall signal strength and decreased variance for numerical representation in memory. Thus, an account of such selective attention to number may best explain why multisensory information improves numerical competence.

Such an attentional account as described previously is consistent with Gibson's (1969) view, as well as the intersensory redundancy hypothesis of

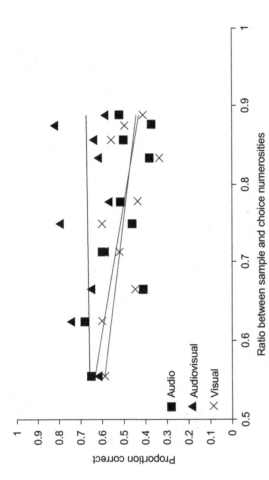

FIGURE 11-4 Uni- and multimodal trial performance from Jordan and Baker's (2011) study of intersensory redundancy and numerical discrimination. Mean accuracy for each ratio presented within each trial type. Overall, children performed more accurately on audiovisual trial types than on either unisensory trial type (bold trendline). Ratio dependent performance, which is predicted by Weber's law, can be seen in the negative slopes of performance on both unisensory trial types. In contrast, the flat slope of performance on audiovisual trials indicates consistently higher accuracy across ratios and thus nonadherence to ratio-dependent performance.

Bahrick and Lickliter (2000). It should be noted that an increase in *general* attention due to concurrent audiovisual stimulation—regardless of whether this stimulation provides *numerical* information—is not the cause for this enhanced numerical competence, as demonstrated by Jordan and colleagues (2008; described previously). While such concurrent audiovisual stimulation may have heightened infants' overall levels of attention to the scene, much the same way as loud noises and a visually dynamic scene may heighten our general attention or awareness, the lack of redundant conceptual content between the auditory and visual streams did not increase infant attention to the property of number per se. Simply pairing concurrent, non-numerical auditory information with visual numerical information was not sufficient to significantly heighten infants' attention to the number of bounces—highlighting the likely need for synchronous, redundant multisensory numerical information to achieve this effect.

These and other findings bolster the claim that an increase in selective attention to the redundantly specified property of number may best describe why multisensory information enhances numerical competence, lending support to the intersensory redundancy hypothesis. However, emergent data suggest that the *intersensory* nature of redundant information may not be necessary for such perceptual benefits to arise. Using a variation of Jordan and colleagues' (2008) bouncing ball paradigm, our lab has demonstrated that multiple simultaneous cues to quantity change occurring within the visual modality alone may similarly enhance infants' discrimination abilities (Baker, Mahamane, & Jordan, 2014, Experiment 1). Here, 6-month-old infants were habituated to a movie sequence of a large ball bouncing 12 times, or a ball 2/3 of the surface area of the large ball bouncing 8 times. Following habituation, each infant was exposed to the novel number/size pairing. Thus, in this experiment infants experienced visual simultaneous 2:3 ratio changes in both number and surface area. As noted earlier, previous studies have demonstrated that when these visual changes in quantity are provided in isolation, infants of this age are not capable of identifying a 2:3 change in either property alone (number: Jordan, Suanda, & Brannon, 2008; surface area: Brannon, Lutz, & Cordes, 2006). However, our data suggest that when visual cues to changes in these quantitative properties are provided synchronously, infants' discrimination abilities are enhanced, resulting in dishabituation to the novel number/size pairing. Furthermore, experiment 2 of Baker and colleagues (2014) provided infants with discordant size and number changes, such that the large ball bounced 8 times and the small ball bounced 12 times. Similar to our findings reported earlier, infants were capable of discriminating a 2:3 change in number and surface area when both properties change simultaneously and in opposite directions.

Taken together, these results suggest that the intersensory nature of simultaneous changes in amodal properties may not be the driving force behind enhanced perceptual discrimination. Instead, simultaneous changes to amodal properties that occur in any sensory modality may be the primary source driving enhanced perceptual discriminations in infancy. For example, a simultaneous change in number and surface area enhances infants' quantitative discrimination, suggesting that such redundancy heightens infant attention to amodal properties of quantity such as number and surface area (Baker et al., 2014).

Emergent findings in neuroscience are also shedding much-needed light on the neural networks that may underlie multisensory integration and cross-modal processing, which together may enhance perceptual discrimination. In a review of such studies, Calvert (2001) reports overlap between brain regions—including the superior temporal sulcus, the inferior parietal sulcus, regions of the frontal cortex, the insula cortex, and claustrum—that consistently play a role in multisensory integration. Moreover, individual neurons, often called "multisensory integrative cells," have been identified in various cortical regions of cats (*Felis catus*; Wallace, Meredith, & Stein, 1992; Wilkinson, Meredith, & Stein, 1996), rats (Barth, Goldberg, Brett, & Di, 1995), and macaques (Duhamel, Colby, & Goldberg, 1998; Graziano & Gross, 1998; Perrett, Harries, Mistlin, & Chitty, 1990), which increase their firing rate when two or more sensory cues from different modalities appear in close temporal and spatial proximity. However, the overwhelming consensus, despite the presence of such multisensory integrative cells, is that entire networks of brain regions are primarily responsible for such integrative abilities (Calvert, 2001). For instance, the superior temporal sulcus tends to play a role in the integration of complex featural information, such as the perception of audiovisual speech (Banati, Goerres, Tjoa, Aggleton, & Grasby, 2000; Callan, Callan, Kroos, & Vatikiotis-Bateson, 2001; Calvert, Hansen, Iversen, & Brammer, 2001; Calvert, Campbell, & Brammer, 2000; Raij, Uutela, & Hari, 2000). The intraparietal sulcus seems specialized for cross-modal spatial coordinate cues and mediating cross-modal links in attention (Banati et al., 2000; Bushara et al., 1999; Bushara, Grafman, & Hallett, 2001; Callan et al., 2001; Calvert, 2001; Eimer, 1999; Lewis, Beauchamp, & DeYoe, 2000; Macaluso, Frith, & Driver, 2000) and has also been highly implicated in numerical processing (see Arsalidou & Taylor, 2011; and Ashkenazi, Black, Abrams, Hoeft, & Menon, 2013, for review). Finally, the posterior insula is often involved with the detection of temporal coincidence between cross-modal stimuli (Banati et al., 2000; Bushara et al., 2001; Calvert, 2001; Hadjikhani & Roland, 1998; Lewis et al., 2000). Because any one multisensory scene is likely to involve many of these individual aspects, these and other brain regions probably work together more often than not when integrating multisensory inputs.

UNANSWERED QUESTIONS AND FUTURE DIRECTIONS

The findings reported in the preceding sections provide important insights into the effects of multisensory information on perception of number. However, the nature of this relation still remains largely unknown, as does the broader role of intersensory redundancy in both the ontogenetic development and evolution of the perception of number. For example, to what degree does such information affect adults' learning and perception? According to theories of perceptual narrowing (Lewkowicz & Ghanzanfar, 2009), the beneficial effects of multisensory stimulation decrease with age, and thus may not be as effective for older populations. However, whether there is an age at which multisensory information ceases to enhance numerical cognition remains to be determined. Based on results from Jordan and Baker (2011), such enhancing effects occur at least through 5 years of age. This finding is especially encouraging when considering the potential value of using intersensory redundancy to enhance tools for teaching basic numerical and arithmetic skills.

A relation has recently been identified between individuals' nonverbal numerical discrimination ability and success in mathematics (Halberda, Mazzocco, & Feigenson, 2008). Furthermore, this discrimination ability has been shown to be a stable predictor of math achievement throughout life (Libertus, Feigenson, & Halberda, 2013). In short, a person who is more sensitive to changes in abstract numerical properties may be more likely to succeed in mathematics than a person who is less sensitive (Star & Brannon, this volume). Moreover, recent evidence suggests that directly training such approximate number abilities improves math proficiency (Park & Brannon, 2013). Because intersensory redundancy effectively increases this sensitivity, a teaching tool that provides redundant inter- or intrasensory information about number may positively influence some aspects of children's performance in math.

The ubiquity of computers in today's elementary school classroom may provide a platform for such teaching tools. For example, virtual manipulatives are computer-based analogs to common teaching tools such as pegboards, fraction tiles, and others. It is possible to embed redundant information about number within computerized tools in the form of audiovisual events that occur at specified moments throughout their use. Thus, virtual manipulatives may provide teachers with a powerful tool that is capable of harnessing the enhancing effects of intersensory redundancy. A recent comparison between virtual manipulatives and more traditional teaching tools has indicated that both are equally effective in teaching third- and fourth-grade students fractions in terms of overall student performance (Moyer-Packenham et al., 2013). One benefit to using virtual manipulatives, though, is that demographic factors such as socioeconomic status did not influence performance in the group of children randomly assigned to the virtual manipulative teaching group in this study, although such demographic factors did predict performance in the group of children randomly assigned to the traditional classroom approach to teaching fractions. Thus, instruction with virtual manipulatives could potentially "level the playing field" between students from different socioeconomic status levels, and may provide a useful real-world future platform in which to test the effects of redundant numerical stimulation on mathematics learning. As these studies are nonetheless in their early phases, we await their replication and extension to other mathematical domains.

It will also be important to better understand the effect of multisensory information about number in atypically developing populations. For example, Turner Syndrome (TS) is a neurogenetic disorder that occurs in females and is associated with a cognitive profile that generally includes deficits in executive, visual-spatial, and mathematical ability (Hong, Scaletta Kent, & Kessler, 2009). However, despite these common deficits, women with TS typically otherwise maintain intact intellectual function and verbal abilities. Given the deficits common to TS, understanding the influence of multisensory information on numerical cognition in this population may clarify the mechanisms through which such information influences perception in neurotypical populations. As argued by Butterworth and Kovas (2013), understanding neurocognitive disorders and how they influence cognition in relation to neurotypical populations has the potential to improve our general understanding of cognition and may even improve education. Future research should strive to clarify these and other outstanding issues.

CONCLUSIONS

Human and nonhuman animals alike possess the ability to represent number without language (Agrillo et al., this volume; Beran et al., this volume; Geary et al., this volume; Pepperberg, this volume; Vallortigara, this volume). This potentially homologous system represents the approximate quantities of collections of objects and shows characteristics that follow predictions made by Weber's law; specifically, this ability has evolutionary origins, providing organisms with a fast and efficient way to quantify a chaotic multisensory environment. One possibility is that the number system evolved to support efficient foraging, among other things, which provides contexts that are likely to produce multisensory cues. As a result, animals that are able to use all available information to guide their foraging would necessarily make more reliable decisions than animals that used only one piece of unimodal information.

In this chapter, we have reviewed recent evidence that multisensory inputs that occur synchronously enhance many aspects of cognition, including the approximate number sense. For example, redundant, multisensory stimuli boost the ability to discriminate quantities relative to abilities found in studies that provide only unimodal cues. From birth, such intersensory stimulation is ubiquitous throughout our environments, likely capturing our attentional priorities and resulting in better encoding of such stimuli relative to unisensory stimuli. We have also reviewed research showing that the beneficial effects of such intersensory redundancy may differ across domains. Furthermore, we reviewed literature demonstrating that perceptual narrowing may occur at different ages for one system (e.g., number) than for other systems (e.g., lips/speech).

In sum, we have reviewed the pertinent literature regarding the ability to perceive number in multiple sensory modalities and the factors that influence this ability. Moreover, we have examined current hypotheses concerning how this multisensory ability emerged, the brain regions that potentially facilitate this ability, and how multisensory integration and intersensory redundancy can influence the amodal processing of number. The latter has potentially important implications for enhancing early mathematics learning and instruction. In the future, it will be important to determine the ages at which benefits accrue from redundant numerical information, and what kind of training might be necessary to sustain any such benefit. Thus, multisensory information may benefit children because they do not effectively focus on the numerical features of what they are encoding (see Hannula, Lepola, & Lehtinen, 2010). Future research should address whether older individuals are capable of extracting the same amount of information unimodally as a younger individual can with multimodal information.

REFERENCES

Arsalidou, M., & Taylor, M. J. (2011). Is $2+2=4$? Meta-analyses of brain areas needed for numbers and calculations. *NeuroImage*, *54*(3), 2382–2393.

Ashkenazi, S., Black, J., Abrams, D., Hoeft, F., & Menon, V. (2013). Neurobiological underpinnings of math and reading learning disabilities. *Journal of Learning Disabilities*, *46*, 549–569.

Bahrick, L. E. (1994). The development of infants' sensitivity to arbitrary intermodal relations. *Ecological Psychology*, *6*(2), 111–123.

Bahrick, L. E., Flom, R., & Lickliter, R. (2002). Intersensory redundancy facilitates discrimination of tempo in 3-month-old infants. *Developmental Psychobiology*, *41*(4), 352–363.

Bahrick, L. E., & Lickliter, R. (2000). Intersensory redundancy guides attentional selectivity and perceptual learning in infancy. *Developmental Psychology*, *36*(2), 190.

Bahrick, L. E., Lickliter, R., & Flom, R. (2004). Intersensory redundancy guides the development of selective attention, perception, and cognition in infancy. *Current Directions in Psychological Science*, *13*(3), 99–102.

Baker, J. M., Mahamane, S. P., & Jordan, K. E. (2014). Multiple visual quantitative cues enhance discrimination of dynamic stimuli during infancy. *Journal of Experimental Child Psychology*, *122*, 21–32.

Baker, J. M., Rodzon, K., & Jordan, K. (2013). The impact of emotion on numerosity estimation. *Frontiers in Psychology*, *4*, 521.

Banati, R. B., Goerres, G. W., Tjoa, C., Aggleton, J. P., & Grasby, P. (2000). The functional anatomy of visual–tactile integration in man: A study using positron emission tomography. *Neuropsychologia*, *38*(2), 115–124.

Barth, D. S., Goldberg, N., Brett, B., & Di, S. (1995). The spatiotemporal organization of auditory, visual, and auditory–visual evoked potentials in rat cortex. *Brain Research*, *678*(1), 177–190.

Barth, H. C., & Paladino, A. M. (2011). The development of numerical estimation: Evidence against a representational shift. *Developmental Science*, *14*(1), 125–135.

Beran, M. J. (2001). Summation and numerousness judgments of sequentially presented sets of items by chimpanzees (*Pan troglodytes*). *Journal of Comparative Psychology*, *115*(2), 181.

Beran, M. J. (2004). Chimpanzees (*Pan troglodytes*) respond to nonvisible sets after one-by-one addition and removal of items. *Journal of Comparative Psychology*, *118*(1), 25.

Beran, M. J., & Beran, M. M. (2004). Chimpanzees remember the results of one-by-one addition of food items to sets over extended time periods. *Psychological Science, 15*(2), 94–99.

Beran, M. J., Beran, M. M., Harris, E. H., & Washburn, D. A. (2005). Ordinal judgments and summation of nonvisible sets of food items by two chimpanzees and a rhesus macaque. *Journal of Experimental Psychology. Animal Behavior Processes, 31*(3), 351.

Beran, M. J., & Rumbaugh, D. M. (2001). "Constructive" enumeration by chimpanzees (*Pan troglodytes*) on a computerized task. *Animal Cognition, 4*(2), 81–89.

Brannon, E. M. (2002). The development of ordinal numerical knowledge in infancy. *Cognition, 83*(3), 223–240.

Brannon, E. M. (2006). The representation of numerical magnitude. *Current Opinion in Neurobiology, 16*(2), 222–229.

Brannon, E. M., Abbott, S., & Lutz, D. J. (2004). Number bias for the discrimination of large visual sets in infancy. *Cognition, 93*(2), B59–B68.

Brannon, E. M., Lutz, D., & Cordes, S. (2006). The development of area discrimination and its implications for number representation in infancy. *Developmental Science, 9*(6), F59–F64.

Brannon, E. M., & Roitman, J. D. (2003). Nonverbal representations of time and number in animals and human infants. *Functional and Neural Mechanisms of Interval Timing, 143–182*.

Brannon, E. M., & Terrace, H. S. (1998). Ordering of the numerosities 1 to 9 by monkeys. *Science, 282*(5389), 746–749.

Brannon, E. M., & Terrace, H. S. (2000). Representation of the numerosities 1–9 by rhesus macaques (*Macaca mulatta*). *Journal of Experimental Psychology. Animal Behavior Processes, 26*(1), 31.

Bushara, K. O., Grafman, J., & Hallett, M. (2001). Neural correlates of auditory–visual stimulus onset asynchrony detection. *The Journal of Neuroscience, 21*(1), 300–304.

Bushara, K. O., Weeks, R. A., Ishii, K., Catalan, M. J., Tian, B., Rauschecker, J. P., et al. (1999). Modality-specific frontal and parietal areas for auditory and visual spatial localization in humans. *Nature Neuroscience, 2*(8), 759–766.

Butterworth, B., & Kovas, Y. (2013). Understanding neurocognitive developmental disorders can improve education for all. *Science, 340*(6130), 300–305.

Callan, D. E., Callan, A. M., Kroos, C., & Vatikiotis-Bateson, E. (2001). Multimodal contribution to speech perception revealed by independent component analysis: A single-sweep EEG case study. *Cognitive Brain Research, 10*(3), 349–353.

Calvert, G. A. (2001). Crossmodal processing in the human brain: Insights from functional neuroimaging studies. *Cerebral Cortex, 11*(12), 1110–1123.

Calvert, G. A., Campbell, R., & Brammer, M. J. (2000). Evidence from functional magnetic resonance imaging of crossmodal binding in the human heteromodal cortex. *Current Biology, 10*(11), 649–657.

Calvert, G. A., Hansen, P. C., Iversen, S. D., & Brammer, M. J. (2001). Detection of audio–visual integration sites in humans by application of electrophysiological criteria to the BOLD effect. *NeuroImage, 14*(2), 427–438.

Cantlon, J. F., & Brannon, E. M. (2006). Shared system for ordering small and large numbers in monkeys and humans. *Psychological Science, 17*(5), 401–406.

Cantlon, J. F., Safford, K. E., & Brannon, E. M. (2010). Spontaneous analog number representations in 3-year-old children. *Developmental Science, 13*(2), 289–297.

Cordes, S., Gelman, R., Gallistel, C. R., & Whalen, J. (2001). Variability signatures distinguish verbal from nonverbal counting for both large and small numbers. *Psychonomic Bulletin & Review, 8*(4), 698–707.

Corso, J. F. (1971). Sensory processes and age effects in normal adults. *Journal of Gerontology,* *26*(1), 90–105.

Dehaene, S., Dupoux, E., & Mehler, J. (1990). Is numerical comparison digital? Analogical and symbolic effects in two-digit number comparison. *Journal of Experimental Psychology: Human Perception and Performance, 16*(3), 626–641.

Duhamel, J. R., Colby, C. L., & Goldberg, M. E. (1998). Ventral intraparietal area of the macaque: Congruent visual and somatic response properties. *Journal of Neurophysiology,* *79*(1), 126–136.

Eimer, M. (1999). Can attention be directed to opposite locations in different modalities? An ERP study. *Clinical Neurophysiology, 110*(7), 1252–1259.

Feigenson, L., Dehaene, S., & Spelke, E. (2004). Core systems of number. *Trends in Cognitive Sciences, 8*(7), 307–314.

Flom, R., & Bahrick, L. E. (2007). The development of infant discrimination of affect in multi-modal and unimodal stimulation: The role of intersensory redundancy. *Developmental Psychology, 43*(1), 238.

Gallistel, C. R., & Gelman, R. (1992). Preverbal and verbal counting and computation. *Cognition,* *44*(1), 43–74.

Gardner, J. M., Lewkowicz, D. J., Rose, S. A., & Karmel, B. Z. (1986). Effects of visual and audi-tory stimulation on subsequent visual preferences in neonates. *International Journal of Behavioral Development, 9*(2), 251–263.

Gibson, E. J. (1969). *Principles of perceptual learning and development.* East Norwalk, CT: Prentice Hall.

Gogate, L. J., & Bahrick, L. E. (1998). Intersensory redundancy facilitates learning of arbitrary relations between vowel sounds and objects in seven-month-old infants. *Journal of Experi-mental Child Psychology, 69*(2), 133–149.

Graziano, M. S., & Gross, C. G. (1998). Spatial maps for the control of movement. *Current Opin-ion in Neurobiology, 8*(2), 195–201.

Hadjikhani, N., & Roland, P. E. (1998). Cross-modal transfer of information between the tactile and the visual representations in the human brain: A positron emission tomographic study. *The Journal of Neuroscience, 18*(3), 1072–1084.

Halberda, J., & Feigenson, L. (2008). Developmental change in the acuity of the "Number Sense": The approximate number system in 3-, 4-, 5-, and 6-year-olds and adults. *Developmental Psy-chology, 44*(5), 1457–1465.

Halberda, J., Ly, R., Wilmer, J. B., Naiman, D. Q., & Germine, L. (2012). Number sense across the lifespan as revealed by a massive Internet-based sample. *Proceedings of the National Academy of Sciences, 109*(28), 11116–11120.

Halberda, J., Mazzocco, M. M., & Feigenson, L. (2008). Individual differences in non-verbal number acuity correlate with maths achievement. *Nature, 455*(7213), 665–668.

Hannula, M. M., Lepola, J., & Lehtinen, E. (2010). Spontaneous focusing on numerosity as a domain-specific predictor of arithmetic skills. *Journal of Experimental Child Psychology,* *107*(4), 394–406.

Hauser, M. D., Tsao, F., Garcia, P., & Spelke, E. S. (2003). Evolutionary foundations of number: Spontaneous representation of numerical magnitudes by cotton-top tamarins. *Proceedings of the Royal Society of London, Series B: Biological Sciences, 270*(1523), 1441–1446.

Hong, D., Scaletta Kent, J., & Kesler, S. (2009). Cognitive profile of Turner syndrome. *Develop-mental Disabilities Research Reviews, 15*(4), 270–278.

Horng, S. H., & Sur, M. (2006). Visual activity and cortical rewiring: Activity-dependent plastic-ity of cortical networks. *Progress in Brain Research, 157,* 3–381.

Huntley-Fenner, G. (2001). Children's understanding of number is similar to adults' and rats': Numerical estimation by 5–7-year-olds. *Cognition, 78*(3), B27–B40.

Huntley-Fenner, G., & Cannon, E. (2000). Preschoolers' magnitude comparisons are mediated by a preverbal analog mechanism. *Psychological Science, 11*(2), 147–152.

Jordan, K. E., & Baker, J. M. (2011). Multisensory information boosts numerical matching abilities in young children. *Developmental Science, 14*(2), 205–213.

Jordan, K. E., & Brannon, E. M. (2006a). A common representational system governed by Weber's law: Nonverbal numerical similarity judgments in 6-year-olds and rhesus macaques. *Journal of Experimental Child Psychology, 95*(3), 215–229.

Jordan, K. E., & Brannon, E. M. (2006b). The multisensory representation of number in infancy. *Proceedings of the National Academy of Sciences of the United States of America, 103*(9), 3486–3489.

Jordan, K. E., & Brannon, E. M. (2006c). Weber's law influences numerical representations in rhesus macaques (*Macaca mulatta*). *Animal Cognition, 9*(3), 159–172.

Jordan, K. E., Brannon, E. M., Logothetis, N. K., & Ghazanfar, A. A. (2005). Monkeys match the number of voices they hear to the number of faces they see. *Current Biology, 15*(11), 1034–1038.

Jordan, K. E., Suanda, S. H., & Brannon, E. M. (2008). Intersensory redundancy accelerates preverbal numerical competence. *Cognition, 108*(1), 210–221.

Judge, P. G., Evans, T. A., & Vyas, D. K. (2005). Ordinal representation of numeric quantities by brown capuchin monkeys (*Cebus apella*). *Journal of Experimental Psychology. Animal Behavior Processes, 31*(1), 79–94.

Kahana-Kalman, R., & Walker-Andrews, A. S. (2001). The role of person familiarity in young infants' perception of emotional expressions. *Child Development, 72*(2), 352–369.

King, A. J. (2009). Visual influences on auditory spatial learning. *Philosophical Transactions of the Royal Society, B: Biological Sciences, 364*(1515), 331–339.

Kuhl, P. K., & Meltzoff, A. N. (1982, December). The bimodal perception of speech in infancy. *Science, 218*(4577), 1138–1141.

Laurienti, P. J., Burdette, J. H., Maldjian, J. A., & Wallace, M. T. (2006). Enhanced multisensory integration in older adults. *Neurobiology of Aging, 27*(8), 1155–1163.

Lewis, J. W., Beauchamp, M. S., & DeYoe, E. A. (2000). A comparison of visual and auditory motion processing in human cerebral cortex. *Cerebral Cortex, 10*(9), 873–888.

Lewkowicz, D. J. (2000). The development of intersensory temporal perception: An epigenetic systems/limitations view. *Psychological Bulletin, 126*(2), 281.

Lewkowicz, D. J., & Ghazanfar, A. A. (2009). The emergence of multisensory systems through perceptual narrowing. *Trends in Cognitive Sciences, 13*(11), 470–478.

Lewkowicz, D. J., & Kraebel, K. S. (2004). The value of multisensory redundancy in the development of intersensory perception. In G. Calvert, et al. (Ed.). *The handbook of multisensory processes* (pp. 655–678). Cambridge, MA: MIT Press.

Lewkowicz, D. J., & Lickliter, R. (1994). Insights into mechanisms of intersensory development: The value of a comparative, convergent-operations approach. *The Development of Intersensory Perception: Comparative Perspectives*, 403–413.

Lewkowicz, D. J., & Schwartz, B. B. (2002). Intersensory perception in infancy: Response to competing amodal and modality-specific attributes. *Poster presented at the biennial international conference for infant studies*. Toronto: ON, Canada.

Lewkowicz, D. J., Sowinski, R., & Place, S. (2008). The decline of cross-species intersensory perception in human infants: Underlying mechanisms and its developmental persistence. *Brain Research, 1242*, 291–302.

Lewkowicz, D. J., & Turkewitz, G. (1980). Cross-modal equivalence in early infancy: Auditory–visual intensity matching. *Developmental Psychology, 16*(6), 597.

Lewkowicz, D. J., & Turkewitz, G. (1981). Intersensory interaction in newborns: Modification of visual preferences following exposure to sound. *Child Development, 827–832.*

Libertus, M. E., Feigenson, L., & Halberda, J. (2013). Is approximate number precision a stable predictor of math ability? *Learning and Individual Differences, 25,* 126–133.

Lichtenstein, M. J. (1992). Hearing and visual impairments. *Clinics in Geriatric Medicine, 8*(1), 173.

Lickliter, R., Bahrick, L. E., & Honeycutt, H. (2002). Intersensory redundancy facilitates prenatal perceptual learning in bobwhite quail (*Colinus virginianus*) embryos. *Developmental Psychology, 38*(1), 15.

Lipton, J. S., & Spelke, E. S. (2003). Origins of number sense large-number discrimination in human infants. *Psychological Science, 14*(5), 396–401.

Lipton, J. S., & Spelke, E. S. (2004). Discrimination of large and small numerosities by human infants. *Infancy, 5*(3), 271–290.

Lovelace, C. T., Stein, B. E., & Wallace, M. T. (2003). An irrelevant light enhances auditory detection in humans: A psychophysical analysis of multisensory integration in stimulus detection. *Cognitive Brain Research, 17*(2), 447–453.

Macaluso, E., Frith, C. D., & Driver, J. (2000). Modulation of human visual cortex by crossmodal spatial attention. *Science, 289*(5482), 1206–1208.

Maurer, D., Mondloch, C. J., & Lewis, T. L. (2007). Sleeper effects. *Developmental Science, 10*(1), 40–47.

Mellon, R. C., Kraemer, P. J., & Spear, N. E. (1991). Development of intersensory function: Age-related differences in stimulus selection of multimodal compounds in rats as revealed by Pavlovian conditioning. *Journal of Experimental Psychology. Animal Behavior Processes, 17*(4), 448–464.

Meredith, M. A., & Stein, B. E. (1983). Interactions among converging sensory inputs in the superior colliculus. *Science, 221*(4608), 389–391.

Moyer, R. S., & Landauer, T. K. (1967). Time required for judgements of numerical inequality. *Nature, 215*(5109), 1519–1520.

Moyer-Packenham, P., Baker, J. M., Westenskow, A., Rodzon, K., Anderson, K., Shumway, J., et al. (2013). Achievement performance: A random assignment study in seventeen third- and fourth-grade classrooms comparing virtual manipulatives with other instructional treatments. *Journal of Education, 193*(2), 25–39.

Nieder, A., Freedman, D. J., & Miller, E. K. (2002). Representation of the quantity of visual items in the primate prefrontal cortex. *Science, 297*(5587), 1708–1711.

Nieder, A., & Miller, E. K. (2004). A parieto-frontal network for visual numerical information in the monkey. *Proceedings of the National Academy of Sciences of the United States of America, 101*(19), 7457–7462.

Park, J., & Brannon, E. M. (2013). Training the approximate number system improves math proficiency. *Psychological Science, 24*(10), 2013–2019.

Patterson, M. L., & Werker, J. F. (2002). Infants' ability to match dynamic phonetic and gender information in the face and voice. *Journal of Experimental Child Psychology, 81*(1), 93–115.

Patterson, M. L., & Werker, J. F. (2003). Two-month-old infants match phonetic information in lips and voice. *Developmental Science, 6*(2), 191–196.

Perrett, D. I., Harries, M., Chitty, A. J., & Mistlin, A. J. (1990). Three stages in the classification of body movements by visual neurons. In H. B. Barlow, C. Blakemore, & M. Weston-Smith (Eds.), *Images and understanding* (pp. 94–108). New York: Cambridge University Press.

Phillips, R. D., Wagner, S. H., Fells, C. A., & Lynch, M. (1990). Do infants recognize emotion in facial expressions?: Categorical and "metaphorical" evidence. *Infant Behavior and Development*, *13*(1), 71–84.

Platt, J. R., & Johnson, D. M. (1971). Localization of position within a homogeneous behavior chain: Effects of error contingencies. *Learning and Motivation*, *2*(4), 386–414.

Raij, T., Uutela, K., & Hari, R. (2000). Audiovisual integration of letters in the human brain. *Neuron*, *28*(2), 617–625.

Rosenblum, L. D., Schmuckler, M. A., & Johnson, J. A. (1997). The McGurk effect in infants. *Perception & Psychophysics*, *59*(3), 347–357.

Rumbaugh, D. M., Savage-Rumbaugh, S., & Hegel, M. T. (1987). Summation in the chimpanzee (*Pan troglodytes*). *Journal of Experimental Psychology. Animal Behavior Processes*, *13*(2), 107.

Sekuler, R., & Mierkiewicz, D. (1977). Children's judgments of numerical inequality. *Child Development*, *630–633*.

Slater, A., Quinn, P. C., Brown, E., & Hayes, R. (1999). Intermodal perception at birth: Intersensory redundancy guides newborn infants' learning of arbitrary auditory – visual pairings. *Developmental Science*, *2*(3), 333–338.

Smith, B. R., Piel, A. K., & Candland, D. K. (2003). Numerity of a socially housed hamadryas baboon (*Papio hamadryas*) and a socially housed squirrel monkey (*Saimiri sciureus*). *Journal of Comparative Psychology*, *117*(2), 217.

Spelke, E. S., Born, W. S., & Chu, F. (1983). Perception of moving, sounding objects by four-month-old infants. *Perception*, *12*(6), 719–732.

Stein, B. E., Huneycutt, W. S., & Meredith, M. A. (1988). Neurons and behavior: The same rules of multisensory integration apply. *Brain Research*, *448*(2), 355–358.

Taylor, R. C., & Ryan, M. J. (2013). Unique interactions of multisensory components perceptually rescue túngara frog mating signals. *Science*, *341*, 273–274.

Temple, E., & Posner, M. I. (1998). Brain mechanisms of quantity are similar in 5-year-old children and adults. *Proceedings of the National Academy of Sciences*, *95*(13), 7836–7841.

Turkewitz, G., & Kenny, P. A. (1982). Limitations on input as a basis for neural organization and perceptual development: A preliminary theoretical statement. *Developmental Psychobiology*, *15*(4), 357–368.

Wagner, S., Winner, E., Cicchetti, D., & Gardner, H. (1981). "Metaphorical" mapping in human infants. *Child Development*, *728–731*.

Walker-Andrews, A. S. (1986). Intermodal perception of expressive behaviors: Relation of eye and voice? *Developmental Psychology*, *22*(3), 373.

Walker-Andrews, A. S., & Lennon, E. (1991). Infants' discrimination of vocal expressions: Contributions of auditory and visual information. *Infant Behavior and Development*, *14*(2), 131–142.

Wallace, M. T., Meredith, M. A., & Stein, B. E. (1992). Integration of multiple sensory modalities in cat cortex. *Experimental Brain Research*, *91*(3), 484–488.

Wallace, M. T., & Stein, B. E. (2007). Early experience determines how the senses will interact. *Journal of Neurophysiology*, *97*(1), 921–926.

Walton, G. E., & Bower, T. G. R. (1993). Newborns form "prototypes" in less than 1 minute. *Psychological Science*, *4*(3), 203–205.

Washburn, D. A., & Rumbaugh, D. M. (1991). Ordinal judgments of numerical symbols by macaques (*Macaca mulatta*). *Psychological Science*, *2*(3), 190–193.

Weber, E. H. (1996). De tactu. Annotationes anatomicae et physiologicae [The sense of touch: Anatomical and physiological aspects]. In H. E. Ross & D. J. Murray (Eds.), *E. H. Weber on the tactile senses.* (2nd ed.). Hove, UK: Taylor & Francis. Original work published in 1834.

Whalen, J., Gallistel, C. R., & Gelman, R. (1999). Nonverbal counting in humans: The psychophysics of number representation. *Psychological Science, 10*(2), 130–137.

Wilkinson, L. K., Meredith, M. A., & Stein, B. E. (1996). The role of anterior ectosylvian cortex in cross-modality orientation and approach behavior. *Experimental Brain Research, 112*(1), 1–10.

Wood, J. N., & Spelke, E. S. (2005). Infants' enumeration of actions: Numerical discrimination and its signature limits. *Developmental Science, 8*(2), 173–181.

Xu, F. (2003). Numerosity discrimination in infants: Evidence for two systems of representations. *Cognition, 89*(1), B15–B25.

Xu, F., & Arriaga, R. I. (2007). Number discrimination in 10-month-old infants. *British Journal of Developmental Psychology, 25*(1), 103–108.

Xu, F., & Spelke, E. S. (2000). Large number discrimination in 6-month-old infants. *Cognition, 74*(1), B1–B11.

Xu, F., Spelke, E. S., & Goddard, S. (2005). Number sense in human infants. *Developmental Science, 8*(1), 88–101.

Part III

Number Judgments: Theoretical Perspectives and Evolutionary Foundations

Chapter 12

The Precision and Internal Confidence of Our Approximate Number Thoughts

Justin Halberda and Darko Odic
Department of Psychological and Brain Sciences, Johns Hopkins University, Baltimore, MD, USA

INTRODUCTION

We all have an approximate sense for numbers. That is, we have experiences of estimating of how many voices we hear or stars we see, and we have experiences of ordinal relation such as judging that the number of voices we hear right now is fewer than the number of stars we see in the sky right now. The sense of number that supports these kinds of experiences is quite immediate (in perception) and primitive (i.e., operating from birth and serving as a foundation on which we learn about the world as understood through numbers). We share this primitive sense of number with other animals (i.e., it has been measured in nonhuman primates, mammals more broadly, birds and fish; e.g., Cantlon & Brannon, 2006; Hauser, Tsao, Garcia, & Spelke, 2003; Meck & Church, 1983; Nieder & Miller, 2004), with babies from birth (Izard, Sann, Spelke, & Streri, 2009), with individuals from every human culture (NB, even those with no written mathematics of any kind; e.g., Dehaene, Izard, Spelke, & Pica, 2008; Frank, Everett, Fedorenko, & Gibson, 2008; Gordon, 2004; Spaepen, Coppola, Spelke, Carey, & Goldin-Meadow, 2011), and with every human-like mind that has ever walked the face of this earth (e.g., the painters of the Chauvet cave and Jesus of Nazareth).[1]

The portion of cognition that generates these experiences of approximate number has been called an "approximate number system" (ANS). It is

---

1. But note that homology has yet to be demonstrated—and, a prudent theorist might bet that the number sense is a case of convergent evolution in fish and humans or insects and humans given that we know of so little at the system's neuroscience scale that would currently count as homologous among these organisms (special thanks to Alvaro Mailhos and Dave Geary for discussions and inspiration surrounding this point).

Mathematical Cognition and Learning, Vol. 1. http://dx.doi.org/10.1016/B978-0-12-420133-0.00012-0

generated by neurons in the intraparietal sulcus (IPS) and area lateral intraparietal (LIP) cortex (Nieder, 2005; Nieder & Miller, 2004; Piazza, Izard, Pinel, Le Bihan, & Dehaene, 2004; Roitman, Brannon, & Platt, 2007). It is an evolutionarily ancient and primitive system for numerical thought, and yet, surprisingly, individual differences in the accuracy and precision of these approximate number representations relate to our performance in symbolic, formal, school mathematics (for review, Feigenson, Libertus, & Halberda, 2013; but see De Smedt, Noël, Gilmore, & Ansari, 2013).

Much of the evidence that supports our current understanding of the ANS is reviewed in other chapters within this volume (e.g., Cantlon, this volume; Starr & Brannon, this volume; vanMarle, this volume), so we try not to duplicate those here. Instead, our aim is to describe the psychophysical model describing ANS representations and their precision (i.e., the Weber fraction and related concepts) and to suggest a reason why this precision may differ across observers. We've organized the chapter into three major sections. We begin by discussing the critical ANS behavioral signatures, including internal confidence, individual differences, and ratio dependence. Subsequently, we review the psychophysical model that accounts for these signatures, including a discussion on the nature of the Weber fraction (w), which we conceptualize as a scaling factor that determines the precision of all ANS representations. Finally, we elaborate on this model and argue that the ANS's key role is in providing "internal confidence" to the observer, rather than the absolute number of items in a scene.

BEHAVIORAL AND NEURAL SIGNATURES OF THE ANS

The ANS has been extensively studied, and there are many well-established signatures of its use (see following sections). Here, we focus on three behavioral signatures that we believe are central to the proper understanding of the ANS: the internal confidence generated by decisions about approximate number; individual differences in ANS performance; and the effect of numerical ratio on accuracy, response time, and internal confidence.

The heart of the ANS (and the psychological experiences of number that it generates) is its ordinal and approximate character. Consider the images in Figure 12-1 (NB, the ANS is multimodal, e.g., Nieder, 2012, and provides a sense of number for both voices heard and stars seen, but our examples focus on vision for ease of demonstration; Baker & Jordan, this volume). The reader will likely find it quite simple to judge that there are more black dots than white dots in Figure 12-1a, even with only a brief glance. It is also likely that deciding whether there are more black dots than white dots in Figure 12-1b is a bit more difficult. Irrespective of the reader's answer to this particular "more" question, we invite you to reflect on your internal confidence for any ordinal guess you might make with respect to Figures 12-1a and 12-1b. For which figure, 12-1a or 12-1b, would you feel more confident about your

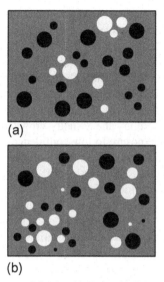

(a)

(b)

FIGURE 12-1 Quick. Are more of the dots black or white? You should find (a) to be easier, quicker, and that your internal confidence in your decision should be higher.

answer, after a brief glance? Which figure are you more likely to make a mistake on?

We expect that the reader feels more confident about his or her guess for Figure 12-1a (i.e., "Black has more!") than for Figure 12-1b (i.e., "Black has more???"). Experiments in multiple labs have found that this difference in internal confidence (and a corresponding increase in errors and response time as trials become more difficult) is not simply a result of there being more dots in Figure 12-1b than 12-1a, nor of missing a few dots or counting a dot twice by accident (Cordes, Gallistel, Gelman, & Latham, 2007; Cordes, Gelman, Gallistel, & Whalen, 2001). As expanded on in the following sections, we believe that a sense of internal confidence for number thoughts (e.g., Figure 12-1a resulting in higher confidence than Figure 12-1b) is the most important psychological experience that the ANS gives rise to—rather than experiences of any particular cardinality (e.g., "that looks like around 18 dots")—and that the ANS and all other magnitude dimensions (e.g., time, length, loudness, brightness; Lourenco, this volume) are in the business of supporting ordinal comparisons and not absolute judgments (i.e., more-black-dots, and not approximately-7-black-dots*).[2]

---

2. *A fortiori,* estimation (e.g., "that looks like around 16 dots") is a case of relative, ordinal comparison (see also Laming, 1997) and probably always involves comparisons to internal standard comparators (see, e.g., Izard & Dehaene, 2008; Sullivan & Barner, 2013). A more thorough defense of this view is beyond the scope of this chapter, but we hope that work in progress, as well as a book in progress, will be able to explore these issues in greater detail over the coming years.

One important lesson that can be drawn from viewing Figures 12-1a and 12-1b is that the internal confidence you feel when viewing them arises incredibly rapidly, probably even before you have determined your answer to the number question (e.g., "Hmm...Black has more!"). As a result, internal confidence, or lack thereof, is felt throughout the "Hmm..." period leading up to your decision, and informs and guides your evidence-gathering procedures (e.g., adjusting the parameters of an internal size normalization algorithm, inspecting with greater scrutiny the visually crowded regions of the image). For example, prolonged exposure to low confidence trials impairs subsequent discrimination performance, whereas exposure to high confidence trials improves it (Odic, Hock, & Halberda, 2012; Wang, Odic, Halberda, & Feigenson, under review). Much of the recent work in our lab demonstrates that internal confidence is continuous and quantitatively rich (i.e., not merely "high, low, or medium"), and may be the primary source for the individual differences in numerical estimation and discrimination performance in tasks that measure our abilities to estimate and compare numerosities (Odic et al., 2012; Wang et al., under review). This sense of internal confidence has yet to be fully explored empirically, and this chapter serves as an introduction to ideas that are currently in development.

The second important lesson to draw from viewing Figures 12-1a and 12-1b is that different observers will experience different levels of internal confidence for these same numerical judgments. If you happen to have children nearby as you read this, we encourage you to ask them to give you their opinion about Figures 12-1a and 12-1b—whether there are more black dots or white dots in each figure. You should find that they take longer to answer than you would take (try to convince them to answer without explicit verbal counting), but that they, like you, answer more quickly for Figure 12-1a than Figure 12-1b.

Individual and developmental differences are also found in ANS discrimination: individuals vary widely in both the speed and accuracy of deciding whether more of the dots are black or white, and ANS speed and accuracy gradually improve over development (Halberda & Feigenson, 2008; Halberda, Ly, Wilmer, Naiman, & Germine, 2012; Odic, Libertus, Feigenson, & Halberda, in press; Piazza et al., 2010). These individual and developmental differences are usually measured through Weber fractions (w), a key concept describing the acuity or precision of an individual's ANS. For example, w has been found to be impaired in individuals who struggle with dyscalculia, or math learning disability (MLD; Mazzocco, Feigenson, & Halberda, 2011; Piazza et al., 2010). In what follows, we discuss the nature of the Weber fraction and the ANS psychophysical model, and subsequently return to unifying it with the concept of internal confidence.

The third important behavioral signature of note is that children and adults should also answer faster and provide more accurate responses for Figure 12-1a than Figure 12-1b. The faster and more accurate answering for Figure 12-1a compared to 12-1b is a universal behavioral signature that every organism shows when relying on approximate number representations (i.e., you, us, children, rats, pigeons). All creatures so far tested have shown this kind of ratio-dependent responding (i.e., Weber's law), where we are both faster and more accurate for "easier" numerical comparisons, and our performance degrades as the ratio between the two numbers being compared moves closer to 1 (e.g., the numerosities in Figure 12-1b are closer to a ratio of 1 where the two sets of dots would be equal in number and there would, therefore, be no correct answer; Figure 12-1a ratio $= 18/8 = 2.25$; Figure 12-1b ratio $= 18/16 = 1.125$).

Internal confidence, individual differences, and ratio dependence are three signatures that are of great value for promoting our understanding of the approximate number system. Research scientists are relying on these patterns to help inform their understanding of the functioning of the ANS. All formal models of ANS representation must strive to provide a detailed account for why these patterns have been observed in every study of ANS performance yet published and in every animal yet tested. Why do these patterns emerge? Like the flow of sunspots across the face of the sun was for Galileo (i.e., and the systematic relationship between the time of year and the angle of their traversal), the elegant systematicity of these response time and error distributions in numerical tasks is a coded key whose proper description will help unlock a door unto our more accurate understanding of numerical cognition and the foundations of our numerical thoughts.

In the remaining sections of this chapter, we review a psychophysical model that accounts for ratio-dependence, internal confidence, and individual and developmental differences in the ANS. But researchers have also identified numerous other signatures of interest; in the end, a sufficient theory of our approximate number representations should be able to provide explanations for all the observed patterns in the data, and not just a subset of them. Due to space constraints, we can only briefly review the list of other relevant signatures:

- Longer response times and higher error rates for younger observers and for observers who struggle with a math learning disability; each discussed in connection with Figure 12-1.
- Multimodal representations of number that include vision, audition, tactation, as well as serial and parallel presentation of collections through these modalities; e.g., the "voices heard and stars seen" mentioned previously (Izard et al., 2009; Lipton & Spelke, 2003; Nieder, 2012). Similarly, cue-combination effects of combining evidence across two or more modalities; e.g., the increased confidence we feel in our estimate of the number of people who are around us when we can both hear and see them talking around the campfire (Jordan & Brannon, 2006; Raposo, Sheppard, Schrater, & Churchland, 2012).
- Relationships between physical parameters of stimulus presentation and a resulting sense for numerosity, including dimensions such as visual

density (Dakin, Tibber, Greenwood, Kingdom, & Morgan, 2011; Durgin, 1995) and physically intermixing or separating sets of stimuli (Gebuis & Reynvoet, 2012; Price, Palmer, Battista, & Ansari, 2012; Zosh, Halberda, & Feigenson, 2011); that is, how do our sensory systems take perceptual evidence and translate this evidence into a numerical thought?

- Relations across various psychological dimensions that may share representational resources or formats with approximate number, such as temporal duration or surface area (Hurewitz, Gelman, & Schnitzer, 2006; Odic et al., in press), as well as effects of forming mappings across these various dimensions, such as longer lines or tone durations mapping naturally to larger numerosities, including the SNARC (spatial–numerical association of response codes) effect wherein observers tend to associate physical space from left to right with a mental number line ordered from small numbers on the left to larger numbers on the right (Gevers, Verguts, Reynvoet, Caessens, & Fias, 2006; Wood, Willmes, Nuerk, & Fischer, 2008),
- Effects of memory and executive control for approximate number representation and adjusting to the varying contexts of different display parameters (Gilmore et al., 2013; Pailian, Libertus, Feigenson, & Halberda, under review).
- The relation between perceptual effects of numerosity and later cognitive effects of numerosity, such as the effects of clustering or Gestalt grouping in a visual display (Im, Zhong, & Halberda, 2013); visual adaptation effects (Burr & Ross, 2008; Ross & Burr, 2010); visual and auditory parsing of individual items that form a collection (Franconeri, Bemis, & Alvarez, 2009; Halberda, Sires, & Feigenson, 2006).
- Mappings between nonverbal approximate number representations and formal math words and symbols; e.g., that "around ten" can be an approximation (Barth, Starr, & Sullivan, 2009; Le Corre & Carey, 2007; Mundy & Gilmore, 2009; Odic, Le Corre, & Halberda, under review; Sullivan & Barner, in press), and mappings between approximate number representations and spatial understandings of a number line; e.g., that "6 > 5" picks out an ordinal direction on the mental number line (Booth & Siegler, 2006; Opfer & Siegler, 2007; Siegler & Booth, 2004).
- Neuropsychological evidence, such as evidence for various deficits that emerge from brain damage, that inform which brain regions support our ANS representation and what other psychological functions may be supported by those regions (Dehaene & Cohen, 1997), as well as imaging studies of the human brain that also help to address these questions; e.g., such studies suggest that overlapping brain regions support our representations of approximate numbers and approximate areas (Castelli, Glaser, & Butterworth, 2006; Pinel, Piazza, Le Bihan, & Dehaene, 2004).
- Neurophysiological recordings that provide data to fuel our theorizing about the representational format and implementations code for numerical

representations as well as information about how such codes may differ across various brain regions (e.g., IPS versus LIP; Nieder, 2005; Piazza et al., 2004; Roitman et al., 2007).

All of these sources of evidence are important and should eventually inform theories of approximate number representation.

A PSYCHOPHYSICAL MODEL FOR ANS REPRESENTATIONS

The key to understanding ratio dependence, individual and developmental differences, and internal confidence (as well as the other signatures described previously) is understanding the psychophysical model of the ANS and correctly understanding what a Weber fraction is.

When we just glance at a picture, even without an explicit task, our experience of Figure 12-1a feels inherently comparative (e.g., "there are more black dots!"); that is, it would be very surprising if someone glanced at Figure 12-1a and reported, "Well, I see one specific dot on the bottom right" (implying that "I see nothing else on the page worth reporting") or "I see approximately 18 black dots and nothing else worthy of note." Displays like Figures 12-1a and 12-1b have been used to measure human and animal numerical discrimination performance (i.e., how accurate we are at determining which color has more dots after just a quick glance); such tasks are called "discrimination tasks."

To model our accuracy (and internal confidence) for judgments that engage the approximate number system (i.e., the "more" judgments we made for Figures 12-1a and 12-1b), we must first specify a model for the underlying ANS representations. It is generally agreed that our internal response to a numerosity in the world is a distribution of activation on a mental "number line." These distributions are inherently variable (sometimes called "noisy") and do not represent number exactly or discretely (Dehaene, 1997; Gallistel & Gelman, 2000). This means that there is some error each time they represent number, and this error can be thought of as a spread of activation around the number being represented. The mental number line is often modeled as having linearly increasing means and linearly increasing standard deviations (Gallistel & Gelman, 2000).[3] In such a format, the representation for, e.g., approximately-7 is a normal (Gaussian) probability density function that has its mean at 7 on the mental number line and a smooth degradation to either side of approximately-7; hence, approximately-6 and approximately-8

---

3. The mental number line has also been conceived of as logarithmically organized with constant standard deviation (Dehaene, 2003). Either this format or the linear one in Figure 12-2a results in the ratio-dependent performance that is the hallmark of the ANS. We rely on the linear format, as it generates fairly intuitive graphs (e.g., Cordes et al., 2001; Gallistel & Gelman, 2000; Meck & Church, 1983, Whalen et al., 1999).

on the mental number line are also highly activated by instances of sevenness in the world.

In Figure 12-2a, we have drawn curves that depict the ANS representations for numerosities 4–10. You can think of these curves as representing the location and spread of activity generated on a mental number line by a particular collection of items in the world with a different bump for each numerosity you might experience (e.g., 4, 5, or 6 black dots). Rather than activating a single discrete value (e.g., 7), the curves are meant to indicate that a range of activity is present each time a collection of (e.g., 7) items is presented.

In fact, the bell-shaped, or Gaussian, ANS representations depicted in Figure 12-2a are more than just a theoretical fantasy; "bumps" like these have been observed in neuronal recordings of the cortex of awake behaving monkeys as they engage in numerical discrimination tasks (e.g., shown an array of 7 dots, neurons that are preferentially tuned to representing approximately-7 are most highly activated, while neurons tuned to approximately-6 and approximately-8 are also fairly active, and those tuned

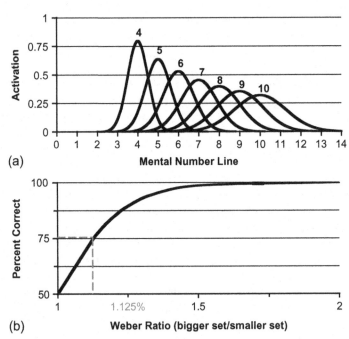

FIGURE 12-2 (a) The psychophysical model describes ANS representations as Gaussian distributions along an ordered number line. As discussed in the text, the Weber fraction is best conceptualized as a scaling factor for how the standard deviation in these distributions linearly increases with the mean. (b) Discrimination performance in the ANS follows a smoothly increasing function with ratio. For this idealized observer with a Weber fraction of 0.125, the ratio at which he or she will perform at about 75% is 1.125.

to approximately-5 and approximately-9 are active only slightly above their resting state; Nieder, 2005; Nieder & Miller, 2004). These neurons are found in the monkey brain in roughly the same region of cortex that has been found to support approximate number representations in human subjects in fMRI studies (Piazza et al., 2004).

It is important to keep in mind that this type of spreading activation is common throughout the cortex, and it is not unique to ANS representations. For example, we (and a rat) will also have neurons in our hippocampi that are preferentially tuned to particular locations in our office/bedroom/cramped-but-well-ventilated-cage that represent our position in space as we move around, with a spreading activation quite similar to the spreading activations depicted in Figure 12-2a (just with the spread occurring in the two-dimensional mental space of our floor plane rather than the one-dimensional space of numerosity; Fyhn, Molden, Witter, Moser, & Moser, 2004; Hafting, Fyhn, Molden, Moser, & Moser, 2005; Moser, Kropff, & Moser, 2008). That is, approximate number representations obey the same principles of "noisy" approximate coding that operate quite broadly throughout the mind/brain.

This point is worth highlighting because it invites you to recognize that, whatever theory you end up preferring for approximate number system representations, that theory must make use of constructs that can apply quite broadly across cortical and subcortical representations. The differences in response times, error rates, and internal confidence that we noted during our discussions of Figures 12-1a and 12-1b have also been observed for the vast majority of the psychological dimensions that humans and other animals represent (e.g., scalar variability and ratio-dependent performance for time, number, distance, flavor concentration, electric shock, perceived weight, density, viscosity; Cantlon, Platt, & Brannon, 2009; Gescheider, 1997; Odic, Im, Eisinger, Ly, & Halberda, under review). Many psychological dimensions rely on coding schemes based on scalar variability and internal confidence (signal fidelity) that operate similarly to ANS representation.

In Figure 12-2a, as the number of items in an array presented to an observer increases from 4 to 10, the standard deviation of the bell-shaped curves that represent the numerosity increases, resulting in a flattening and spreading of the activations (note the peakier curve for approximately-4 and the broader curve for approximately-9 in Figure 12-2a). This increase in spread with increasing number is the basis for the hallmark properties of the ANS and, as discussed previously, is similar to discrimination in many other dimensions (e.g., brightness, loudness), discrimination dependent on ratio and not their absolute number (i.e., scalar variability, or Weber's law, described later). When you are trying to discriminate one numerosity from another using the Gaussian representations in Figure 12-2a, the more overlap there is between the two Gaussians being compared, the less accurately they can be discriminated.

Critically, the overlap between Gaussian distributions is also the source for the differences in accuracy, RT, and internal confidence we experienced when viewing Figures 12-1a and 12-1b. Numerosities that are closer together have more overlap in their curves on the mental number line, making them harder to separate from each other to determine which collection is more numerous. Ratios that are closer to 1, where the two numbers being compared are closer (e.g., Figure 12-1b), give rise to Gaussian ANS representations with greater overlap, resulting in poorer and slower discrimination (i.e., "ratio-dependent performance")—e.g., it feels easier to decide that there are more black dots than white dots when looking at Figure 12-1a than at Figure 12-1b (and observers would make fewer errors, and decide faster, when shown Figure 12-1a than Figure 12-1b). Looking at the curves, and their overlap, in Figure 12-2a helps you to picture why errors, response times, and internal confidence may change as the numerosities being compared become larger and closer in proportion.

To see how the bell-shaped representations of the ANS in Figure 12-2a can predict differences in errors, response times, and internal confidence, consider that the curve for approximately-5 in Figure 12-2a is broader than the curve representing approximately-4 (i.e., approximately-5 has a larger standard deviation than approximately-4). These two curves are fairly easy to visually tell apart in Figure 12-2a. But, as one increases in number (i.e., as one moves right in Figure 12-2a), the curves become more and more similar looking (e.g., is curve 9 higher and skinnier than curve 10, or do they look pretty much the same?). As the ANS representations become more similar— i.e., as there is more overlap between the representations of the two numerosities to be discriminated—discrimination becomes harder, is more error-prone, and takes longer.[4] These bell-shaped representations predict that discrimination should smoothly become more and more difficult as the two numerosities become more and more similar.

In the ANS, it is not simply that larger numbers are harder to discriminate across the board. For example, an observer's performance at discriminating approximately-16 from approximately-20 (not shown in Figure 12-2) is predicted to be identical (in error rate, response time, and internal confidence) to the observer's performance at discriminating approximately-8 from approximately-10—as both of these trials would involve the same ratio (i.e., $10/8 = 1.125 = 20/16$). Although the curves for approximately-16 and approximately-8 do not have the same overall shape (e.g., the curve representing approximately-16 would be broader and flatter than the curve representing

---

4. This example based on the height and skinniness of the curves is simply to generate the intuition that discrimination becomes harder as the curves become more similar. Actual discrimination in the ANS is not based on the heights of the curves, but on the similarity of the activations elicited by the two sets of quantities (shown graphically in Figure 12-2a) and the amount of overlap between the two curves representing these numerosities.

approximately-8), it is the amount of overlap between the curves being compared that determines error rates, response times, and internal confidence. Because the standard deviation (SD) of the curves increases linearly with the mean (SD $= \mu$ * w), the curves representing approximately-8 and approximately-10 will overlap in area to the same extent that approximately-16 overlaps with approximately-20.[5]

Behavioral performance in tasks that engage the ANS is richly textured and exquisitely well structured. What do we mean by this? Observers don't simply "do a bit worse" as the numbers become more similar; nor do they feel "just a bit less confident." Rather, each observer's error rate, response time, and internal confidence are exquisitely well predicted by the bell-shaped representations in Figure 12-2a, and the changes in observers' performance as a function of trial difficulty is very systematic. This systematicity is what any candidate theory of approximate number representations must account for.

There are, however, numerous misunderstandings about Weber's law (i.e., ratio-dependent performance) and especially the Weber fraction (w), which indexes individual differences in ANS accuracy and internal confidence. In what follows, we elaborate on the nature of the Weber fraction and go through some of the most common misconceptions about it, including that the Weber fraction indexes just-noticeable differences, that it is defined as 75% accuracy, etc.

How to Think of a Weber Fraction (w) in the Approximate Number System (ANS)

What is a Weber fraction (w), and what does it tell us about an observer's approximate number system (ANS) representations? Some common misunderstandings of a Weber fraction include that it is (1) the fraction by which a stimulus with numerosity n would need to be increased in order for a subject to detect and report the direction of this change resulting in 75% correct performance across trials (i.e., that it is the "difference threshold" or the "just noticeable difference," JND), (2) the smallest ratio at which subjects will be significantly above chance in a numerical discrimination task, and (3) the midpoint between subjective equality of two collections and asymptotic performance in numerical discrimination. Rather, the Weber fraction is all of these things, and it is also simpler, more abstract, and more basic than any of these. After illustrating some problems with the above views, we sketch a proposal that the Weber fraction can be understood as an internal scaling factor that

---

5. Note also that it is the numerical similarity between the sets that is important for determining how difficult a trial might be, and not their absolute size. Bigger is not always harder; it depends on the numerical distances involved. For example, 7 black versus 8 gray dots is a *harder* trial than 17 black versus 30 gray dots—because the ratio 30:17 is larger than the ratio 8:7. This is sometimes called the "size effect."

indexes the amount of internal precision (i.e., signal fidelity) of every approximate number representation, and that the Weber fraction, so understood, can be used to determine the standard deviation of every numerosity representation within the ANS, and can turn knowledge of any one approximate number representation into any other approximate number representation.

Consider Figure 12-2a to represent the ANS number representations for a particular individual who has a Weber fraction = 0.125. In the following sections, we describe what role this number (0.125) is taken to play by each of the four conceptualizations listed earlier. In the end, we suggest that understanding a Weber fraction to be an internal scaling factor indexing the internal precision, confidence, or signal fidelity of a person's approximate number thoughts is the most valuable and true conceptualization. We suggest that this number (0.125), so understood, tells us how imprecise, or "noisy," a person's approximate number thoughts are.

The Weber Fraction Is Not a Just Noticeable Difference (JND)

If you present the hypothetical subject (whose ANS representations are depicted in Figure 12-2a) with the tasks we did with Figures 12-1a and 12-1b (i.e., the task of determining which of two collections has the greater number of dots) on a trial where there are 16 gray dots, this subject would require an increase of 2 dots from this standard ($n_1 = 16$; $16 \bullet .125 = 2$; $n_2 = 16 + 2 = 18$) in order to respond that black ($n_2 = 18$) is more numerous than gray ($n_1 = 16$) on 75% of the trials that present these two numerosities.[6] That is, a subject's Weber fraction can be used to determine the amount by which you would need to change a particular stimulus in order for the subject to correctly determine which number was larger on 75% of the trials (where chance = 50%). Conceived in this way, the Weber fraction describes a relationship between any numerosity and the numerosity that will consistently be discriminated from this standard. This gives one way of understanding why you might choose 75% correct performance; however, to specify what "consistently discriminated from" might mean, you could also choose some other standard (e.g., 66.7% correct, or any other percent above 50%). From this point of view, which is often the dominant one taught in psychophysics, the point is to estimate the steepness of the linear portion of the psychometric function, or the slope of the linear rising portion of this function (depicted in Figure 12-2b), and 66.7% would work for such purposes just as well as 75% or 80%.

However, as we will see below, the seemingly arbitrary reasons for choosing 75% correct as an index of performance are somewhat justified once we understand the mathematical relationship that holds between correct discrimination performance, the Weber fraction (w), and the standard deviations of the underlying Gaussian representations.

---

6. Note, we use "$\bullet$" throughout to indicate multiplication.

The Weber Fraction Is Not the Smallest Discriminable Ratio

Some readers, more familiar with research on the acuity of the ANS in infants (Izard et al., 2009; Libertus & Brannon, 2009; Lipton & Spelke, 2003; Xu & Spelke, 2000; Xu, Spelke, & Goddard, 2005) and less familiar with the literature on adult psychophysics, may have come to believe that a Weber fraction describes the ratio below which a subject will fail to discriminate two numerosities (e.g., 6-month-olds succeed with a 1:2 ratio and fail with a 2:3 ratio; Xu et al., 2005). This suggests a categorical interpretation of the Weber fraction (e.g., a threshold where you will succeed if a numerical difference is "above threshold" and fail if it is "below threshold"). That is, some may have come to believe that performance should be near perfect with ratios easier than a subject's Weber fraction and at chance for ratios harder than a subject's Weber fraction.

Categorical performance, however, is not observed in typical performance where a large number of trials test a subject's discrimination abilities across a wide range of ratios (Halberda & Feigenson, 2008; Halberda et al., 2012; Piazza et al., 2010). In such cases, behavioral performance shows a smooth improvement from a ratio of 1 (where $n_1 = n_2$ and there is no correct answer) toward increasing accuracy; and not a "step function" from at-chance performance below the Weber fraction to above-chance performance above the Weber fraction.

Consider again the simple task of being briefly shown a display that includes some black and white dots and being asked to determine on each flash if there were more black or more white dots. Percent correct on this numerical discrimination task is not a step function with poor performance "below threshold" and good performance "above threshold," but rather is a smoothly increasing function from near-chance performance to consistent success. This performance and the range of individual differences, gathered from more than 10,000 subjects between the ages of 8 and 85 years of age participating in this type of numerical discrimination task, can be seen in Figures 12-3a and 12-3b.

The actual behavioral data from subjects seen in Figure 12-3a, and the modeled ideal behavior seen in Figure 12-2b, suggest that the subjects will always be above chance no matter how small the difference between n_1 and n_2 (e.g., in theory, even a baby will be "above chance" at seeing that 10,001 black dots is numerically more than 10,000 gray dots; see Green & Swets, 1966); what changes is not whether an observer will succeed or fail to make a discrimination but rather the number of trials an experimenter would have to run in order to find a statistically significant difference in performance on the most difficult trials. Consider that the region nearest to equality (a ratio of 1) is the region of most rapid improvement in every observer's performance (e.g., Figures 12-3a and 12-2b). That is, subjects' performance shows more improvement when the ratios being tested increase from 1.01 to 1.1 than they show when the ratios increase from 1.33 to 1.4. There are two take-home points that we'd like to stress: (1) even a baby should be able to tell that 21 black dots is numerically more than 20 gray dots (what changes is the number of trials we'd have to run to be able to show that

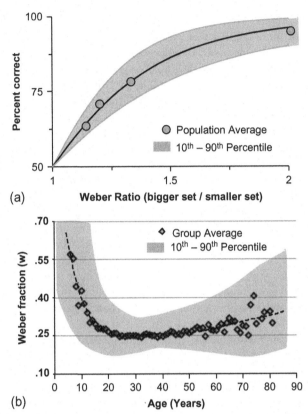

FIGURE 12-3 ANS performance from more than 10,000 participants. (a) ANS discrimination smoothly increases with ratio; (b) Weber fractions initially become better (i.e., become lower) with age, and eventually plateau around age 30, and subsequently slowly become worse. Note that these graphs present previously unpublished data collected on testmybrain.org and panamath.org during 2008 (see Halberda et al., 2012).

infants detect this difference), and (2) it is at the hardest ratios (e.g., 1.01 versus 1.1) that we see the most rapid improvements in numerical discrimination performance (and not at some "threshold" or fraction that changes from "at chance" to "above chance").[7]

---

7. However, we would also like to note that, within the practical limits of testing real babies, the infant literature's method of looking for a change from at-chance performance to above-chance performance is a quite reasonable approach. It allows one to *roughly* locate the Weber fraction of subjects who, like infants, cannot participate in the large number of trials it takes to achieve the smooth data seen in Figure 12-3a. We have published papers that use this kind of approach, and it is a fine thing to do. But we're suggesting that it would be best if we do not allow such practical concerns to inspire a faulty foundation on which to grow our theory-based intuitions of what is possible and impossible for the ANS and other magnitude representations.

Even for infants, then, and untrained observers, performance in numerical discrimination tasks should not shoot up from "at chance" for harder ratios to "significantly above chance" at easier ratios. Melissa Libertus and colleagues have ingeniously demonstrated that infants looking time to numerically varying stimuli can reveal a smoothly graded function of increasing looking that is quite similar to what is seen in Figure 12-3a (Libertus & Brannon, 2009, 2010). In these tasks, infants are shown a display of dots at a particular ratio, and their looking time to the display is measured; the amount of time infants spend looking at the display is a continuous function dependent on the ratio: as ratios get harder and harder, infants gradually look less and less. This work highlights, in dramatic fashion, that misunderstanding a Weber fraction to indicate something about a change from chance performance (or an "inability to distinguish") to a sudden ability to distinguish numerically varying stimuli will generate the wrong intuitions (i.e., in theorists, teachers, and students).

The Weber Fraction Is Not the Midpoint between Subjective Equality and Asymptotic Performance

One common approach to localize the Weber fraction at some point along the smoothly increasing curve in Figure 12-3a is at 75% accuracy—the midpoint between subjective equality of the two numerosities being compared (without biases, occurring at a ratio = 1, where $n_1 = n_2$) and asymptotic performance (typically occurring nearing 100% correct, although asymptotic performance could be lower in unskilled subjects, resulting in a midpoint that falls at a percent correct lower than 75%; for example, see Halberda & Feigenson, 2008).[8] Hence, to calculate a Weber fraction, a researcher may take the ratio at which the observer performed at 75% and then subtract 1 (e.g., if 75% performance is at a ratio of 1.25, then the w is estimated at 0.25).

If observers behave optimally and if the Weber fraction is within a particular range, this shorthand does produce the correct value. In Figure 12-2b, we have drawn the expected percent correct for the ideal subject in Figure 12-2a whose Weber fraction $(w) = 0.125$ as derived by a model from classical psychophysics. This idealized subject would perform at 75% around ratio 1.128.

There are two challenges, however, with conceptualizing the Weber fraction in this manner. First, the mathematical relation between w (conceptualized as a scaling factor) and the ratio at 75% is not constant: whereas an observer with a w of 0.125 will perform at about 75% around ratio 1.125, an observer with a w of 0.5 will perform at around 75% around a ratio of 2.0, not 1.5!

---

8. Typically, behavioral performance will cross 50% at ratio = 1 for an observer who has no bias to choose black or white and who is simply guessing at chance = 50% when $n_1 = n_2$ (#black = #gray); and it may never reach 100% no matter how easy the trials become (e.g., because everyone has some tendency to make a miss-hit on the response keys from time to time, even if merely from sheer boredom with all those dots).

Second, understanding of the Weber fraction as the midpoint between sub-jective equality and asymptotic performance misses the deeper continuous nature of discrimination within the ANS. For example, this focus on 75% has led many researchers throughout history to believe that accuracy is the theoretical variable they are hoping to measure. For example, many readers may be familiar with "staircase" methods or adaptive procedures that adjust trial difficulty in response to the subject's performance in an attempt to focus the majority of trials on a position where the subject is at 75% correct (as if this point were of special importance). As we see in the next section, there is nothing at all special about 75% correct performance. The Weber fraction (w), properly understood as a scaling factor for determining internal variabil-ity for every approximate number representation, perfectly predicts perfor-mance at 75% correct, or 85%, or 51.3756% correct, and everything in between; and it determines performance along the entire smoothly improving curve seen in Figures 12-2b and 12-3b.[9]

Weber Fraction Conceptualized as a Scaling Factor

Let us consider a fourth way of understanding the Weber fraction: as a scaling factor that indexes the amount of "noisiness" surrounding every numerical representation of the ANS.

Consider again the Gaussian curves in Figure 12-2a. The spread of each successive numerosity from 4 to 10 is steadily wider than the numerosity before it. This means that the discriminability of any two numerosities is a smoothly varying function, dependent on the ratio between the two numeros-ities to be discriminated. In theory, such discrimination is never perfect because any two numerosities—no matter how distant from one another—will always share some overlap. At the same time, discrimination will never be entirely impossible, so long as the two numerosities are not identical, because any two numerosities, no matter how close (e.g., 67 and 68), will always have some nonoverlapping area where the larger numerosity is detectably larger (Green & Swets, 1966). Correct discrimination may occur on only a small percentage of trials if the two sets are very close in number, but it will never be impossible (up to the limits of the sensory detector). This motivates the intuition that percent correct in a numerical discrimination task should be a smoothly increasing function from the point of subjective equality to asymp-totic performance. The smooth increase in percent correct as a function of

9. For those interested in practical concerns, the most reliable and stable performance for human subjects, where trials are neither too easy nor too hard, occurs at around 86% correct performance (let's call it the "Goldilocks position") and not at 75% correct (this factoid garnered from model-ing work in our lab, and our practical experiences testing subjects across a wide range of ability levels and ages, and informed by conversations with the great and stimulating Zhong-Lin Lu). So, even for practical reasons (beyond theoretical concerns) we should not focus on 75% as something special.

ratio is no accident. It is the smoothly increasing spread in the underlying Gaussian representations depicted in Figure 12-2a that is the source of the smoothly increasing "Percent Correct" ideal performance in Figure 12-2b.

Noting the smoothly increasing spread of the Gaussian representations in Figure 12-2a might motivate you to ask what is the parameter that determines the rate of increase in standard deviation with numerosity, and what determines the amount of spread in each Gaussian representation on the mental number line? In fact, it is the Weber fraction that determines the spread of every single representation on the mental number line by the following formula ($SD_{n1} = \bar{x}_{n1} \bullet w$). The standard deviation (SD) of the Gaussian bell-shaped curve representing any particular numerosity on the mental number line is the central tendency for that representation ($\bar{x}_{n1}$) multiplied by the Weber fraction (w).

Why is this the case? Well, intuitively, it is the standard deviations of the underlying Gaussian representations that determine the amount of overlap between the curves that represent any two numerosities, and it is the amount of overlap between the numerosities that determines how well any two numerosities can be discriminated. The categorical views of the Weber fraction as a kind of threshold between successful discrimination and failure, or as the midpoint between subjective equality and asymptotic performance, choose to focus on only one particular point of what is actually a continuous and smooth function of increasing success at discrimination. As a result, this entire function is determined by the Weber fraction because this parameter describes the standard deviations of every single numerosity representation in the ANS— and therein the degree of overlap between any two numerosities on the mental number line.

The Weber fraction (w) is the constant that describes the amount of precision for each observer's ANS number representation. It is a scaling factor by which you could take any one of the curves in Figure 12-2a and turn it into any of the other curves in Figure 12-2a in an accordion-like fashion. In the linear model depicted in Figure 12-2a, the analog representation for any numerosity (e.g., $n=7$) is a Gaussian random variable with a mean at n (e.g., $n = 7$) and a standard deviation of ($n \bullet w$).[10,11] This means that for a subject who has a Weber fraction of 0.125, the ANS representation for $n = 7$ will be a bell-shaped normal curve with a mean of 7 on the mental number line and a standard deviation of $0.875 = 0.125 \bullet 7$. By substituting any number you like for n, you can easily determine the shape of the underlying ANS representation without ever having the observer engage in a numerical

---

10. Note that signal compression or expansion is also important because it can change the position of representations along the mental number line (for detailed discussion, see Odic et al., under review)—a detail that does not concern us for the present moment.
11. Note also that the relationship of the Weber fraction (w) to internal confidence is also true for a logarithmic model of numerosity representation, with any differences in details not relevant for the present discussion.

discrimination task that compares two numbers. This illustrates the power of understanding the Weber fraction as an index of signal fidelity, internal confidence, or internal noise. Rather than simply telling us something about how well a subject will do at discriminating two numbers "near their just noticeable difference," the Weber fraction (w) tells us the shape and overlap of every single number representation along a mental number line. The Weber fraction is about all of the representations, not just the ones "near threshold."

Understood in this way, the Weber fraction is not even specific to the task of numerical discrimination; indeed, it is wholly independent and prior to discrimination. An animal that, bizarrely, could only represent a single numerical value in its ANS (e.g., could represent only *approximately-7* and no other numbers) and could therefore never discriminate 12 from any other number (i.e., could not even perform a numerical discrimination task) would nonetheless have a Weber fraction, and we could measure it!

Meet Justin The Rat

In this section, we want to briefly discuss the beautifully limited mind of an animal named "Justin The Rat," which, strangely, can only represent the number *approximately-7* and no other number. Justin The Rat can represent all the other things that we represent (e.g., dots, colors), but for numbers, he has only one thought, and that is the thought *approximately-7*.

Question: Hey, Justin The Rat, how many food pellets did you just eat?
Answer: *Approximately-7*.
Question: On another topic, Justin The Rat, surely you do not believe in God?
Answer: Well, not in an interventionist Christian god, if that's what you mean.
Question: Dear Justin The Rat, on a scale from 1 to 10, with 10 being "smoking hot" and 1 being "let's not talk about this," how sexy am I really?
Answer: *Approximately-7*.

You get the idea.

How well would Justin The Rat do if we asked him to choose which array has more dots while showing him 16 gray dots and 18 black dots? It may seem predestined that Justin The Rat would be terrible at a numerical discrimination task involving two sets of dots, each with more than 7 dots—owing to his unique brain abnormality that limits his numerical thoughts to *approximately-7* and nothing else. For instance, how would we teach him such a task or measure his performance? And, is it even possible to have a living creature with numerical cognitive abilities so impaired? Does Justin The Rat *have* a Weber fraction (w), and how would we measure it? To answer these questions, we invite you to take our earlier, technical, sections as a point of departure.

As we have seen, a Weber fraction can be understood to be a scaling factor that determines the standard deviation of the bell-shaped curves representing each and every number representation in a subject's approximate number system (ANS). Numerical discrimination tasks (e.g., Figures 12-1a and 12-1b) are not the only way of measuring this type of internal scaling parameter.

Although production tasks (such as the "tap your finger n times" task; Whalen, Gallistel, & Gelman, 1999) and discrimination tasks (such as the "who has more" task; e.g., Figure 12-1) have often been discussed separately, they measure theoretically identical aspects of ANS representations. For a "tap your finger n times" task, researchers generate a measure of the coefficient of variation (CoV or CV), which is the standard deviation of the number of presses divided by the mean number of presses. For example, ask a subject to press a button 9 times too quickly for explicit counting while saying the word "the" to further block verbal counting. The result will be a bell-shaped distribution of responses; the subject will most often press 9 times, but will also sometimes press 8 or 7, and sometimes press 10 or 11 or 12, etc. Graphing the number of instances where the subject presses 7, 8, 9, 10, 11, etc., times when requested to press 9 times will reveal a smooth bell-shaped curve centered around 9 (similar to the curve for 9 in Figure 12-2a). Take the standard deviation of this bell-shaped curve and divide it by the mean of this curve to return the CV for this subject (i.e., $CV_{n9} = SD_{n9}/\bar{x}_{n9}$). For this task, which also engages the ANS, CV is expected to be constant across all numbers probed. That is, ask the same subject to do the study again pressing, e.g., $n = 14$ times, build a similar-looking (but fatter) bell-shaped curve centered around 14, divide the standard deviation of this curve by the mean of this curve, and you should get the same CV that you got for the version of the task in which the requested number of presses was 9 (Cordes et al., 2001; Whalen et al., 1999); $CV_{n14} = SD_{n14}/\bar{x}_{n14}$; and, $CV_{n14} \approx CV_{n9}$.

The source of the bell-shaped curves in a numerical production task is not, in theory, simply mispresses or mistakes (a curve built out of mispresses would not be a bell-shaped Gaussian, but a more narrow, binomial, non-Gaussian, curve; Cordes et al., 2001). The source of the bell-shaped curves, and the fatness of the curves, is the variability in the underlying representations of the ANS. And this leads to a little-remarked-upon identity: an observer's CV and his or her Weber fraction (w) should, theoretically, be identical numbers.

Note, $CV_n = SD_n/\bar{x}_n$. And, as we mentioned previously, the SD of the underlying ANS representation for any number can be determined using a subject's Weber fraction (w) as an internal scaling parameter, i.e., $SD_n = \bar{x}_n \bullet w$. Rearrange this equation and you get $SD_n/\bar{x}_n = w$, the same equation that we use to calculate CV (i.e., $SD_n/\bar{x}_n = CV_n$). That is, $CV = w$.

This identity makes intuitive sense upon reflection. As subjects try to tap their finger quickly in a numerical production task, they give up on verbal counting and allow their ANS to assess when the target number of taps has been reached. But these ANS representations are "noisy," leading observers to sometimes tap too many times and sometimes tap too few times over the course of many trials. In this way, the source of the errors in a "tap your finger" task is the noisy representations of the ANS. The source of the errors in a numerical discrimination task (e.g., Figures 12-1a and 12-1b) is also the noisy representations of the ANS. The coefficient of variance (CV) and Weber fraction (w) are two ways of estimating the imprecision, or variability, in these underlying representations. And so, CV and w are two ways of measuring the same thing.

Thus, to estimate the Weber fraction for Justin The Rat—an animal that, strangely, can represent only *approximately-7*—train him to press a button 7 times, run many trials, calculate CV, and CV$=w$ (for an alternative method, see Odic, Im, et al., under review, and visit www.panamath.org/psimle). Here, you have found the Weber fraction for an animal without ever having that animal compare two numbers or see two collections. This is an illustration of the inductive power of understanding the Weber fraction (w) to be an internal scaling factor. A Weber fraction need not require an understanding of failure or success at numerical discrimination, nor even the ability to make a numerical discrimination. Rather a Weber fraction (w) is simply a way of indexing the internal precision (aka signal fidelity, internal confidence, noise) in a person's ANS approximate number representations.[12]

How a Weber Fraction (*w*) Indexes Individual Differences in ANS Precision

The inductive power of understanding the Weber fraction (w) to be an internal scaling factor is further highlighted when we compare the Weber fractions of different individuals. Individuals differ in the precision of their ANS representations. Some people have less precise approximate number representations, and some people have more precise representations (Halberda & Feigenson, 2008; Halberda et al., 2012). In Figure 12-4a, we have illustrated some idealized curves that display the underlying ANS representations for a subject whose $w=0.125$ and, in Figure 12-4b, for a subject whose $w=0.20$. Crucially, you can see that the subject in Figure 12-4b has a greater degree of overlap between the bell-shaped curves of their ANS representations than the subject in Figure 12-4a (recall, a bigger Weber fraction means more noise and fatter curves). It is this overlap that leads to differences in internal

---

12. In full disclosure, production tasks and discrimination tasks may not always be measuring the same thing, because in fact, it is unlikely that any psychological task is measuring only one thing. No matter how simple you make the task, it is likely that many different psychological factors are required for encoding, response generation, and decision making. As such, in practice, measured CV will not perfectly predict measured w. A scientifically productive question CV might be, "How might the differences in measured estimates of CV and w help us determine the variety of psychological variables these tasks have in common and those that they have distinctly?"

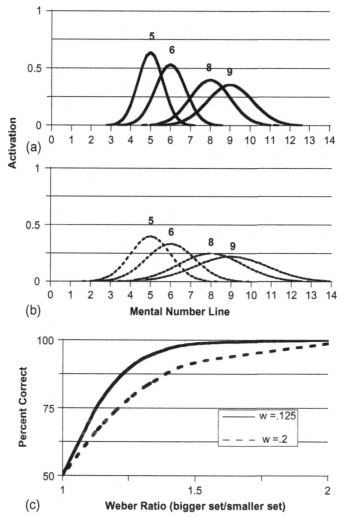

FIGURE 12-4 Individual differences in the ANS. The difference between an individual with a better (lower) Weber fraction in (a) and an individual with a worse (higher) Weber fraction in (b) is entirely in the variability of the ANS representations: higher overlap between the representations results in lower accuracy (illustrated in [c]) and lower confidence and higher RT.

confidence, error rates, and response times—and to the difficulty in discriminating two stimuli that are close in numerosity. The hypothetical subject in Figure 12-4b would have poorer performance than the subject in Figure 12-4a in a numerical discrimination task (e.g., Figures 12-1a and 12-1b).

The ideal performance in Figure 12-4c also shows the smooth gradual increase in percent correct as a function of ratio that we have been discussing. In Figures 12-3a and 12-3b, we saw data from more than 10,000 individuals who played a numerical discrimination task online. Every one of the more than 10,000 observers in this sample obeyed this kind of gradual increase in percent correct (seen in Figure 12-3a) from a ratio of 1 (where the number of black and gray dots are equal) to easier ratios like 2 (where there might be 20 black dots and 10 gray dots; $20/10 = 2$). What changes from observer to observer is how steep the left side of the performance curve is (NB, you can see this difference in Figure 12-4c for the two hypothetical subjects who differ in their Weber fraction).

In Figures 12-3a and 12-3b, individual differences are shown by indicating the range of performance from the lower 10th to the upper 90th percentile rank of the more than 10,000 observers (i.e., the lower bound of the gray-shaded region in Figure 12-3b indicates the average performance of the 90th percentile group [i.e., best], and the upper bound of the gray-shaded region indicates the average performance of the 10th percentile group [i.e., worse]; note that the upper and lower bounds are reversed for Figure 12-3a because higher percent correct translates to lower Weber fraction (w), i.e., lower internal noise). Figure 12-3b shows how the average Weber fraction improves over development.

A steeper, quicker rise in the psychometric function (Figures 12-4c and 12-3a) indicates better sensitivity, better discrimination abilities, more precise ANS representations (e.g., sharper bell-shaped humps in the ANS, with less "noise"; smaller standard deviations for each hump), and this is indexed by the subject having a smaller Weber fraction (Figures 12-4a, 12-4b, and 12-3b) (i.e., a smaller Weber fraction indicates less noise in the underlying ANS representations).

The values for the Weber fractions in Figure 12-4 have been chosen so as to illustrate another value of understanding the Weber fraction to be an internal scaling factor: it empowers comparisons across individuals and formal models of individual differences. Converting the Weber fraction for each of these subjects into the nearest whole number fraction reveals that the Weber fraction for the subject in Figure 12-4a is 8:9 and for the subject in Figure 12-4b is 5:6 (i.e., $9/8 = 1.125$; $6/5 = 1.20$). Investigating the Gaussian curves in Figure 12-4a and 12-4b reveals that the bell-shaped curves for the numerosities 8 and 9 for the subject in Figure 12-4a are identical in shape to the bell-shaped curves for the numerosities 5 and 6 for the subject in Figure 12-4b. This too is no accident. The only parameter that has been changed in the construction of Figures 12-4a and 12-4b is the Weber fraction for the subject. This single parameter determines the spread in the curves that represent every possible numerosity in the ANS of each subject.

In this way, understanding the Weber fraction to be an internal scaling factor that determines the spread of every ANS number representation not only empowers us to compare one number representation to another within a particular subject (e.g., the lesson we learned from Justin The Rat), but also empowers us to compare across individuals and to create mathematically tractable predictions about how the ANS representations of one person (e.g., the subject in Figure 12-4a) relate to the ANS representations of another (e.g., the subject in Figure 12-4b). This is not the case for any other estimate of individual differences that you might prefer to use (e.g., percent correct, average response time, the slope of error rate as a function of ratio, the slope of response time as a function of ratio), although these may be used as rough approximations (e.g., just like subtracting 1 from the 75% ratio can, in some circumstances, be used as an approximation).

Two important goals of psychology are to measure and to understand the sources of individual differences in a wide range of social behaviors and cognitive abilities. A valuable approach to approaching these challenges is to develop a formal model of the particular aspect of the psychological system that you hypothesize is different from one person to another—such as the precision and accuracy of the bell-shaped representations of the ANS. When we understand a Weber fraction (w) to be an internal scaling factor that indexes the precision of each person's approximate number thoughts, we find that this allows us to directly translate (in a formal sense, seen graphically in Figures 12-4a to 12-4c) from one individual's ANS precision to another individual's ANS precision; and to build specific proposals for the shape and activation of each numerical representation within each individual's ANS.

There remains important work to be done, both practical and theoretical, to ensure that we are correctly measuring subjects' Weber fractions (e.g., how much display time is optimal? Does the Weber fraction change if we present auditory stimuli rather than visual stimuli?). Also, we must strive to ensure that our formal models of a Weber fraction reflect the actual behavior and neuronal activity of our subjects of interest. This is an ongoing process for our research field, and we do believe there are major discoveries still to be made. But, we also believe that understanding a Weber fraction as a scaling factor is an important foundation for studying individual differences, and for beginning a journey of making new discoveries that will help us build more appropriate and accurate models of cognition.

THE RELATION BETWEEN THE WEBER FRACTION AND INTERNAL CONFIDENCE

One further lesson we can draw from our experiences with Figures 12-1a and 12-1b is the power of internal confidence to inform how we search and interact with the world. If you happen to still have children nearby, you might try asking which picture (Figure 12-1a or 12-1b) they think looks like the easier picture to answer without actually making the judgment. Which picture will it be easier to figure out "who has more?" We imagine that you would find that even without counting, children will judge that Figure 12-1a will be the easier trial. We believe that this ability to respond which figure would be easier to answer emerges from our sense of internal confidence for ANS questions and displays, which directly stems from the internal variability in the

Gaussian (bell-shaped) representations (whose value is scaled by the Weber fraction). In this way, the psychophysical model outlined above of the Weber fraction as a scaling parameter provides a unified explanation for the ratio signatures, individual differences, and the source of internal confidence in our approximate number decisions.

Importantly, notice that you and the child can answer this "easiness" question (and the "which are you more likely to make an error on" question asked earlier) *even before you figure out the correct answer.* That is, even without ever being told which color has more dots (i.e., before telling the child that "black has more" for both Figure 12-1a and Figure 12-1b), we seem to be able to tell that Figure 12-1a will be the easier image to answer which has more. Because our feeling of internal confidence occurs prior to our decision, we can use it in several ways. For example, it is a signal to "slow down" and be more careful about answering the question for Figure 12-1b. We might also have a sense that we should "look more closely at those dots on the lower left corner of Figure 12-1b before we answer to see if there are more black or gray dots down in that visually crowded region of the display." This means that our sense of internal confidence (and trial difficulty), generated by our ANS representations, can help us decide how to *approach* answering a question (e.g., where to allocate our attention, or when to be careful and take a second glance). Thus, the ANS is not simply in the business of giving us an answer to a numerical question; it is, perhaps more importantly, involved in helping us direct our limited attention and memory resources to help us make more effective decisions (Odic et al., 2012).

In this way, far from amounting to counterproposals to the importance of the ANS, some recent results revealing that observers are affected by stimulus factors such as size-conflicting stimuli (Dakin et al., 2011; Gebuis & Reynvoet, 2012; Szucs, Nobes, Devine, Gabriel, & Gebuis, 2013), spatially intermixed stimuli (Gebuis & Reynvoet, 2012; Price et al., 2012), or briefly flashed stimuli (Inglis & Gilmore, 2014) are all beautiful demonstrations of the importance of internal confidence, generated by ANS representations for guiding our numerical decisions. We hypothesize that observers rely on internal confidence from the ANS (which is sensitive to the context of stimulus presentation) to marshal their cognitive control abilities in order to respond more effectively to numerical stimuli, which vary wildly in their mode of presentation across contexts (e.g., sounds heard, objects seen, all at once or serially presented). The effects of stimulus presentation and, e.g., size/duration-controlled or size/duration-confounded stimuli are beautiful demonstrations of the importance of approximate number representations and internal confidence for numerical cognition.

Finally, internal confidence, in being fundamentally related to Weber fractions, is also an important individual difference that may be related to other cognitive abilities. An observer who has less precise ANS representations for number will feel somewhat confident that "black has more" for

Figure 12-1a, and may feel very low confidence that "black has more" for Figure 12-1b (and all observers are likely to take longer and to make more errors for Figure 12-1b than 12-1a). We theorize that this difference in internal confidence has a major impact on how we feel about mathematics across our entire lives and may be *the dominant source* for what we experience as, "I am (am not) a math person." Ongoing work in our lab is testing the relation between internal confidence and school math performance. We believe that an understanding of internal confidence can be a major unifying force for the study of approximate number representations. All of the work being done in this exciting field is valuable and relevant; e.g., every empirical paper reports findings that can help to refine our theories of approximate number representation. Rather than defending older ideas about how internal representations of approximate number might affect our mathematical thinking, we are excited to look to the future for new constructs that can unify across our older distinctions.

CONCLUSION

In this chapter, we have tried to promote understanding a Weber fraction (w) as a scaling factor that enables any ANS number representation to be turned into any other; or, equivalently, as an index of the amount of internal confidence a person experiences in his or her approximate number thoughts. Understood in this way, a Weber fraction does not require the commonsense notion of a "threshold" (i.e., a change from failure to success), and it does not generate the same kinds of confusions that this commonsense notion gives rise to. Additionally, this psychophysical model integrates with a variety of signatures that have been experimentally observed.

We believe that thinking about a Weber fraction as JNDs, critical ratios, of 75% performance has given rise to some confusions and that it is currently limiting our theorizing (e.g., the confusion that performance should change from chance performance at difficult ratios to above-chance performance at easy ratios, while, as shown in Figure 12-3, the actual performance of subjects does not look this way at all, but instead is a smoothly increasing function). It also does not promote the kind of understanding of the approximate number system (ANS) that highlights the systematic nature of variability throughout the system. Understanding a Weber fraction (w) to be a scaling factor—i.e., an estimate of signal fidelity across all possible ANS representations— promotes our understanding that variability inherent in ANS representations is not merely a bug but is rather a feature of our approximate number system.

The heart of the ANS (and the psychological experiences of number that it generates) is its ordinal and approximate character. Because the ANS displays scalar variability in its coding of numerosity, understanding the variability of any one ANS representation (e.g., through measuring CV) can be easily translated into an understanding of the internal variability, and internal confidence, for every single ANS number representation. Furthermore, understanding a Weber fraction (w) as a scaling factor also promotes our understanding of the systematic relationships that exist across individuals (e.g., the comparison of the two subjects in Figure 12-4).

When we understand the Weber fraction (w) to be an internal scaling parameter that indexes the amount of precision and internal confidence in each person's approximate number thoughts, we can begin to see many new doors for research begin to open—e.g., connections to math anxiety, stereotype threat, the feeling of "I'm just not a number person." Connections to executive functioning and the possibilities for interventions to improve number sense also come into a sharper focus. All these avenues have yet to be fully explored, and we are excited to be able to play just a small role in testing out some of these ideas.

REFERENCES

Barth, H., Starr, A., & Sullivan, J. (2009). Children's mappings of large number words to numerosities. *Cognitive Development, 24,* 248–264.

Booth, J. L., & Siegler, R. S. (2006). Developmental and individual differences in pure numerical estimation. *Developmental Psychology, 42*(1), 189.

Burr, D. C., & Ross, J. (2008). A visual sense of number. *Current Biology, 18*(6), 425–428.

Cantlon, J. F., & Brannon, E. M. (2006). Shared system for ordering small and large numbers in monkeys and humans. *Psychological Science, 17*(5), 401–406. http://dx.doi.org/10.1111/j.1467-9280.2006.01719.x.

Cantlon, J. F., Platt, M., & Brannon, E. M. (2009). Beyond the number domain. *Trends in Cognitive Sciences, 13*(2), 83–91. http://dx.doi.org/10.1016/j.tics.2008.11.007.

Castelli, F., Glaser, D., & Butterworth, B. (2006). Discrete and analogue quantity processing in the parietal lobe: A functional MRI study. *Proceedings of the National Academy of Sciences, 103*(12), 4693–4698. http://dx.doi.org/10.1073/pnas.0600444103.

Cordes, S., Gallistel, C. R., Gelman, R., & Latham, P. (2007). Nonverbal arithmetic in humans: Light from noise. *Attention, Perception, & Psychophysics, 69*(7), 1185–1203. http://dx.doi.org/10.3758/BF03193955.

Cordes, S., Gelman, R., Gallistel, C. R., & Whalen, J. (2001). Variability signatures distinguish verbal from nonverbal counting for both large and small numbers. *Psychonomic Bulletin & Review, 8*(4), 698–707.

Dakin, S. C., Tibber, M. S., Greenwood, J. A., Kingdom, F. A. A., & Morgan, M. J. (2011). A common visual metric for approximate number and density. *Proceedings of the National Academy of Sciences, 108*(49), 19552–19557. http://dx.doi.org/10.1073/pnas.1113195108.

De Smedt, B., Noël, M.-P., Gilmore, C., & Ansari, D. (2013). How do symbolic and non-symbolic numerical magnitude processing skills relate to individual differences in children's mathematical skills? A review of evidence from brain and behavior. *Trends in Neuroscience and Education, 2*(2), 48–55.

Dehaene, S. (1997). *The number sense.* New York: Oxford University Press.

Dehaene, S. (2003). The neural basis of the Weber–Fechner law: A logarithmic mental number line. *Trends in Cognitive Sciences, 7*(4), 145–147.

Dehaene, S., & Cohen, L. (1997). Cerebral pathways for calculation: Double dissociation between rote verbal and quantitative knowledge of arithmetic. *Cortex, 33,* 219–250.

Dehaene, S., Izard, V., Spelke, E., & Pica, P. (2008). Log or linear? Distinct intuitions of the number scale in Western and Amazonian indigene cultures. *Science, 320*(5880), 1217–1220. http://dx.doi.org/10.1126/science.1156540.

Durgin, F. H. (1995). Texture density adaptation and the perceived numerosity and distribution of texture. *Journal of Experimental Psychology. Human Perception and Performance, 21*(1), 149.

Feigenson, L., Libertus, M. E., & Halberda, J. (2013). Links between the intuitive sense of number and formal mathematics ability. *Child Development Perspectives, 7*(2), 74–79.

Franconeri, S. L., Bemis, D. K., & Alvarez, G. A. (2009). Number estimation relies on a set of segmented objects. *Cognition, 113*(1), 1–13.

Frank, M. C., Everett, D. L., Fedorenko, E., & Gibson, E. (2008). Number as a cognitive technology: Evidence from Pirahã language and cognition. *Cognition, 108*(3), 819–824.

Fyhn, M., Molden, S., Witter, M. P., Moser, E. I., & Moser, M.-B. (2004). Spatial representation in the entorhinal cortex. *Science, 305*(5688), 1258–1264.

Gallistel, C. R., & Gelman, R. (2000). Non-verbal numerical cognition: From reals to integers. *Trends in Cognitive Sciences, 4*(2), 59–65.

Gebuis, T., & Reynvoet, B. (2012). The interplay between nonsymbolic number and its continuous visual properties. *Journal of Experimental Psychology. General, 141*(4), 642.

Gescheider, G. A. (1997). *Psychophysics: The fundamentals* (3rd ed.). Mahwah: Lawrence Erlbaum Associates.

Gevers, W., Verguts, T., Reynvoet, B., Caessens, B., & Fias, W. (2006). Numbers and space: A computational model of the SNARC effect. *Journal of Experimental Psychology. Human Perception and Performance, 32*(1), 32–44. http://dx.doi.org/10.1037/0096-1523.32.1.32.

Gilmore, C., Attridge, N., Clayton, S., Cragg, L., Johnson, S., Marlow, N., et al. (2013). Individual differences in inhibitory control, not non-verbal number acuity, correlate with mathematics achievement. *PLoS ONE, 8*(6), e67374.

Gordon, P. (2004). Numerical cognition without words: Evidence from Amazonia. *Science, 306*(5695), 496.

Green, D. M., & Swets, J. A. (1966). *Signal detection theory and psychophysics.* Los Altos Hills, CA, USA: Peninsula.

Hafting, T., Fyhn, M., Molden, S., Moser, M.-B., & Moser, E. I. (2005). Microstructure of a spatial map in the entorhinal cortex. *Nature, 436*(7052), 801–806.

Halberda, J., & Feigenson, L. (2008). Developmental change in the acuity of the "number sense": The approximate number system in 3-, 4-, 5-, and 6-year-olds and adults. *Developmental Psychology, 44*(5), 1457–1465. http://dx.doi.org/10.1037/a0012682.

Halberda, J., Ly, R., Wilmer, J. B., Naiman, D. Q., & Germine, L. (2012). Number sense across the lifespan as revealed by a massive Internet-based sample. *Proceedings of the National Academy of Sciences, 109*(28), 11116–11120. http://dx.doi.org/10.1073/pnas.1200196109.

Halberda, J., Sires, S. F., & Feigenson, L. (2006). Multiple spatially overlapping sets can be enumerated in parallel. *Psychological Science, 17*(7), 572–576. http://dx.doi.org/10.1111/j.1467-9280.2006.01746.x.

Hauser, M. D., Tsao, F., Garcia, P., & Spelke, E. S. (2003). Evolutionary foundations of number: Spontaneous representation of numerical magnitudes by cotton-top tamarins. *Proceedings of the Royal Society of London, Series B: Biological Sciences, 270*(1523), 1441–1446. http://dx.doi.org/10.1098/rspb.2003.2414.

Hurewitz, F., Gelman, R., & Schnitzer, B. (2006). Sometimes area counts more than number. *Proceedings of the National Academy of Sciences, 103*(51), 19599.

Im, H. Y., Zhong, S., & Halberda, J. (2013). Biases in human number estimation are well-described by clustering algorithms from computer vision. *Journal of Vision, 13*(9), 829.

Inglis, M., & Gilmore, C. (2014). Indexing the approximate number system. *Acta Psychologica, 145*, 147–155.

Izard, V., & Dehaene, S. (2008). Calibrating the mental number line. *Cognition, 106*(3), 1221–1247. http://dx.doi.org/10.1016/j.cognition.2007.06.004.

Izard, V., Sann, C., Spelke, E. S., & Streri, A. (2009). Newborn infants perceive abstract numbers. *Proceedings of the National Academy of Sciences, 106*(25), 10382–10385. http://dx.doi.org/10.1073/pnas.0812142106.

Jordan, K. E., & Brannon, E. M. (2006). The multisensory representation of number in infancy. *Proceedings of the National Academy of Sciences of the United States of America, 103*(9), 3486.

Laming, D. (1997). *The measurement of sensation.* Oxford University Press. Retrieved from, http://www.oxfordscholarship.com/view/10.1093/acprof:oso/9780198523420.001.0001/acprof-9780198523420.

Le Corre, M., & Carey, S. (2007). One, two, three, four, nothing more: An investigation of the conceptual sources of the verbal counting principles. *Cognition, 105*(2), 395–438.

Libertus, M. E., & Brannon, E. M. (2009). Behavioral and neural basis of number sense in infancy. *Current Directions in Psychological Science, 18*(6), 346.

Libertus, M. E., & Brannon, E. M. (2010). Stable individual differences in number discrimination in infancy. *Developmental Science, 13*(6), 900–906. http://dx.doi.org/10.1111/j.1467-7687.2009.00948.x.

Lipton, J. S., & Spelke, E. S. (2003). Origins of number sense large-number discrimination in human infants. *Psychological Science, 14*(5), 396–401. http://dx.doi.org/10.1111/1467-9280.01453.

Mazzocco, M. M. M., Feigenson, L., & Halberda, J. (2011). Impaired acuity of the approximate number system underlies mathematical learning disability (Dyscalculia). *Child Development, 82*(4), 1224–1237.

Meck, W. H., & Church, R. M. (1983). A mode control model of counting and timing processes. *Journal of Experimental Psychology. Animal Behavior Processes, 9*(3), 320–334.

Moser, E. I., Kropff, E., & Moser, M.-B. (2008). Place cells, grid cells, and the brain's spatial representation system. *Neuroscience, 31*(1), 69.

Mundy, E., & Gilmore, C. K. (2009). Children's mapping between symbolic and nonsymbolic representations of number. *Journal of Experimental Child Psychology, 103*(4), 490–502.

Nieder, A. (2005). Counting on neurons: The neurobiology of numerical competence. *Nature Reviews. Neuroscience, 6*(3), 177.

Nieder, A. (2012). Supramodal numerosity selectivity of neurons in primate prefrontal and posterior parietal cortices. *Proceedings of the National Academy of Sciences, 109*(29), 11860–11865.

Nieder, A., & Miller, E. K. (2004). Analog numerical representations in rhesus monkeys: Evidence for parallel processing. *Journal of Cognitive Neuroscience, 16*(5), 889–901.

Odic, D., Hock, H., & Halberda, J. (2012). Hysteresis affects approximate number discrimination in young children. *Journal of Experimental Psychology. General, 143*(1), 255–265. http://dx.doi.org/10.1037/a0030825.

Odic, D., Im, H. Y., Eisinger, R., Ly, R., & Halberda, J. (under review). PsiMLE: A maximum-likelihood approach to estimating psychophysical scaling and variability more reliably, efficiently, and flexibly.

Odic, D., Le Corre, M., & Halberda, J. (under review). Children's mappings between number words and the approximate number system.

Odic, D., Libertus, M. E., Feigenson, L., & Halberda, J. (in press). Developmental change in the acuity of approximate number and area representations. *Developmental Psychology.*

Opfer, J. E., & Siegler, R. S. (2007). Representational change and children's numerical estimation. *Cognitive Psychology, 55*(3), 169–195. http://dx.doi.org/10.1016/j.cogpsych.2006.09.002.

Pailian, H., Libertus, M. E., Feigenson, L., & Halberda, J. (under review). Development of visual working memory in childhood.

Piazza, M., Facoetti, A., Trussardi, A. N., Berteletti, I., Conte, S., Lucangeli, D., et al. (2010). Developmental trajectory of number acuity reveals a severe impairment in developmental dyscalculia. *Cognition, 116*(1), 33–41. http://dx.doi.org/10.1016/j.cognition.2010.03.012.

Piazza, M., Izard, V., Pinel, P., Le Bihan, D., & Dehaene, S. (2004). Tuning curves for approximate numerosity in the human intraparietal sulcus. *Neuron, 44*(3), 547–555. http://dx.doi.org/10.1016/j.neuron.2004.10.014.

Pinel, P., Piazza, M., Le Bihan, D., & Dehaene, S. (2004). Distributed and overlapping cerebral representations of number, size, and luminance during comparative judgments. *Neuron, 41*(6), 983–993.

Price, G. R., Palmer, D., Battista, C., & Ansari, D. (2012). Nonsymbolic numerical magnitude comparison: Reliability and validity of different task variants and outcome measures, and their relationship to arithmetic achievement in adults. *Acta Psychologica, 140*(1), 50–57. http://dx.doi.org/10.1016/j.actpsy.2012.02.008.

Raposo, D., Sheppard, J. P., Schrater, P. R., & Churchland, A. K. (2012). Multisensory decision-making in rats and humans. *The Journal of Neuroscience, 32*(11), 3726–3735.

Roitman, J. D., Brannon, E. M., & Platt, M. L. (2007). Monotonic coding of numerosity in macaque lateral intraparietal area. *PLoS Biology, 5*(8), e208.

Ross, J., & Burr, D. C. (2010). Vision senses number directly. *Journal of Vision, 10*(2), 10.

Siegler, R. S., & Booth, J. L. (2004). Development of numerical estimation in young children. *Child Development, 75*(2), 428–444.

Spaepen, E., Coppola, M., Spelke, E. S., Carey, S., & Goldin-Meadow, S. (2011). Number without a language model. *Proceedings of the National Academy of Sciences, 108*(8), 3163.

Sullivan, J., & Barner, D. (2013). How are number words mapped to approximate magnitudes? *The Quarterly Journal of Experimental Psychology, 66*(2), 389–402.

Sullivan, J., & Barner, D. (in press). Inference and association in children's early numerical estimation. *Child Development.* doi:10.1111/cdev.12211.

Szucs, D., Nobes, A., Devine, A., Gabriel, F. C., & Gebuis, T. (2013). Visual stimulus parameters seriously compromise the measurement of approximate number system acuity and comparative effects between adults and children. *Developmental Psychology, 4*, 444. http://dx.doi.org/10.3389/fpsyg.2013.00444.

Wang, J., Odic, D., Halberda, J., & Feigenson, L. (under review). A five-minute intervention improves preschoolers' intuitive sense of number with significant transfer to symbolic mathematics performance.

Whalen, J., Gallistel, C. R., & Gelman, R. (1999). Nonverbal counting in humans: The psychophysics of number representation. *Psychological Science, 10*(2), 130–137.

Wood, G., Willmes, K., Nuerk, H.-C., & Fischer, M. H. (2008). On the cognitive link between space and number: A meta-analysis of the SNARC effect. *Psychology Science Quarterly, 50*(4), 489.

Xu, F., & Spelke, E. S. (2000). Large number discrimination in 6-month-old infants. *Cognition, 74*(1), B1–B11. http://dx.doi.org/10.1016/S0010-0277(99)00066-9.

Xu, F., Spelke, E. S., & Goddard, S. (2005). Number sense in human infants. *Developmental Science, 8*(1), 88–101.

Zosh, J. M., Halberda, J., & Feigenson, L. (2011). Memory for multiple visual ensembles in infancy. *Journal of Experimental Psychology. General, 140*(2), 141.

Chapter 13

The Evolution of Number Systems

David C. Geary[1], Daniel B. Berch[2] and Kathleen Mann Koepke[3]
[1]Psychological Sciences, University of Missouri, Columbia, MO, USA
[2]Curry School of Education, University of Virginia, Charlottesville, VA, USA
[3]Eunice Kennedy Shriver National Institute of Child Health and Human Development (NICHD), Bethesda, MD, USA

INTRODUCTION

All of the chapters in this volume provide evidence that a wide range of non-human animals (Agrillo et al.; Beran et al.; Pepperberg; Vallortigara) and human infants and very young children (Baker & Jordan; Cantlon; Lourenco; McCrink; Posid & Cordes; vanMarle) have a number sense (Dehaene, 1997). They are sensitive to variation in quantities of many different types of objects, sounds, and behaviors, and many can learn to associate symbols with these representations. Number sense is not simply the ability to represent discrete numbers of objects and other magnitudes (e.g., area, volume, length), but to operate on them in ways that are analogous in some respects to arithmetic (Gallistel & Gelman, 2000; McCrink, this volume; vanMarle, this volume).

As illustrated by the chapters in this volume, the behavior of a wide range of nonhuman species and human infants and young children indicates the existence of mechanisms that implicitly represent, process, and integrate information about number and other types of magnitudes (Gallistel, 1990). Indeed, a series of recently published experiments with great apes has convincingly demonstrated for the first time that Bonobos (*Pan paniscus*), Chimpanzees (*P. troglodytes*), Gorillas (*Gorilla gorilla*), and Orangutans (*Pongo pygmaeus*) can all draw inferences about the probability of obtaining a preferred (banana slices) or nonpreferred (carrot slices) food item based on the relative amounts of these foods mixed together in transparent buckets. Individuals from all of these species chose the experimenter's hand that held a single concealed item from the bucket with proportionally more banana slices than carrot slices, suggesting that intuitive statistical reasoning, at least in its most fundamental mode, is not an exclusively human ability (Rakoczy et al., 2014).

Mathematical Cognition and Learning, Vol. 1. http://dx.doi.org/10.1016/B978-0-12-420133-0.00013-2
335

All of the chapters in this volume describe laboratory-based studies of number sense, although many of them note that this is an evolved competence, and several integrate their laboratory studies using situations that are similar to those found in natural contexts (Agrillo et al., this volume). We believe that a full appreciation of the significance of number sense also requires a deeper analysis of its evolutionary function and the contexts in which it is expressed. This would require another volume in and of itself; here we provide the reader with a foothold on these issues and ultimately a greater appreciation of the importance of number sense for nonhuman species in the first section. In the second section, we provide a few examples of anthropological and archaeological studies of number, and how models of number and magnitude evolution may contribute to a fuller understanding of human decision making, as well as an introduction to recent work capitalizing on computational modeling of the evolution of number systems and the cultural evolution of number systems. We conclude with a final, brief call to expand on these perspectives and integrate them across domains.

THE EVOLUTIONARY FUNCTIONS OF NUMBER SENSE IN NONHUMAN SPECIES

Evolution is about staying alive long enough to find a mate, reproduce, and for some species, invest in offspring (Darwin, 1859, 1871). Two important aspects of staying alive involve finding food (foraging) and not becoming food (predator avoidance). These are, of course, important features of natural selection that are noted in many of the chapters of this volume. In this first section, we provide a brief analysis of foraging—predator avoidance is the other side of the foraging coin—following Gallistel (1990), and illustrate key points using several field studies of how animals use their number sense in these contexts. Although mentioned by Agrillo (this volume), sexual selection—competition for mates and mate choice—is not often discussed in this literature, but is yet another dynamic that can influence the evolution and expression of number sense and sensitivity to magnitude more generally (see Lourenco, this volume). In fact, as we illustrate in the second section, discrimination of competitors and potential mates using various types of magnitude information is quite common.

Foraging

Predator and prey relations are fundamental to natural selection (Darwin, 1859). As Darwin described, the struggle for existence is exacting, and small differences in the traits that are at the forefront of this struggle can mean the difference between who lives and who dies.

> Owing to this struggle for life, any variation, however slight and from whatever cause proceeding, if it be in any degree profitable to an individual of any species, in its infinitely complex relations to other organic beings and to external

nature, will tend to the preservation of that individual, and will generally be inherited by its offspring.... [And yet] we behold the face of nature bright with gladness, we often see superabundance of food; we do not see, or we forget, that the birds which are idly singing round us mostly live on insects or seeds, and are thus constantly destroying life; or we forget how largely these songsters, or their eggs, or their nestlings, are destroyed by birds and beasts of prey; we do not always bear in mind, that though food may be now superabundant, it is not so at all seasons of each recurring year. (Darwin, 1859, pp. 61–62)

In other words, the physical and cognitive traits that support an individual's competence at foraging and at not being foraged by other species are under constant and strong selection pressures. Evolution has not simply resulted in traits that enable locating and capturing prey, but to do so efficiently (e.g., Krebs, Ryan, & Charnov, 1974). Efficiency, in turn, requires searching in locations or patches that will provide the best payoff per unit of foraging time [termed optimal foraging theory (OFT); for recent overviews, see Hamilton (2010) and Ydenberg (2010)]. As detailed in Gallistel's (1990) seminal *The Organization of Learning,* the payoff will necessarily be dependent on the discrete number of prey and their size (i.e., magnitude) at a particular location, as well as the time needed to capture them and the number of conspecifics (others of the same species) foraging in the same place. Selection on foraging behaviors and decisions (e.g., when to switch to a new patch) will thus favor the evolution of brain and cognitive systems that can represent and operate on discrete number and size of prey, and integrate these with a representation of time and the number of competitors for the same prey.

Many of the chapters in this volume describe a myriad of clever experimental studies with a wide range of species that support the existence of brain and cognitive systems that represent number (Agrillo et al., this volume; Beran et al., this volume; Pepperberg, this volume; Vallortigara, this volume). There is debate as to whether these competencies are due to two distinct systems—the object tracking system (OTS) and the approximate number system (ANS)—or only a single system (i.e., the ANS) that can account for a wide range of numbers. The OTS individuates and tracks objects but is limited to the three to four items that can be maintained in visual attention. As vanMarle (this volume) discusses, it is also possible that the object tracking system can be used to represent small quantities, even if it did not evolve for this purpose. The ANS is an evolved system that represents the approximate number of collections of objects, with precision varying inversely with collection size (see Halberda, this volume). Many of the nonhuman studies of the OTS and ANS involve considerable training, and thus, it is possible that the observed competencies are influenced to some extent by this training. Field studies with wild animals in natural contexts would address this issue. Moreover, many of the experimental studies implicitly assume efficient foraging (e.g., when the behavior involves choosing the larger of two sets of food) but do not fully consider how this would operate in natural contexts. Field studies with untrained animals in these contexts provide critical convergent evidence and illustrate the pressures that contributed to the evolution of systems for representing number and magnitude more generally.

A series of experimental field studies—an experimental manipulation that involved wild animals in their natural habitat—of the New Zealand robin (*Petroica australis*) illustrates our point (Garland, Low, & Burns, 2012; Hunt, Low, & Burns, 2008). These robins are hoarders; that is, they forage for insects and cache these in various locations on their territory for later consumption. They also monitor the caching of their mate (who shares the same territory) and will pilfer these if an opportunity to do so arises. Moreover, these birds have few natural predators, and as a result, they do not avoid humans and readily accept food from them. These features make them an ideal species for studying numerical competencies in a natural setting. Watching others cache—an experimenter in these studies—is a natural behavior, and foraging would be most efficient if these birds visited caches based on the overall amount of food stored in them.

Hunt et al. (2008) constructed artificial cache sites in a tree branch, with access to the caches requiring the flipping over of a cover that is similar to the robin's leaf flipping during prey searches. The caches were filled one item at a time with discrete food items, and the birds were allowed to choose and eat the contents of one cache. Across a variety of manipulations, these birds consistently discriminated among caches with 1 to 4 items (e.g., 1 vs. 2 or 3 vs. 4) but not caches that contained larger numbers of items (e.g., 5 vs. 8). It is not clear how the birds were making these discriminations, but given the range in which they were accurate, the OTS is a strong contender. Garland et al. (2012) conducted a similar study but used larger numbers of food items (e.g., 32 vs. 64) that are more likely to consistently engage the ANS than the combinations used in Hunt et al. These birds chose the larger cache when the ratio was 1:8 (e.g., 2 vs. 16, 4 vs. 32), 1:4 (e.g., 8 vs. 32) and 1:2 (e.g., 16 vs. 32) but did not discriminate among caches with larger ratios (e.g., 3:4, 24 vs. 32). Garland et al.'s findings are consistent with the ratio-dependent choices that are a signature of the ANS (e.g., Halberda, this volume; Star & Brannon, this volume), but do not rule out the use of the object tracking system for discriminating among smaller sets.

In another experimental field study, Farnsworth and Smolinski (2006) presented Northern Mockingbirds (*Mimus polyglottos*) with feeders that required birds to remove sticks from one of two locations to obtain food (again, an easily learned behavior that is not too different than natural behaviors); food was available once all of the sticks on one or the other side were removed. The number of sticks varied across the two locations (1 vs. 6, 2 vs. 5, and 3 vs. 4), and thus if these birds can discriminate number and are biased to forage efficiently, then they should remove the smaller set of sticks. Indeed, these birds did so 84% of the time for the 1 stick vs. 6 sticks condition, and 79% of the time in the 2 vs. 5 condition but, as found for the New Zealand robin, showed no preference in the 3 vs. 4 condition.

Foraging based on relative payoffs is not just for the birds. The buff-tailed bumble bee (*Bombus terrestris*), for instance, forages for nectar from the plant *Alcea setosa*. The plant typically grows five flowers that produce nectar and pollen, and thus efficient foraging would involve visiting five and only five flowers for each plant and not returning to any previously visited flower (Bar-Shai, Keasar, & Shmida, 2011). Bar-Shai et al. recorded 516 such foraging episodes in the field and found that only 1% of the bees revisited a previously foraged flower. Very few bees departed after visiting one to three flowers, but 25% departed after visiting four flowers and 92% of the remaining bees departed after visiting five flowers, even when the plant had six or seven total flowers. Other analyses indicated that departure was driven by the number of visited flowers, but not by other potential cues (e.g., flower location on the plant). A follow-up experimental study showed that these bees had not simply evolved a strategy that involved five foraging bouts per plant, but rather adjusted their foraging behavior based on the number of flowers that produced nectar (for a review of numerical competence in invertebrates see Pahl, Si, & Zhang, 2013).

Of course, species that forage for one prey item at a time would be expected to make decisions based on other types of magnitude information, such as relative size. And, as one might predict, foraging is not simply based on number and size of prey, but how easily prey are detected and caught in a particular ecology (e.g., Almenar, Aihartza, Goiti, Salsamendi, & Garin, 2013) and the intensity of the competition with conspecifics for the same prey (e.g., Suraci & Dill, 2011).

In other words, animals use number and magnitude information when foraging and in many cases also need to track the number of conspecifics foraging for the same prey (e.g., Nelson & Jackson, 2012). For many group-living species, social competition is typically about control of specific resources (e.g., a fruit tree) or a larger territory, that is, competition over access to food. A very consistent pattern across species is that larger groups have a competitive advantage over smaller groups, and in fact, conflicts with larger groups can often be fatal (Wrangham, 1999). These dynamics set the stage for the evolution of mechanisms for comparing the size of ones' own group to that of competing groups.

We see this dynamic and sensitivity to relative group size in female lions (*Panthera leo*; Grinnell, 2002; McComb, Packer, & Pusey, 1994), as just one example. In this species, groups of related females (prides) maintain a territory; females will stay on this territory across generations, whereas males will migrate and compete for access to other prides. The territories have a limited amount of prey, and thus, females fiercely defend it against other prides. Females signal their territorial boundaries and group size through a roaring chorus, whereby groups of females will roar together. In a field experiment, McComb et al. used recordings of either one intruder or three intruders to simulate intrusion into the territories of 21 different prides and monitored defenders' approach or avoidance of the intruders. The ratio of defenders to intruders significantly affected approach behaviors, with "female adult defenders without dependent offspring preferring a 2:1 ratio before approaching" (McComb et al., 1994, p. 383); females with cubs almost always approached regardless of the ratio of intruders to defenders. With three intruders and three defenders, the probability of approach was about 20%, but this jumped to about 50% with four defenders and more than 90% with six defenders. Their behavior suggested that even though they preferred a 2:1 ratio, they were sensitive to smaller differences (i.e., 3 vs. 4). Such findings, with ratio-dependent sensitivity, seemingly support the hypothesis of a single number system, i.e., the ANS.

Mating

It is common across species of insect, fish, bird, and mammal for males and sometimes females to have evolved traits that signal genetic and physical health and reproductively relevant behaviors (Andersson, 1994). These are traits involved in competition for access for mates (typically male–male competition) and discriminative choice of mating partners (typically female choice; Darwin, 1871). Many of these traits are exaggerated (e.g., peacock tail; *Pavo cristatus*; Petrie, 1994), and their muted expression can compromise the health and sometimes the life of unfit individuals and their offspring (Zahavi, 1975). Males can evaluate the relative fighting ability, for instance, of other males by the size or number of these traits, and females can use the same or similar traits to discriminate fit from marginally less fit would-be mates.

To be functional in terms of assessing competitive ability or mate fitness, animals necessarily have to be able to make some type of magnitude-based comparisons of at least two individuals on these traits (at minimum, requiring an implicit understanding of more vs. less; Halberda, this volume). Moreover, comparisons can involve visiting and assessing multiple dispersed mates, implying some mechanisms for rank ordering mates, as well as integration of multiple cues across sensory modalities, implying an abstract representation of magnitude. The corresponding magnitudes can be based on continuous extent, such as area, perimeter, or length, or discrete features of the trait (Lourenco, this volume).

Assessments of competitors or mates based on continuous extent are very common. These traits include the size of colorful plumage or beak patches in a wide variety of bird species, such as collard flycatchers (*Ficedula albicollis*; Garamszegi et al., 2004), the American goldfinch (*Spinus tristis*; McGraw & Hill 2000), and red jungle fowl (*Gallus gallus*; Zuk, Thornhill, & Ligon, 1990), among many others. Analogous traits are also common in fish, such as guppies (*Poecilia reticulata;* Sheridan & Pomiankowski 1997), and in insects, such as many species of damselfly (e.g., *Hetaerina Americana;* Contreras-Garduño, Buzatto, Abundis, Nájera-Cordero, & Córdoba-Aguilar, 2007) and stalk-eyed fly (e.g., *Diasemposis meigenii*; Bellamy, Chapman,

Fowler, & Pomiankowski, 2013). With a few exceptions, such as mandrills (*Mandrillus sphinx;* Setchell & Dixson, 2001), most mammals do not have these colorful badges, but they do have other size-related traits that signal fitness, such as horn size in bighorn sheep (*Ovis canadens;* Coltman, Festa-Bianchet, Jorgenson, & Strobeck, 2002) or antler size in *cervids* (e.g., red deer, *Cervus elaphus;* Bartoš & Bahbouh 2006).

More complex signals include rate of courtship or dominance displays. Alatalo, Glynn, and Lundberg (1990), for instance, manipulated the song rates of male collared flycatchers by provisioning some of them but not others. Provisioned males were in better physical condition and produced, on average, 34 song bouts every 5 minutes compared to 16 bouts for nonprovisioned males, roughly a 2:1 ratio. Females chose the provisioned males as mates 5.5 times more frequently than the nonprovisioned males. Male wolf spiders (*Hygrolycosa rubrofasciata*) signal their condition by drumming their lower abdomen on dry leaves (Kotiaho, Alatalo, Mappes, & Parri, 1996). In natural conditions, females will be courted by up to six males and prefer the male with the highest drum rate. The extent of female sensitivity to variation in drum rate was not detailed, but a 1.5:1 ratio of drum rates predicted male survivability; thus, it is likely that females are sensitive to at least this degree of difference (Kotiaho et al., 1996).

Clutton-Brock and Albon (1979) assessed the relation between male red deer roaring rates and their fighting ability and overall fitness. As is common in species in which competition can escalate to risky and sometimes deadly physical contests (Clutton-Brock, Albon, Gibson, & Guinness, 1979), male red deer use a variety of signals to first assess the fighting ability of potential rivals. Roaring is one of the last of these evaluative signals that is assessed before escalation; roaring rate is highly correlated with fighting ability ($r = .75$ to $.80$; Clutton-Brock & Albon, 1979). Bouts occur in sequences of 1 to 10 discrete roars per minute. These field studies did not allow for a precise estimate of males' sensitivity to roar number, but it appears that they can discriminate bouts that differ by about 2 roars, whether they are in short sequences (1 to 3 roars) or long sequences (7 to 10 roars). Females also prefer males with longer roar bouts and can also discriminate bouts that differ by 2 roars, at least for short sequences (one-roar differences and longer sequences were not assessed; McComb, 1991).

Analogous patterns are found in several species of tree frog (*Hyla chrysoscelis; H. Versicolor*). In these species, females make mate choices based on male calls (Gerhardt, Dyson, & Tanner, 1996). These calls are emitted as a sequence of sound pulses, and males that produce more pulses per call sire healthier offspring than do males with fewer pulses (Welch, Semlitsch, & Gerhardt, 1998). The number of pulses per call can be experimentally simulated, and female preference for one call over another provides information on sensitivity to pulse number, controlling overall duration of the call. Although females are highly discriminating among males with a below-average number of pulses and less discriminating among males with an above-average number of pulses, they can nevertheless discriminate among males whose calls differ by a only a single pulse (Gerhardt, Tanner, Corrigan, & Walton, 2000).

Although few of these studies were explicitly designed to assess the limits of animals' sensitivity to continuous or discrete magnitudes, it is clear that many mating-related signals convey magnitude information. Given the wealth of information known about these signals and their apparent ubiquity across sexually reproducing species (Andersson, 1994), they provide an untapped opportunity to study the evolution and proximate expression of systems for processing and representing magnitudes.

QUANTITATIVE PROCESSING IN THE EVOLUTION OF HUMAN DECISION MAKING

In this section, we discuss several topics pertaining to the evolution and proximate development of various kinds of quantitative skills in humans. They include (a) how archaeological approaches might inform our understanding of the evolutionary emergence of symbolic representations in *Homo*; (b) the role of quantitative processing in the evolution of human decision making; (c) recent evidence from laboratory experiments and computer simulations concerning how a primitive system for perceiving and evaluating continuous dimensions such as size and amount may have been a precursor to the development of the discrete numerical system; and (d) how anthropological and cognitive archaeological approaches might inform our understanding of the ways in which *Homo sapiens* was able to make the transition from approximate judgments of quantity to exact numerical thinking.

The Evolutionary Emergence of Symbolic Representation

Archaeological studies suggest that hominids (our bipedal ancestors) survived like many others in the animal kingdom, via gathering and hunting in small familial groups, perhaps with the aid of primitive tools (Foley, 1987). Because these hominids were largely migratory and produced very little materially, there are few artifacts on which to base conclusions of their cognitive capacities, but most scientists accept that number appreciation and capacity were at least equal to that of contemporary nonhuman animals. As we briefly noted earlier, OFT (recall, optimal foraging theory) has provided a valuable approach for determining how animals can best maximize their efficiency in locating and capturing prey. These implicit, cost-benefit decisions are known to be based at least in part on the processing of quantitative information. Indeed, as Gallistel (2011) and Hagen et al. (2012) have pointed out, the implicit arithmetical competencies of the ANS (McCrink, this volume) support representations that can be elaborated during decision making. For more than 30 years, anthropologists have extensively used the fundamental assumptions, principles, and modeling techniques of OFT to great advantage in making inferences about the foraging behaviors and decision making in Late Pleistocene hunter-gatherers (Zeder, 2012).

But what pushed *Homo* toward symbolic thinking? Although still vigorously debated, local demographic (Kuhn, 2012; Powell, Shennan, & Thomas, 2009) and geographic/climatic (Barton & d'Errico, 2012) pressures are considered the most likely factors to have played a role in the emergence of anatomically modern humans. One possibility is that as our early *Homo* ancestors became increasingly effective at reducing the external sources of mortality (e.g., through building of shelters, tools, fire, etc.; Foley, 1987); population sizes increased and triggered a within-species arms race that helped to drive substantial increases in brain size (Alexander, 1989; Flinn, Geary, & Ward, 2005; Geary, 2005). Alternative explanations include the complexity of hunting in complex ecologies (Kaplan, Hill, Lancaster, & Hurtado, 2000), and climatic variation (Potts, 1998). Whatever drove the increases in brain size—including modest increases in the parietal cortex above and beyond the overall change in brain volume (Holloway & de la Coste-Lareymondie, 1982)—since the emergence of *Homo*, these changes laid the foundation for what archaeologists term culture in early modern humans, including the emergence of symbolic communication (Belfer-Cohen, & Goring-Morris, 2011; Powell, Shennan, & Thomas, 2009; Rossano, 2010).

Symbols, as defined by Peirce (1978/1931) include three hierarchical levels of increasing complexity and abstractness: iconic (i.e., the sign is physically similar to the referent; e.g., clay models of two pigs represent two actual pigs), indexical (i.e., indicates the temporal or spatial presence or existence of the referent; e.g., vultures flying overhead indicates carrion nearby, or the approach of three members of an enemy group indicates a larger hidden number of enemies), and symbolic (where the relation between the symbol and the referent exists purely by convention and is arbitrary; e.g., the Arabic digit "2" to represent "two-ness"). Number and magnitude cognitions are represented across all three types of symbols, but especially iconic and symbolic. Almost by definition, the use of symbols generates archaeological artifacts (in the form of markings, pottery, art works, built tools, etc.) that enable the archaeologist to study and fix in historical time cognitive transitions of humans. The order and timing of the emergence of symbols is still unclear but would be expected to include those that represent exact number.

In a few places, large sedentary communities $(25 < N < 100)$ such as the Mediterranean Natufians (Late Epipaleolithic, 15,000–11,600 cal BP) developed around naturally occurring abundant resources. Even before agriculture, the complex social culture that emerged around these food resources necessitated the use of symbolic representations for social organization and communication. Rossano (2010) provides evidence that the domestication of agriculture enabled ever larger social networks, including those with distant trading communities. Symbolic representations are particularly useful for managing some aspects of social life within growing communities, and numerical representations may have been especially important for managing trade and other forms of commerce.

The Evolution of Numerical Heuristics and Decision Making

It is important to appreciate the contributions that an evolutionary perspective can make to the contemporary study of decision making in general and the basic role that quantitative reasoning plays in the confluence of cognitive operations that give rise to specific choices. What follows here is a brief discussion of some of the prevailing approaches most relevant to these topics.

As Hammerstein and Stevens (2012) have pointed out, despite many advances in the study of decision making stemming from research in such fields as cognitive psychology, experimental economics, and neuroscience, among other disciplines, "the explanatory power of evolutionary theory has been neglected" (p. 1). In outlining what they call "Darwinian Decision Theory," these authors offer six reasons for framing decision making from an evolutionary perspective. Here, we describe only the first reason they present: *adaptive specialization*, where a given species' evolved decision mechanisms should operate very efficiently under the same environmental conditions that drove the evolution of these mechanisms. As Hammerstein and Stevens note, this concept is analogous to Gigerenzer and colleagues' (Gigerenzer, 2008; Gigerenzer, Todd, & ABC Research Group, 1999) "adaptive toolbox" of simple rules of thumb that are selected for implicit decision making in different environmental (physical or social) contexts—what Gigerenzer refers to as *ecological rationality*. More specifically, the adaptive toolbox, a "Darwinian-inspired theory," is hypothesized to contain an assortment of heuristics—experience-based strategies used for rapidly solving problems or making decisions under conditions of uncertainty that serve to "satisfice" rather than optimize. Thus equipped, one can arrive at a solution that is good enough, rather than the best one; the assumption is that optimal choices are especially costly, and thus trade-offs allow for good enough decisions that result in the best marginal returns.

Most important for our purposes is that at least 4 of the 10 heuristics that Gigerenzer (Gigerenzer, 2008; Gigerenzer et al., 1999) purports to be in the human adaptive toolbox clearly depend on basic quantitative or magnitude comparison skills: (1) Tallying—i.e., to estimate a criterion, do not estimate weights of the utilities of choice outcomes by their probabilities, but simply *count the number* of favoring cues; (2) Satisficing—i.e., search through alternatives and choose the first one that *exceeds your aspiration level* (*comparing magnitudes*); (3) 1/N equality heuristic—i.e., allocate resources *equally* to each of N alternatives (*division*); and (4) Imitate the majority—i.e., look at a *majority of people* in your peer group and imitate their behavior (*numerical comparison based on proportion*).

Finally, it is also worth pointing out here that the same kinds of decision rules used by nonhuman animals when foraging, described earlier, are also used by humans when engaging in both physical and cognitive foraging activities, such as searching for food in a simulated fishing task (Hutchinson, Wilke, & Todd, 2008) and "fishing" through memory for meaningful words made from random letter sequences (Wilke, Hutchinson, Todd, & Czienskowski, 2009). As Hammerstein and Stevens (2012) have pointed out, however, rather than serving as general-purpose rules, these algorithms operate in a manner that is tuned to the statistical structure of particular contexts and environments.

From Continuous Magnitude Perception to Discrete Numerical Representation

Recently, it has been proposed that a system that evolved to support the perception and evaluation of continuous, noncountable dimensions such as size and amount may have been the evolutionary precursor to the discrete numerical system or the ANS (Henik, Leibovich, Naparstek, Diesendruck, & Rubinsten, 2012). As Henik et al. correctly point out, however, it is not yet clear whether this system was co-opted for making judgments along a new dimension such as number (Cantlon, Platt, & Brannon, 2009; Geary, 1995), was exploited by cultural inventions (Dehaene & Cohen, 2007), or was able to be accessed as a consequence of evolution (Rozin, 1976). In order to test the evolutionary account, Katz, Benbassat, Diesendruck, Sipper, and Henik (2013) employed an evolutionary computational approach entailing the use of *genetic algorithms*—i.e., search heuristics that mimic the process of natural selection. Implemented in order to generate computational models inspired by biological neuronal systems referred to as *artificial neural networks* (ANNs), ANNs are capable of learning, perception, and pattern recognition. This approach revealed that ANNs trained to perceive size were significantly better at evolving the ability to count discrete items than were networks trained to evolve counting skills de novo. The authors interpreted these findings as support for the idea that counting skills evolved from the more primitive size perception system.

Taking a different approach to testing the Henik et al. (2012) evolutionary perspective, Gabay, Leibovich, Henik, and Gronau (2013) recently proposed that being able to represent the *conceptual size* of animals—long-term knowledge of their real-world size, regardless of their actual retinal size—may have been indispensable for an organism's survival. To wit, Gabay et al. reasoned that such a representational ability could aid in differentiating predators from prey, among other adaptive behaviors. And as numerals correspondingly denote long-term knowledge of exact quantities and amounts, Gabay et al. maintain that conceptual size could have conceivably acted as a link between continuous and numerical (discrete) magnitude representations.

To test this hypothesis, these investigators examined the influence of conceptual sizes of real-world animal and object primes on adult humans' processing of the magnitudes of Arabic numerals. The procedure involved visually displaying a picture of a large or small (in real life) animal or object, such as HORSE or a CAT, respectively, but that were of the same retinal size; immediately afterward they presented a numerically large (e.g., 9) or small (e.g., 2) Arabic numeral that had to be quickly judged as either odd or even. Gabay et al. found that congruent primes (e.g., a large animal preceding a large numeral) led to faster parity (odd/even) judgments than did incongruent primes (e.g., a large animal preceding a small numeral), leading them to conclude that conceptual size may have mediated the transition from continuous magnitude representations to a discrete numerical system.

From Approximation to Numerical Thinking

"How did *Homo sapiens* alone ever move beyond approximation?" (Dehaene, 1997, p. 91). Anthropologists have sought to answer this question by studying the use of number across various cultures, especially traditional cultures' use of language to express and manipulate number. As language's relationship to the development of exact number concepts is the subject of a later volume, we will not attempt an extended review here. However, these studies have proved vitally important to understanding the historical development of numerical thinking in humans, and as such, a few highlights will be mentioned.

Documenting how and when people in traditional cultures use spoken or drawn or assembled representations of number (e.g., bundles of sticks), researchers have revealed both the diversity of and common features of numerical thinking across cultures. The awareness and use of exact number appears to be universal in all known extant cultures; all have some language for exact numbers,[1] whether the language is spoken, written, manually signed, clicked, or whistled, and all cultures show a working capacity to compare magnitudes and manipulate numbers, at least in approximate numerosities. In contrast, the range of exact numbers counted (e.g., from one to one billion), the system of counting (e.g., body-tally systems that use the names of predefined body parts to represent a specific number), the use of and consistency of a numeric base for counting larger numbers (e.g., the industrial world's use of base 10; Comrie, 2005), and the language of numeric operations vary widely.

---

1. There is some evidence, still disputed, that two distinct precontact indigenous cultures (i.e., the Xilixana Yanomami and the Pirahã of Brazil) both lacked a word for exactly one (see Nevins, Pesetsky, & Rodrigues, 2009). As both groups now have had significant contact with the dominant Brazilian culture and the Western world, it is impossible to meaningfully further assess the validity of these claims. Regardless, early investigations of individuals in both cultures precontact showed clear evidence of individuation of objects in both small and large sets and behaved in ways suggesting their recognition of the differences in set sizes of 1, 2, and 3 objects (i.e., up to the subitizing capacity).

Some cultures even have different language reserved for counting in culturally relevant circumstances (e.g., Wuvulu for counting coconuts[2]; Hafford, 1999). The only three universals in human number systems appear to be: (1) all humans have some exact number system, (2) the number systems are representational and therefore abstract, and (3) all exact number systems at minimum count unity (i.e., one, but note that not all number systems include the null concept or zero).

Some of the new approaches to answering the questions surrounding *Homo sapiens'* leap to the manipulation of large exact numbers have emerged from the work of *cognitive archaeologists*, who attempt to reconstruct the ways in which the human mind operated in the past and illuminate how it has changed over the course of human history (Abramiuk, 2012). For example, the archaeological record suggests that the earliest known numerical signs are dated to approximately 30,000 years ago; these include external supports and devices such as notched sticks and bones, knotted strings, and clay tokens, which may have been used for recording kills from a hunt, among other purposes (Coolidge & Overmann, 2012). It remains unclear as to the specific role that such material objects may have played in the development of symbolic number systems during our recent history, or in the proximate development of people's mapping of the quantities represented by the ANS or object file system to numerical symbols. Nevertheless, a provocative account of how this may have occurred has been put forward by Malafouris in his comprehensive Material Engagement Theory (Malafouris, 2004, 2010, 2012, 2013; Malafouris & Renfrew, 2010).

With respect to the early development of symbolic numerical cognition in particular, Malafouris (2013) contends that before there were linguistic quantifiers or other symbols available to express numerical quantities, the ability to conceptualize discrete numbers was acquired through engagement with material artifacts—a process he refers to as *enactive discovery and signification*. In describing these processes, he emphasizes that artifacts, particularly small clay tokens, were not symbols that directly stood for numbers. Rather, it was the visible and tangible properties of these tokens that "brought forth" numbers by inducing connections in the brain's intraparietal region through a process of cortical reorganization and (partial) neuronal recycling (Dehaene & Cohen, 2007). According to Malafouris, it was through such an embodied "conceptual integration" that the extant neurological links between spatial and numerical processing were modified, at least in the initial emergence of number-like symbols in human culture. In other words, the parietal areas that had previously evolved to support approximate numerosity could subsequently support the representation of exact number. Finally, as Overmann (2013) puts it in a succinct characterization of Malafouris's position: "Inchoate concepts of quantity, once externalized

---

2. As described by Hafford (1999), 37–39, in Hammarström (2010).

in material form, become tangible; in becoming tangible, they also become more explicit, more manipulable and more readily shared between individuals and generations" (p. 21).

These are, of course, fascinating questions that touch on several issues addressed in this volume, including Baker and Jordan's discussion of multi-sensory integration, and Canton's, Lourenco's, and Star and Brannon's discussion of the relation between the ANS and children's learning of symbolic mathematics. Neither Malafouris (2013) nor any of the authors in this volume address the related question of how the emergence of material culture and symbol systems shaped subsequent evolution of the human mind. In other words, did those individuals who easily mapped symbols onto the ANS to construct more precise and flexible representations of quantity have advantages in some cultural contexts (e.g., trade) that resulted in survival and reproductive advantages that in turn led to the evolutionary refinement of this system? At this time, we have no way of knowing whether or not this has occurred, but genomic evidence of evolutionary change since the emergence of material culture leaves open the possibility that it has (Voight, Kudaravalli, Wen, & Pritchard, 2006).

CONCLUSIONS

The field of numerical and mathematical cognition has expanded enormously in the past several decades and now includes scientists from fields as disparate as entomology and archaeology, much less psychology and neuroscience. As illustrated by the chapters in this volume, the influx of scientists into the area has resulted in substantial advances in our understanding of the evolution of the brain and cognitive systems that support our sense of discrete quantity and magnitude more generally, as well as how these systems are expressed and shaped during development. Our goal with this chapter was to illustrate the broad importance of sensitivity to number and magnitude to survival and reproduction in the wild, and to simultaneously point out new avenues for future study of these competencies in humans. For example, an evolved system for processing numerical information, presumably driven by foraging demands, has been recognized for decades (Gallistel, 1990). In stark contrast, the importance of sensitivity to number and magnitude as related to sexual selection (Andersson, 1994; Darwin, 1871), that is, competition for mates and mate choices, has not been fully explored. As sexual selection is currently a very active area of research in evolutionary biology, it provides a potentially rich source of collaborative, interdisciplinary study of how number and magnitude systems evolved and how they change with development.

Human decision making is another area that can inform and be informed by our understanding of basic systems for representing number and magnitude. As noted, many of the fast, implicit decisions that people make day in and day out are dependent in part on the outputs of the number and magnitude systems (Gigerenzer, 2008; Gigerenzer et al., 1999). The study of how number and magnitude information become integrated into implicit and explicit decision making is an area that remains to be fully exploited and an area of considerable practical importance (Reyna, Nelson, Han, & Dieckmann, 2009). The study of the evolved number and magnitude systems can also be expanded with consideration of simulations of how these systems may have evolved (Henik et al., 2012). These approaches cannot provide strong evidence, but they do provide useful feasibility checks and can generate new ways of conceptualizing the evolutionary and developmental processes that shape the number and magnitude systems.

Finally, the merging of archaeological and anthropological research with basic research on number and magnitude representation can help to better inform our understanding of the form and function of historically early symbolic number systems and press us to consider whether the emergence of social practices, such as large-scale trade, may have influenced the recent evolution of these systems. What do these cultural influences tell us about the ontological pressures that drive children's number development today? Do they suggest sensitive periods for interventions? We believe knowledge of our evolutionary history of number provides us with a compass pointing to a myriad of new questions, tools, and directions for research on numerical cognition.

REFERENCES

Abramiuk, M. A. (2012). *The foundations of cognitive archaeology.* Cambridge, MA: The MIT Press.

Alatalo, R. V., Glynn, C., & Lundberg, A. (1990). Singing rate and female attraction in the pied flycatcher: An experiment. *Animal Behaviour, 39,* 601–603.

Alexander, R. D. (1989). Evolution of the human psyche. In P. Mellars & C. Stringer (Eds.), *The human revolution: Behavioral and biological perspectives on the origins of modern humans* (pp. 455–513). Princeton, NJ: Princeton University Press.

Almenar, D., Aihartza, J., Goiti, U., Salsamendi, E., & Garin, I. (2013). Hierarchical patch choice by an insectivorous bat through prey availability components. *Behavioral Ecology and Sociobiology, 67,* 311–320.

Andersson, M. (1994). *Sexual selection.* Princeton, NJ: Princeton University Press.

Bar-Shai, N., Keasar, T., & Shmida, A. (2011). The use of numerical information by bees in foraging tasks. *Behavioral Ecology, 22,* 317–325.

Barton, R. N. E., & d'Errico, F. (2012). North African origins of symbolically mediated behaviour and the Aterian. In S. Elias (Ed.), *Origins of human innovation and creativity developments in quaternary science: Vol. 16.* (pp. 23–34). Amsterdam: Elsevier.

Bartoš, L., & Bahbouh, R. (2006). Antler size and fluctuating asymmetry in red deer (*Cervus elaphus*) stags and probability of becoming a harem holder in rut. *Biological Journal of the Linnean Society, 87,* 59–68.

Belfer-Cohen, A., & Goring-Morris, A. N. (2011). Becoming farmers. *Current Anthropology, 52,* S209–S220.

Bellamy, L., Chapman, N., Fowler, K., & Pomiankowski, A. (2013). Sexual traits are sensitive to genetic stress and predict extinction risk in the stalk-eyed fly, *Diasemposis meigenii*. *Evolution, 67,* 2662–2673.

Cantlon, J. F., Platt, M. L., & Brannon, E. M. (2009). Beyond the number domain. *Trends in Cognitive Sciences, 13,* 83–91.

Clutton-Brock, T. H., & Albon, S. D. (1979). The roaring of red deer and the evolution of honest advertisement. *Behaviour, 69,* 145–170.

Clutton-Brock, T. H., Albon, S. D., Gibson, R. M., & Guinness, F. E. (1979). The logical stag: Adaptive aspects of fighting in red deer (*Cervus elaphus*). *Animal Behaviour, 27,* 211–225.

Coltman, D. W., Festa-Bianchet, M., Jorgenson, J. T., & Strobeck, C. (2002). Age-dependent sexual selection in bighorn rams. *Proceedings of the Royal Society of London B, 269,* 165–172.

Comrie, B. (2005). Numeral bases. In M. Haspelmath, M. S. Dryer, D. Gil & B. Comrie (Eds.), *The world atlas of language structures* (pp. 530–533). Oxford: Oxford University Press.

Contreras-Garduño, J., Buzatto, B. A., Abundis, L., Nájera-Cordero, K., & Córdoba-Aguilar, A. (2007). Wing colour properties do not reflect male condition in the American rubyspot (*Hetaerina americana*). *Ethology, 113,* 944–952.

Coolidge, F. I., & Overmann, K. A. (2012). Numerosity, abstraction, and the emergence of symbolic thinking. *Current Anthropology, 53,* 204–225.

Darwin, C. (1859). *On the origin of species by means of natural selection*. London: John Murray.

Darwin, C. R. (1871). *The descent of man, and selection in relation to sex*. London: John Murray.

Dehaene, S. (1997). *The number sense. How the mind creates mathematics*. New York: Oxford University Press.

Dehaene, S., & Cohen, L. (2007). Cultural recycling of cortical maps. *Neuron, 56,* 384–398.

Farnsworth, G. L., & Smolinski, J. L. (2006). Numerical discrimination by wild Northern mockingbirds. *The Condor, 108,* 953–957.

Flinn, M. V., Geary, D. C., & Ward, C. V. (2005). Ecological dominance, social competition, and coalitionary arms races: Why humans evolved extraordinary intelligence. *Evolution and Human Behavior, 26,* 10–46.

Foley, R. (1987). Hominid species and stone-tool assemblages: How are they related. *Antiquity, 61,* 380–392.

Gabay, S., Leibovich, T., Henik, A., & Gronau, N. (2013). Size before numbers: Conceptual size primes numerical value. *Cognition, 129,* 1–23.

Gallistel, C. R. (1990). *The organization of learning*. Cambridge, MA: Bradford Books/MIT Press.

Gallistel, C. R. (2011). Mental magnitudes. In S. Dehaene & E. M. Brannon (Eds.), *Space, time, and number in the brain: Searching for the foundations of mathematical thought* (pp. 3–12). London: Academic Press.

Gallistel, C. R., & Gelman, R. (2000). Non-verbal numerical cognition: From reals to integers. *Trends in Cognitive Sciences, 4,* 59–65.

Garamszegi, L. Z., Møller, A. P., Török, J., Michl, G., Péczely, P., & Richard, M. (2004). Immune challenge mediates vocal communication in a passerine bird: An experiment. *Behavioral Ecology, 15,* 148–157.

Garland, A., Low, J., & Burns, K. C. (2012). Large quantity discrimination by North Island robins (*Petroica longipes*). *Animal Cognition, 15,* 1129–1140.

Geary, D. C. (1995). Reflections of evolution and culture in children's cognition: Implications for mathematical development and instruction. *American Psychologist, 50,* 24–37.

Geary, D. C. (2005). *The origin of mind: Evolution of brain, cognition, and general intelligence*. Washington, DC: American Psychological Association.

Gerhardt, H. C., Dyson, M. L., & Tanner, S. D. (1996). Dynamic properties of the advertisement calls of gray tree frogs: Patterns of variability and female choice. *Behavioral Ecology, 7*, 7–18.

Gerhardt, H. C., Tanner, S. D., Corrigan, C. M., & Walton, H. C. (2000). Female preference functions based on call duration in the gray tree frog (*Hyla versicolor*). *Behavioral Ecology, 11*, 663–669.

Gigerenzer, G. (2008). Why heuristics work. *Psychological Science, 3*, 20–29.

Gigerenzer, G., Todd, P. M. & The ABC Research Group. (1999). *Simple heuristics that make us smart.* New York: Oxford University Press.

Grinnell, J. (2002). Modes of cooperation during territorial defense by African lions. *Human Nature, 13*, 85–104.

Hafford, J. A. (1999). *Elements of Wuvulu grammar. MA thesis.* Arlington: University of Texas at Arlington.

Hagen, E. H., Chater, N., Gallistel, C. R., Houston, A., Kacelnik, A., Kalenscher, T., et al. (2012). Decision making: What can evolution do for us? In P. Hammerstein & J. R. Stevens (Eds.), *Evolution and the mechanisms of decision making* (pp. 97–126). Cambridge, MA: The MIT Press.

Hamilton, I. M. (2010). Forging theory. In D. F. West & C. W. Fox (Eds.), *Evolutionary behavioral ecology* (pp. 177–193). New York: Oxford University Press.

Hammarström, H. (2010). Rarities in numeral systems. In J. Wohlgemuth & M. Cysouw (Eds.), *Rethinking universals: How rarities affect linguistic theory: Vol. 45.* (pp. 11–53). Berlin: Walter de Gruyter.

Hammerstein, P., & Stevens, J. R. (2012). Six reasons for invoking evolution in decision theory. In P. Hammerstein & J. R. Stevens (Eds.), *Evolution and the mechanisms of decision making* (pp. 1–17). Cambridge, MA: The MIT Press.

Henik, A., Leibovich, T., Naparstek, S., Diesendruck, L., & Rubinsten, O. (2012). Quantities, amounts, and the numerical core system. *Frontiers in Human Neuroscience, 5*, 186.

Holloway, R. L., & de al Coste-Lareymondie, M. C. (1982). Brain endocast asymmetry in pongids and hominids: Some preliminary findings on the paleontology of cerebral dominance. *American Journal of Physical Anthropology, 58*, 101–110.

Hunt, S., Low, J., & Burns, K. C. (2008). Adaptive numerical competency in a food-hoarding songbird. *Proceedings of the Royal Society B, 275*, 2373–2379.

Hutchinson, J. M. C., Wilke, A., & Todd, P. M. (2008). Patch leaving in humans: Can a generalist adapt its rules to dispersal of items across patches? *Animal Behaviour, 75*, 1331–1349.

Kaplan, H., Hill, K., Lancaster, J., & Hurtado, A. M. (2000). A theory of human life history evolution: Diet, intelligence, and longevity. *Evolutionary Anthropology, 9*, 156–185.

Katz, G., Benbassat, A., Diesendruck, L., Sipper, M., & Henik, A. (2013). From size perception to counting: An evolutionary computation point of view. In C. Blum (Ed.), *Proceedings of the Fifteenth Annual Conference on Genetic and Evolutionary Computation Companion* (pp. 1675–1678). New York: ACM.

Kotiaho, J., Alatalo, R. V., Mappes, J., & Parri, S. (1996). Sexual selection in a wolf spider: Male drumming activity, body size, and viability. *Evolution, 1977–1981*.

Krebs, J. R., Ryan, J. C., & Charnov, E. L. (1974). Hunting by expectation or optimal foraging? A study of patch use by chickadees. *Animal Behaviour, 22*, 953–964.

Kuhn, S. L. (2012). Emergent patterns of creativity and innovation in early technologies. In S. Elias (Ed.), *Origins of human innovation and creativity developments in quaternary science: Vol. 16.* (pp. 69–87). Amsterdam: Elsevier.

Malafouris, L. (2004). The cognitive basis of material engagement: Where brain, body and culture conflate. In E. DeMarrais, C. Gosden, & C. Renfrew (Eds.), *Rethinking materiality: The engagement of mind with the material world* (pp. 53–62). Cambridge, UK: McDonald Institute for Archaeological Research.

Malafouris, L. (2010). Grasping the concept of number: How did the sapient mind move beyond approximation? In I. Morley & C. Renfrew (Eds.), *The archaeology of measurement: Comprehending heaven, earth and time in ancient societies.* Cambridge, UK: Cambridge University Press.

Malafouris, L. (2012). Peer comment on journal article 'Numerosity, abstraction, and the emergence of symbolic thinking.' *Current Anthropology, 53,* 204–225.

Malafouris, L. (2013). *How things shape the mind: A theory of material engagement.* Cambridge, MA: The MIT Press.

Malafouris, L., & Renfrew, C. (2010). The cognitive life of things: Archaeology, material engagement, and the extended mind. In L. Malafouris & C. Renfrew (Eds.), *The cognitive life of things: Recasting the boundaries of the mind* (pp. 1–12). Cambridge, UK: McDonald Institute for Archaeological Research.

McComb, K. E. (1991). Female choice for high roaring rates in red deer, *Cervus elaphus. Animal Behaviour, 41,* 79–88.

McComb, K., Packer, C., & Pusey, A. (1994). Roaring and numerical assessment in contests between groups of female lions, *Panthera leo. Animal Behaviour, 47,* 379–387.

McGraw, K. J., & Hill, G. E. (2000). Differential effects of endoparasitism on the expression of carotenoid- and melanin-based ornamental coloration. *Proceedings of the Royal Society of London B, 267,* 1525–1531.

Nelson, X. J., & Jackson, R. R. (2012). The role of numerical competence in a specialized predatory strategy of an araneophagic spider. *Animal Cognition, 15,* 699–710.

Nevins, A., Pesetsky, D., & Rodrigues, C. (2009). Pirahã exceptionality: A reassessment. *Language, 85,* 355–404.

Overmann, K. A. (2013). Material scaffolds in number and time. *Cambridge Archaeological Journal, 23,* 19–39.

Pahl, M., Si, A., & Zhang, S. (2013). Numerical cognition in bees and other insects. *Frontiers in Psychology, 4,* 1–9.

Peirce, C. S. (1978/1931). In C. Hartshorne & P. Weiss (Eds.), *The collected papers of Charles Sanders Peirce.* Cambridge, MA: Harvard University Press.

Petrie, M. (1994). Improved growth and survival of offspring of peacocks with more elaborate trains. *Nature, 371,* 598–599.

Potts, R. (1998). Variability selection in hominid evolution. *Evolutionary Anthropology, 7,* 81–96.

Powell, A., Shennan, S., & Thomas, M. G. (2009). Late Pleistocene demography and the appearance of modern human behavior. *Science, 324,* 1298–1301.

Rakoczy, H., Clüver, A., Saucke, L., Stoffregen, N., Gräbener, A., Migura, J., et al. (2014). Apes are intuitive statisticians. *Cognition, 131,* 60–68.

Reyna, V. F., Nelson, W. L., Han, P. K., & Dieckmann, N. F. (2009). How numeracy influences risk comprehension and medical decision making. *Psychological Bulletin, 135,* 943–973.

Rossano, M. J. (2010). Making friends, making tools, and making symbols. *Current Anthropology, 51,* S89–S98.

Rozin, P. (1976). The evolution of intelligence and access to the cognitive unconscious. In J. M. Sprague & A. N. Epstein (Eds.), *Progress in psychobiology and physiological psychology* (pp. 245–280). New York: Academic Press.

Setchell, J. M., & Dixson, A. F. (2001). Changes in the secondary sexual adornments of male mandrills (*Mandrillus sphinx*) are associated with gain and loss of alpha status. *Hormones and Behavior, 39,* 177–184.

Sheridan, L., & Pomiankowski, A. (1997). Fluctuating asymmetry, spot asymmetry and inbreeding depression in the sexual coloration of male guppy fish. *Heredity, 79,* 515–523.

Suraci, J. P., & Dill, L. M. (2011). Energy intake, kleptoparasitism risk, and prey choice by glaucous-winged gulls (*Larus glaucescens*) foraging on sea stars. *Auk, 128,* 643–650.

Voight, B. F., Kudaravalli, S., Wen, X., & Pritchard, J. K. (2006). A map of recent positive selection in the human genome. *PLoS Biology, 4*(3), e72.

Welch, A. M., Semlitsch, R. D., & Gerhardt, H. C. (1998). Call duration as an indicator of genetic quality in male gray tree frogs. *Science, 280,* 1928–1930.

Wilke, A., Hutchinson, J. M. C., Todd, P. M., & Czienskowski, U. (2009). Fishing for the right words: Decision rules for human foraging behavior in internal search tasks. *Cognitive Science, 33,* 497–529.

Wrangham, R. W. (1999). Evolution of coalitionary killing. *Yearbook of Physical Anthropology, 42,* 1–30.

Ydenberg, R. (2010). Decision theory. In D. F. West & C. W. Fox (Eds.), *Evolutionary behavioral ecology* (pp. 131–147). New York: Oxford University Press.

Zahavi, A. (1975). Mate selection—A selection for a handicap. *Journal of Theoretical Biology, 53,* 205–214.

Zeder, M. A. (2012). The Broad Spectrum Revolution at 40: Resource diversity, intensification, and an alternative to optimal foraging explanations. *Journal of Anthropological Archaeology, 31,* 241–264.

Zuk, M., Thornhill, R., & Ligon, J. D. (1990). Parasites and mate choice in red jungle fowl. *American Zoologist, 30,* 235–244.

Index

Note: Page numbers followed by "*f*" indicate figures, and "*t*" indicate tables and "*np*" indicate footnotes.

Printed in the United States
By Bookmasters